The Seismoelectric Method

The Seismoelectric Method

Theory and applications

André Revil

Associate Professor, Colorado School of Mines, Golden, CO, USA
Directeur de Recherche at the National Centre for Scientific Research (CNRS),
ISTerre, Grenoble, France

Abderrahim Jardani

Associate Professor, Maître de Conférence, Université de Rouen, Mont-Saint-Aignan, France

Paul Sava

Associate Professor, Colorado School of Mines, Golden, CO, USA

Allan Haas

Senior Engineering Geophysicist, hydroGEOPHYSICS, Inc., Tuscon, AZ, USA

WILEY Blackwell

Registered Office
John Wiley & Sons, Ltd, The Atrium, Southern Gate, Chichester, West Sussex, PO19 8SQ, UK

Editorial Offices
9600 Garsington Road, Oxford, OX4 2DQ, UK
The Atrium, Southern Gate, Chichester, West Sussex, PO19 8SQ, UK
111 River Street, Hoboken, NJ 07030-5774, USA

For details of our global editorial offices, for customer services and for information about how to apply for permission to reuse the copyright material in this book please see our website at www.wiley.com/wiley-blackwell.

Library of Congress Cataloging-in-Publication Data

Revil, André, 1970–
  The seismoelectric method : theory and applications / André Revil, associate professor, Colorado School of Mines, Golden CO, USA [and] Directeur de Recherche at the National Centre for Scientific Research (CNRS), ISTerre, Grenoble, France, Abderrahim Jardani, associate professor, Maître de Conference, Université de Rouen, Mont-Saint-Aignan, France, Paul Sava, associate professor, Colorado School of Mines, Golden CO, USA, Allan Haas, senior engineering geophysicist, HydroGEOPHYSICS, Inc., Tuscon, AZ, USA.
    pages cm
  Includes index.
  ISBN 978-1-118-66026-3 (cloth)
1. Seismic prospecting. 2. Prospecting–Geophysical methods. I. Jardani, Abderrahim. II. Sava, Paul. III. Haas, Allan. IV. Title.
  TN269.8.R48 2015
  622′.1592–dc23

                                    2015002784

A catalogue record for this book is available from the British Library.

Set in 8.5/12pt Meridien by SPi Publisher Services, Pondicherry, India

Printed in Singapore by C.O.S. Printers Pte Ltd

1 2015

This book is dedicated to two Russian scientists who developed the electrokinetic concepts attached to the seismoelectric effects.

Andrey Germogenovich Ivanov (1907–1972). Son of a teacher of geography, Andrey worked at the Institute of Earth Physics (SU Academia of Sciences) in 1930s and up to mid-1950s. His work was mostly concerned with the seismoelectrical method, but he worked also on low-frequency electromagnetic methods. Andrey wrote two handbooks. The first book was concerned with geophysical methods applied to the detection of mineral deposits. It was published in 1961 and in collaboration with Feofan Bubleinikov. The second book was entitled *Physics in Investigations of Earth Interior* (1971). Andrey Germogenovich Ivanov is usually credited to have been the first scientist to record seismoelectric effects in field conditions.

Yakov Il'ich Frenkel (1894–1952). Frenkel was born on February 10, 1894, in the southern Russian city of Rostov-on-Don. He was a very influential Russian scientist during the first half of the 20th century. Geophysics was a field of Frenkel's early interest. In 1944, Frenkel visited the Institute of Theoretical Geophysics in Moscow. There, he became interested in the work of Andrey Germogenovich Ivanov. As mentioned above, Ivanov was the first to discover, in 1939, that the propagation of seismic waves in soils was accompanied by the appearance of an electrical field. Ivanov recognized that this new phenomenon was caused by the pressure difference between two points in wet soil resulting from the propagation of longitudinal (P-)waves. Frenkel modeled the wet soil as a two-phase material, and he formulated the first continuum hydromechanical theory for wave propagation in porous media. Frenkel discovered the existence of the second compressional P-wave (usually named later the Biot slow P-wave), but he dismissed the electrical effects associated with this type of wave as unimportant because of the strong damping of this slow P-wave. Frenkel was the first to understand that the seismoelectric effect recorded by Ivanov could be electrokinetic in nature. Indeed, the presence of water in a porous material is responsible for the formation of an electric double layer on the mineral surface. The relative movement of the excess charge of the electrical diffuse layer (the external part of the electrical double layer) due to the passage of a seismic wave is responsible for the generation of a source current density. These currents are responsible in turn for the generation of electromagnetic disturbances. Frenkel's 1944 paper "On the theory of seismic and seismoelectric phenomena in moist soil" is the first to theoretically describe wave propagation in porous media. A complete theory was however produced in 1956 by Maurice Biot. The linear poroelasticity theory should generally be referred to as the Biot–Frenkel theory rather the Biot theory as done classically in the literature. His life and contributions are described in the book *Yakov Ilich Frenkel: His Work, Life and Letters* by Frenkel, V. Ya., and Birkhäuser Verlag (1996).

Contents

Foreword by Bernd Kulessa

At least as far back as the early 1930s, geophysicists were intrigued by the small electrical field disturbances that accompany propagating seismic waves and their potential utility in subsurface exploration. The first ever volume of the Society of Exploration Geophysicists' flagship journal, *Geophysics*, was R. R. Thompson report, in 1936, on "The Seismic Electric" effect and its potential value in recording seismic waves. Geophysicists have of course confirmed since then that there are better ways of recording seismic signals, and the "seismic electric" effect recurrently came into and went out of fashion as dictated by a healthy dose of skepticism that persists to this day. However, over the past three decades, the body of seismoelectric (electrical fields induced by seismic wave propagation) and electroseismic (seismic waves induced by electrical current flow) literature has been growing ever faster, reflecting ongoing academic intrigue and, in my mind, perhaps also the romantic notion that one day the Earth might reveal its innermost secrets by tiny electrical fields when it is gently prodded with seismic waves.

Whatever the underlying motivation, the fundamental principles of the seismoelectric method have been identified and refined over time, and in early chapters are spelled out succinctly by André Revil and his coauthors as they pertain to saturated or unsaturated, clay-bearing or clay-free earth materials. The authors' agenda-setting seismoelectric research in recent years has been grounded firmly in these principles and their implementation in elaborate finite-element forward and stochastic and deterministic inverse modeling schemes is described in detail in Chapter 4. When presenting my own seismoelectric research, I am often asked the obvious question: so what exactly is the point of recording electrical along with seismic signals? In the future, I shall refer the engaging colleagues to the stimulating answer in Chapter 5! A rather niggling irritation in the processing and interpretation of seismoelectric data is the separation of weak interfacial conversions from stronger coseismic arrivals and electrical noise. I am therefore particularly inspired by the authors' offer of enhancing such conversions through seismoelectric beamforming in Chapter 6, thus promising to lessen the frustration. The book concludes with a pertinent case study in Chapter 7, supporting the exciting hypothesis that seismoelectric data can reflect the water content of the vadose zone.

I have the pleasure to congratulate André Revil and his colleagues on seizing the moment to publish a pioneering and timely book that recognizes the global renaissance of the seismoelectric method and provides a milestone in its development. It will inspire academic researchers, advanced-level students, and practitioners alike and challenges us to contribute to future advances in the acquisition, processing, modeling, and interpretation of seismoelectric data.

Bernd Kulessa
Swansea, October 12, 2013

Foreword by Niels Grobbe

Societal challenges regarding environmental issues and the quest for natural resources demand a continuous need for improved imaging techniques. In recent years, quite some research has been performed investigating the potential of seismoelectric phenomena for geophysical exploration, imaging, and monitoring.

The seismoelectric effect describes the coupling between seismic waves and electromagnetic fields. It is a promising technique as it is complementary to conventional seismics. The seismoelectric signals enable seismic resolution and electromagnetic sensitivity at the same time. In addition, it can provide us with high-value information like porosity and permeability of the medium. However, like other geophysical methods, the seismoelectric technique also has its own drawbacks. One of its main challenges is the very low signal-to-noise ratio of the coupled signals, especially the second-order seismoelectric conversion (or interface response fields). Nevertheless, even if the signals would be stronger, the fact remains that the seismoelectric physical phenomenon is a very complex phenomenon, making it hard to fully understand. In addition, existing acoustic geophysical processing, imaging, and inversion techniques are often not directly applicable or not so easily extended to, for example, elastodynamic systems, let alone seismoelectric systems.

André Revil and coauthors present in this book a unique and pioneering overview of the seismoelectric method, thereby addressing the two main challenges as described above.

Starting from the scale of mineral grain surfaces and ions, the authors introduce in Chapter 1 the concepts of the electrical double layer and streaming current density, the driving mechanisms behind seismoelectric phenomena. Extensive theoretical discussions combined with illustrative experimental laboratory results provide the reader with a thorough understanding of the fundamentals of the seismoelectric phenomenon in both saturated (Chapter 2) and unsaturated media, including two-phase flow scenarios (Chapter 3).

In the seismoelectric theory as described by Pride in "Governing equations for the coupled electromagnetics and acoustics of porous media" (1994, *Physical Review B*, **50**, 15678–15696), Biot's poroelasticity equations are coupled to Maxwell's electromagnetic equations. In this case, full coupling between the electric and magnetic fields is considered. In Section 2.1.2, Revil and colleagues introduce a quasistatic approach for the electromagnetic part of the system. This quasistatic approach makes use of the fact that at low frequency, the electric and magnetic parts are not coupled. The electric field is then rotation-free and can hence be written as minus the gradient of an electrical potential.

Since the seismoelectric effect is a complex physical phenomenon where a huge amount of parameters are involved, being able to simplify the system, for example, in this way, might be beneficial for both our understanding of the phenomenon and for further developing the technique toward imaging and inversion.

Perfect examples of this are provided in Chapter 4, where the authors present, besides effective forward modeling of the seismoelectric effect using the finite-element method, both stochastic and deterministic inversion algorithms and seismoelectric inversion results, something that researchers did not expect to be possible in the next 10 or 20 years.

In Chapter 5, the authors take the inversion algorithms one step further in a wonderful attempt to use seismoelectric signals for source characterization. In addition to the numerical results, laboratory experiments are presented for further insight.

In Section 1.4.2, Revil and colleagues introduce an acoustic approximation for effective modeling of seismoelectric phenomena. This elegant approximation turns out to be highly effective for the development and computationally expensive numerical modeling tests of the seismoelectric beamforming technique described in Chapter 6. Using this technique, seismic energy is focused at particular locations in an attempt to improve the weak signal-to-noise ratio of the seismoelectric conversion.

Chapter 7 then finalizes with field data experiments focused on the vadose zone, one of the areas of the Earth's subsurface where the application of seismoelectric methods seems to be very promising.

It is a true honor to be one of the first to congratulate André Revil, Abderrahim Jardani, Paul Sava, and Allan Haas with this wonderful pioneering work describing the seismoelectric method. The completeness and out-of-the-box approaches of the authors not only show a thorough understanding by the authors of this complex physical phenomenon, but they also provide the reader with a perfect guidebook into the fascinating world of seismoelectrics.

Niels Grobbe
Delft University of Technology
Delft, the Netherlands
August 15, 2014

Preface

The seismoelectric method describes the generation of electrical and electromagnetic disturbances associated with the occurrence of seismic sources and seismic wave propagation in partially or totally water-saturated porous media. The existence of these disturbances has been known for over 75 years. However, the development of rigorous and experimentally testable theories to interpret these effects has been done only in the last few decades, especially with the seminal work of Steve Pride in 1994. In parallel, experimental observations have demonstrated that these effects can be recorded in the field, which makes the seismoelectric method much more than just another exotic geophysical method to put on the shelf. Our goal with this book is to present an overview of the seismoelectric method and some of its potential applications in geophysics.

Chapter 1 introduces some of the key concepts required to understand the seismoelectric theory that is developed for the saturated case in Chapter 2 and for the partially saturated case in Chapter 3. These key concepts include the electrical double layer theory (for silica sands and clayey materials) and the reasons why an electrical (streaming) current density is produced when the pore water moves relatively to the skeleton formed by the solid grains. In the context of the seismoelectric theory, the propagation of seismic waves is responsible for such a relative flow of the pore water, and the associated current is responsible for electromagnetic disturbances that can be measured remotely. The streaming current acts as a source term in the Maxwell equations, which can be used to analyze the related electromagnetic disturbances. We provide in Chapter 1 a short history of the seismoelectric method highlighting the pioneering works of Thompson in the United States, the electroacoustic experiments by Hermans (1938), the first field observation by Ivanov (1939), and the first model proposed by Frenkel (1944). The first chapter also includes a simplification of the seismoelectric theory for the case of acoustic waves. Such simplified theory can be very useful when we

are only interested by the kinetics (travel time) of the seismoelectric problem and not interested by the amplitude of the seismoelectric conversions.

In Chapter 2, we present a complete theory for the generation of seismoelectric effects in the quasistatic limit of the Maxwell equations and for various types of rheological constitutive laws for the porous material and the pore fluid. We start with a description of the poroelastic wave propagation in a poroelastic material filled by a viscoelastic fluid that can sustain shear stresses (extended Biot theory). This represents the general case of wave propagation discussed in this chapter for porous media. This case is interesting since the Biot theory appears symmetric in terms of its constitutive equations and, as a result, four waves (two P-waves and two S-waves) can be determined. Then we present the equations describing the propagation of the seismic waves in a linear poroelastic material saturated by a Newtonian fluid (classical Biot theory) as a special case of the more general theory. We describe in this context the properties of the most important parameter linking the seismic and electric phenomena, the so-called streaming potential coupling coefficient.

In Chapter 3, we apply two extensions of the fully saturated case investigated in Chapter 2 to two cases. The first extension concerns unsaturated conditions for which the material is partially saturated with water and the second fluid is very compressible. Water is considered to be the wetting phase for the solid grains. In this case, the nonwetting phase (air) is at the atmospheric pressure and is highly compressible. The unsaturated seismoelectric theory can be used to describe seismoelectric conversions in the vadose zone (i.e., the partially saturated portion of soils, above unconfined aquifers). The second extension corresponds to the case where there are two immiscible Newtonian fluids present in the pore space. In this more general case, we need to explicitly account for the capillary pressure and the different types of P-waves generated in the porous material. Finally, we extend the acoustic model discussed in Chapter 1 to the partially saturated case, and we end this chapter by comparing some predictions of our model with available experimental data.

Then, based on the developed field equations developed to model the seismoelectric effect in saturated and unsaturated conditions, we proceed to discuss how to implement these equations using the finite-element method. This is done so that we can forward-model the occurrence of seismoelectric signals for various applications in earth sciences. This development is presented in Chapter 4. As geophysicists, we are also interested in going one step further, to solve the so-called inverse problem. Indeed, the solution of the inverse problem is needed to determine the degree of information contained in the seismoelectric signals relative to the more classical seismic signals. In Chapter 4, we present both stochastic and deterministic algorithms to invert seismoelectric signals in terms of key material properties or in terms of the location of the boundaries between geological formations.

In Chapter 5, we study the electromagnetic disturbances associated directly with a seismic source. First, we consider a seismic source in a water-saturated linear poroelastic material. For this case, the source itself is characterized by a moment tensor. Our goal with this analysis is to determine what the advantages are, in terms of information content, in collecting electromagnetic signals in addition to the seismic signals. We apply the deterministic and stochastic mathematical approaches discussed in Chapter 4 to combine the information content of electrograms, magnetograms, and seismograms in terms of seismic source characterization. We also present a laboratory experiment showing, at the scale of a cement block, what types of electrical disturbances can be observed during a hydraulic fracturing experiment. We continue with an example of laboratory data showing clear bursts in the electrical field associated with the occurrence of Haines jumps (corresponding to jumps of the meniscus between the two fluid phases) during the drainage of a sandbox. Finally, we present a field experiment, at a small scale, showing how we can use seismoelectric information to localize a burst in water injection in a well.

As seen in Chapter 4, the seismoelectric conversion can be rather weak with respect to the coseismic electrical fields. In Chapter 6, we develop a new technique called "seismoelectric beamforming" with a goal to enhance the electrical field associated with seismoelectric conversions over the spurious and relatively less informative coseismic electrical disturbances. We present the basic ideas underlying this new method and some numerical tests in piecewise constant and heterogeneous materials. Finally, we discuss how this new method can be used to improve cross-well resistivity tomography and can potentially be a breakthrough to provide high-resolution geophysical images for cross-well tomography using a principle called image-guided inversion.

In Chapter 7, we analyze field seismoelectric data related to the vadose zone, that is, the unsaturated portion of the ground. In addition to a short literature review, we show how seismoelectric data can be gathered for such shallow applications. We present field data from a case study in United Kingdom and apply the numerical model discussed in Chapter 4 to this case study to reproduce the field data. We show that our model can match fairly well the observations and that we can infer the water content of the vadose zone using the seismoelectric method.

André Revil, Abderrahim Jardani, Paul Sava,
and Allan Haas
January 2015

References

Hermans, J. (1938) Charged colloid particles in an ultrasonic field. *Philosophical Magazine*, **25**, 426.

Ivanov, A.G (1939) Effect of electrization of earth layers by elastic waves passing through them. *Proceedings of the USSR Academy of Sciences (Dokl. Akad. Nauk SSSR)*, **24**, 42–45.

Frenkel, J. (1944) On the theory of seismic and seismoelectric phenomena in a moist soil. *Journal Physics (Soviet)*, **8**(4), 230–241.

Acknowledgments

We thank also our colleagues and students who have helped us with this book. Our deep appreciations go to Guillaume Barnier, Bernd Kulessa, Harry Mahardika, and Philippe Leroy, who have helped us through the preparation of figures, stimulating discussions, and some numerical simulations, and Christian Dupuis for sharing two figures from one of his papers. Without their work and help, the making of the book would not have been possible.

CHAPTER 1
Introduction to the basic concepts

The goal of the first chapter is to introduce some of the key concepts required to understand the seismoelectric theory that will be developed for the saturated case in Chapter 2 and for the partially saturated and two-phase flow cases in Chapter 3. These key concepts include the electrical double layer theory and the reasons why an electrical (streaming) current density is produced when the pore water flows relative to the skeleton formed by the solid grains. In the context of the seismoelectric theory, the propagation of seismic waves will be responsible for the relative flow of pore water, and the resulting source current density will be responsible for electromagnetic (EM) disturbances. We will provide a short history of the seismoelectric method as well as its basic concepts. We will also give an introduction to wave propagation theory. At the end of this chapter, we will also provide some simulations using a simplified version of the seismoelectric theory that is based on the acoustic approximation. These models will illustrate, in a simple way, the key concepts behind the seismoelectric method, especially the difference between coseismic signals and seismoelectric conversions. Finally, we will present a preliminary model of seismoelectric phenomena pertaining to the Biot–Frenkel theory of linear poroelasticity.

1.1 The electrical double layer

As discussed later in Section 1.4, the existence of seismoelectric effects is closely related to the existence of the electrical double layer at the interface between the pore water and the skeleton (made of the elastic minerals). In the presence of several immiscible fluids in the pore space, seismoelectric effects can be also associated with the existence of an electrical double layer at the interface between the pore water and these other fluids such as air or oil. Therefore, we believe that it is important to start this book with an extensive description of what the electrical double layer is for silica and clay minerals that are in contact with an electrolyte composed of water molecules and ions. We will focus on silica and clays but the electrical double layer theory has been also developed for carbonates (Cicerone et al., 1992; Strand et al., 2006; Hiorth et al., 2010) and other types of aluminosilicates such as zeolites (van Bekkum et al., 2001).

The electrical double layer is a generic name given to electrochemical disturbances existing at the surface of minerals in contact with water containing dissolved ions. The electrical double layer comprises (1) the Stern layer of sorbed ions on the mineral surface (Stern, 1924) and (2) the diffuse layer of ions bound to the surface through the coulombic force associated with the deficiency or excess of electrical charges on the mineral surface and the Stern layer (Gouy, 1910; Chapman, 1913). The sorbed ions of the Stern layer possess a specific affinity for the mineral surface in addition to the coulombic interaction (specific is usually used to include all types of interactions that are not purely coulombic). In the case of the diffuse layer, the ions are interacting with the mineral surface only through the coulomb interaction.

The readers that are interested to understand the seismoelectric effect but that are not interested by the

The Seismoelectric Method: Theory and Applications, First Edition. André Revil, Abderrahim Jardani, Paul Sava and Allan Haas.

interfacial electrochemistry can skip Sections 1.1.1 and 1.1.2 and can go directly to Section 1.1.3 of this chapter.

1.1.1 The case of silica
1.1.1.1 A simplified approach

Figure 1.1 sketches the surface of a silica grain coated by an electrical double layer. When a mineral like silica is in contact with water, its surface becomes charged due to chemical reactions between the available surface bonding and the pore water as shown in Figure 1.2. For instance, the silanol groups, shown by the symbol >SiOH, of the surface of silica (where > refers to the mineral crystalline framework), behave as weak acid–base (amphoteric sites). This means that they can lose a proton when in contact with

water to generate negative surface sites (>SiO⁻). They can also gain protons to become positive sites (>SiOH₂⁺). Putting water in contact with a fresh silica surface leads to a slight acidification of the pore water, as shown in Figure 1.2, which explains why silica is considered to be an acidic rock. At the opposite end, a mineral like carbonate will generate a basic pH (>7.0) in the pore water.

It follows that the mineral surface charge of silica appears to be pH dependent. It is typically negative at near-neutral pH values (pH 5–8) and possibly positive or neutral for very acidic conditions (pH <3). The simplest complexation reactions at the surface of silica can be summarized as (e.g., Wang & Revil, 2010, and references therein)

$$> SiOH + H^+ \Leftrightarrow\ > SiOH_2^+ \quad K_{(+)}, \qquad (1.1)$$

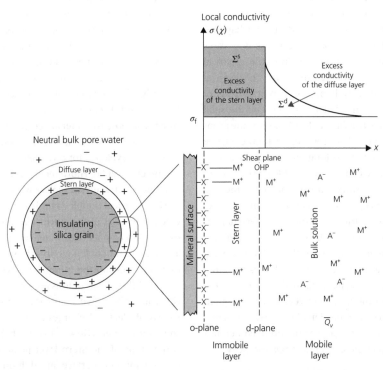

Figure 1.1 Sketch of the electrical double layer at the pore water–mineral interface coating a spherical grain (modified from Revil & Florsch, 2010). The local conductivity $\sigma(\chi)$ depends on the local distance χ from the charged surface of the mineral. The pore water is characterized by a volumetric charge density $\bar{Q}_V$ corresponding to the (total) charge of the diffuse layer per unit pore volume (in coulombs (C) m⁻³). The Stern layer is responsible for the excess surface conductivity Σ^S (in siemens, S) with respect to the conductivity of the pore water σ_f, while the diffuse layer is responsible for the excess surface conductivity Σ^d. These surface conductivities are sometimes called specific surface conductance because of their dimension, but they are true surface conductivities. The Stern layer is comprised between the o-plane (mineral surface) and the d-plane, which is the inner plane of the electrical diffuse layer (OHP stands for outer Helmholtz plane). The diffuse layer extends from the d-plane into the pores. The element M^+ stands for the metal cations (e.g., sodium, Na⁺), while A^- stands for the anions (e.g., chloride, Cl⁻). In the present case (negatively charged mineral surface), M^+ denotes the counterions, while A^- denotes the coions. The fraction of charge contained in the Stern layer with respect to the total charge of the double layer is called the partition coefficient f.

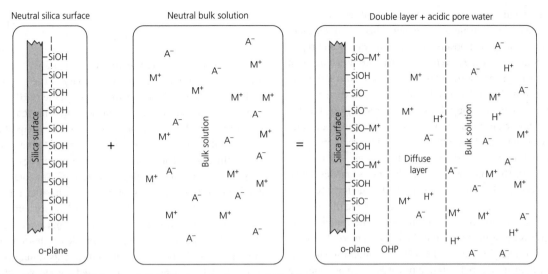

Figure 1.2 Formation of the electrical double layer in the case of silica. In the present case, a neutral silica surface is brought in contact with a neutral pore water solution composed of cations M^+ and anions A^-. The silanol surface groups at the surface of silica release a certain number of protons in the pore water, making the solution slightly acidic. Some of the cations from the pore water are adsorbed in the Stern layer. The surface charge density and the Stern layer charge density are compensated in the diffuse layer. In a sandstone, the bulk pore water is neutral (no net charge density), and only the diffuse layer is not neutral and more precisely characterized usually by an excess of (positive) charges.

$$> SiOH \Leftrightarrow > SiO^- + H^+ \quad K_{(-)}, \qquad (1.2)$$

where $K_{(\pm)}$ are the two equilibrium constants associated with the surface sorption and desorption of protons. This 2-pK model considers that two charged surface species, namely, $>SiO^-$ and $>SiOH_2^+$, are responsible for the surface charge density of silica. That said, the reaction in Equation (1.1) is often neglected in a number of studies because the occurrence of the positive sites, $>SiOH_2^+$, can only happen at low pH values (typically below pH <3 as mentioned briefly previously).

We also assume that the pore water contains a completely dissociated monovalent salt (e.g., NaCl providing the same amount of cations Na^+ and anions Cl^-). In the following, a "counterion" is an ion that is characterized by a charge opposite to the charge of the mineral surface, while a "coion" has a charge of the same sign as the mineral surface. The typical case for silica is to have a negative surface charge, and therefore, the counterions are the Na^+ cations and the coions are the Cl^- anions. Note however that the sorption of cations is characterized by a high valence and a strong affinity for the silica surface (for instance, Al^{3+}) and can reverse the charge of the mineral surface (surface and Stern later together) and therefore can reverse the sign of the charge of the diffuse layer. The sorption is described by the following reaction:

$$> SiOH + M^+ \Leftrightarrow > SiO^-M^+ + H^+, \quad K_M, \qquad (1.3)$$

where K_M corresponds to the equilibrium constant for this reaction. Sorption is distinct from precipitation, which involves the formation of covalent bonds with the mineral surface. This sorption can be strong (formation of an inner-sphere complexes with no mobility along the mineral surface) or weak. In the "weak case," the formation of the Stern layer is a kind of condensation effect demonstrated by molecular dynamics. A weak sorption example is the case of a hydrated sodium. In this example, the sorbed counterion Na^+ keeps its hydration sphere, and it forms a so-called outer-sphere complex with the mineral surface (e.g., Tadros & Lyklema, 1969). Such counterions are expected to keep some mobility along the mineral surface, responsible (as briefly explained in Section 1.3) for a low-frequency polarization of the mineral grains in an alternating electrical field. The layer of ions formed by the sorption of these counterions directly on the mineral surface is called the Stern layer. The Stern layer is therefore located between the o-plane (mineral surface) and the d-plane, which is the inner plane of the electrical diffuse layer (Figures 1.1 and 1.2). The sorption of counterions occurs at the "β-plane" which is located in between the o- and d-planes shown in Figure 1.1.

As stated earlier, at near-neutral pH values, the surface charge of silica is generally negative. This negative surface attracts the ions of positive sign (counterions) and repels the ions of the same sign (coions). The surface charge that is not balanced by the sorption of some counterions in the Stern layer is balanced further away in the so-called diffuse layer. In normal conditions, the diffuse layer is therefore characterized by an excess of counterions and a depletion of coions with respect to the free pore water located in the central part of the pores (Figures 1.1 and 1.2). This concept of a diffuse layer was first developed by Gouy (1910) and Chapman (1913). The term "electrical double layer" is a generic name describing this electrochemical system coating the surface of the minerals and comprising of the Stern and the diffuse layers. The term electrical "triple layer model" (TLM) is often used in electrochemistry when different types of sorption phenomena are considered at the level of the Stern layer. In this case and as briefly discussed previously, electrochemists use the term "inner-sphere complexes" for ions strongly bound to the mineral surface (e.g., Cu^{2+} or NH_4^+ on the surface of silica). The term "outer-sphere complex" is used to characterize ions that are weakly bound to the mineral surface (e.g., K^+, Na^+) and generally keep their hydration layer and a certain mobility along the mineral surface. We will return later in this section to the idea of a strong sorption mechanism.

Electrokinetic properties are defined by measurable macroscopic effects associated with the relative displacement of the diffuse layer with respect to the solid phase, with the Stern layer attached to it (e.g., von Smoluchowski, 1906). One of the key parameters to define electrokinetic properties is the zeta potential. For simplicity, we assume that the zeta potential is the inner potential of the diffuse layer. Our goal is to define a simple model to determine the value of the zeta potential as a function of the pore water salinity for a simple 1:1 solution like NaCl or KCl. The availability of the different sites is obtained by solving one continuity equation for the surface sites and two constitutive equations based on reactions (1.2) and (1.3) earlier. These three equations are given by

$$\Gamma_S^0 = \Gamma_{SiOH}^0 + \Gamma_{SiO^-}^0 + \Gamma_{SiOM}^0, \qquad (1.4)$$

$$K_{(-)} = \frac{\Gamma_{SiO}^0 \cdot \alpha_{H^+}^0}{\Gamma_{SiOH}^0}, \qquad (1.5)$$

$$K_M = \frac{\Gamma_{SiOM}^0 \alpha_{H^+}^0}{\Gamma_{SiOH}^0 \alpha_{M^+}^0}, \qquad (1.6)$$

where Γ_i^0 corresponds to the number of sites i per surface area and $\alpha_{H^+}^0$ and $\alpha_{M^+}^0$ denote the activity of the protons and cations M^+ on the outer Helmholtz plane (OHP; see Figure 1.1). Equation (1.4) expresses the fact that the sum of all the different types of surface sites is equal to the crystalline site density Γ_S^0 of silanol groups on the surface of the grains. The value of this quantity can be determined from crystallographic considerations. The total site density Γ_S^0 is typically between 5 and 10 sites per nm^2.

Equations (1.5) and (1.6) represent the balance between species associated with the constitutive chemical reactions (1.2) and (1.3) assuming thermodynamic equilibrium for these reactions (kinetics is neglected) and assuming that reaction (1.1) can be safely neglected for near-neutral pH values. According to Revil et al. (1999a), we have $pK_{(-)} = -\log_{10} K_{(-)}$ is typically around 7.4–7.5 at 25°C and $pK_{Na^+} = -\log_{10} K_{Na^+}$ is typically close to 3.3 at 25°C, while $pK_{K^+} = -\log_{10} K_{K^+}$ is close to 2.8 at 25°C.

The solution of Equations (1.4)–(1.6) is straightforward and given by

$$\Gamma_{SiO^-}^0 = \frac{K_{(-)}\Gamma_S^0}{A\alpha_{H^+}^0}, \qquad (1.7)$$

$$\Gamma_{SiOM}^0 = \frac{K_M\Gamma_S^0 \alpha_{M^+}^0}{\alpha_{H^+}^0 A}, \qquad (1.8)$$

where the quantity A is defined by

$$A = 1 + \frac{K_{(-)}}{\alpha_{H^+}^0} + \alpha_{M^+}^0 \frac{K_M}{\alpha_{H^+}^0}. \qquad (1.9)$$

Since, in this simplified model, there is only one type of charged site on the surface of the mineral, the surface charge density (charge per surface are of the mineral surface) is simply given by

$$Q_0^S = -e\Gamma_{SiO^-}^0, \qquad (1.10)$$

where e is the elementary charge (1.6×10^{-19} C, C for coulomb).

The next step is to determine the equivalent surface charge density of the diffuse layer, which is characterized by an electrostatic potential that decreases away from the mineral surface $\varphi(\chi)$ where χ denotes the local distance away from the mineral surface assuming to be locally smooth. The activity or concentrations of the ions in the electrical diffuse layer are determined through the use of Poisson–Boltzmann statistics. To understand these distributions, we need to define the so-called electrochemical potentials of cations (+) and anions (−). These electrochemical potentials are defined by (e.g., Gouy, 1910; Hunter, 1981)

$$\mu_i = \mu_i^0 + k_b T \ln \alpha_i + q_i \varphi, \tag{1.11}$$

where $\mu_{(\pm)}^0$ is the chemical potential of the ions in a reference state (a constant), k_b is the Boltzmann constant, T is temperature (in degrees K, Kelvin), α_i is the activity of species i (equal to the concentrations for dilute solutions), q_i is the charge of species i (in C; for instance, $q_{(+)} = e$ for Na$^+$ where e denotes the elementary charge 1.6×10^{-19} C), and φ is the electrostatic potential (in volts, V).

Local thermodynamic equilibrium between the electrical diffuse layer and the bulk pore water is given by the equality of the electrochemical potentials. We can consider equilibrium between a position χ away from the OHP (see position in Figure 1.1) and an arbitrary position in the bulk pore water for which the local potential of the electrical diffuse layer φ vanishes ($\varphi(\infty) = 0$). For monovalent ions, the condition (Hunter, 1981)

$$\mu_i(\chi) = \mu_i(\infty) \tag{1.12}$$

yields

$$\mu_i^0 + k_b T \ln \alpha_i(\chi) \pm e \varphi(\chi) = \mu_i^0 + k_b T \ln \alpha_i^f. \tag{1.13}$$

In Equation (1.13), $\alpha_{(\pm)}^f$ denotes the activity of the cations (+) or anions (−) far from the mineral surface and taken in the bulk pore water (in the bulk pore fluid, characterized by superscript f). It follows that the ionic activity of species i at the position of the OHP itself, α_i^0, is given as a function of the activity in the bulk pore water α_i^f by

$$\alpha_i^0 = \alpha_i^f X^{2q_i}, \tag{1.14}$$

$$X = \exp\left(-\frac{\varphi_d}{2k_b T}\right), \tag{1.15}$$

where φ_d denotes the electrical potential at the OHP (i.e., the inner plane of the electrical diffuse layer). The charge in the diffuse layer is given by averaging the concentrations over the thickness of the electrical diffuse layer.

In the general case, the charge density in the diffuse layer is given by

$$Q_S = \int_0^\infty \sum_{i=1}^N q_i C_i(\chi) d\chi \tag{1.16}$$

$$Q_S = \sum_{i=1}^N q_i \int_0^\infty C_i^f \exp\left(-\frac{q_i \varphi(\chi)}{k_b T}\right) d\chi. \tag{1.17}$$

We have also the useful property (Pride, 1994)

$$\int_0^\infty \exp\left(-\frac{q_i \varphi(\chi)}{k_b T}\right) d\chi = 2\chi_d \exp\left(-\frac{q_i \varphi_d}{2k_b T}\right), \tag{1.18}$$

where $2\chi_d$ represents an average thickness for the diffuse layer ($\chi_d = (\epsilon_f k_b T / 2e^2 C_f)^{1/2}$ where e denotes the elementary charge 1.6×10^{-19} C, k_b denotes the Boltzmann constant, and ϵ_f denotes the dielectric constant of water). The length scale χ_d is called the Debye screening length in electrical double layer theory (e.g., Gouy, 1910, Chapman, 1913). From Equations (1.17) and (1.18), we obtain

$$Q_S = 2\chi_d \sum_{i=1}^N q_i C_i^f X^{q_i}. \tag{1.19}$$

The potential in the diffuse layer is approximately given by the Debye formula $\varphi(\chi) = \varphi_d \exp(-\chi/\chi_d)$ (e.g., Pride, 1994) where φ_d denotes the local potential on the OHP. For a binary symmetric 1:1 electrolyte, the expression of the charge density of the diffuse layer reduces to (using Eq. 1.15)

$$Q_S = \sqrt{aC_f} \sinh\left(\frac{-\varphi_d}{2k_b T}\right), \tag{1.20}$$

$$a = 8 \times 10^3 \epsilon_f k_b T N, \tag{1.21}$$

where N denotes the Avogadro number (6.0221×10^{23} mol^{-1}). We can rewrite the charge density of the diffuse layer as

$$Q_S = \frac{1}{2} \sqrt{aC_f} \left(X - \frac{1}{X}\right). \tag{1.22}$$

The electrical double layer problem can be finally solved by using a final condition in the form of a global electroneutrality condition for the electrical double layer and the mineral surface. This condition implies that the charge density on the mineral surface is exactly counterbalanced by the charge density in the Stern layer and the charge density in the diffuse layer. In order to get an analytical solution for the zeta potential, we are going to omit the charge density in the Stern layer (a fair approximation for silica but not for clays). It follows that the total electroneutrality condition can be written as

$$Q_S + Q_0^S = 0. \tag{1.23}$$

Using Equations (1.7), (1.10), (1.14), and (1.22) into Equation (1.23), the potential of the Stern layer φ_d is the solution of the following equation:

$$\alpha\left(X - \frac{1}{X}\right)\left(1 + \beta X^2\right) - 1 = 0, \tag{1.24}$$

where

$$\alpha = \frac{\sqrt{aC_f}}{2e\Gamma_S^0}, \tag{1.25}$$

$$\beta = \frac{10^{-pH} + K_M C_f}{K_{(-)}}, \tag{1.26}$$

and where X is defined by Equation (1.15) and a by Equation (1.21). At low salinities, we have $X \gg (1/X)$. With this assumption, Equation (1.24) simplified to the following cubic equation:

$$X^3 + pX + q = 0, \tag{1.27}$$

with $p = 1/\beta$ and $q = -1/(\alpha\beta)$. The real root of this cubic equation is given by

$$X = \left(-\frac{q}{2} + \sqrt{\Delta}\right)^{1/3} + \left(-\frac{q}{2} - \sqrt{\Delta}\right)^{1/3}, \tag{1.28}$$

where

$$\Delta = \left(\frac{q}{2}\right)^2 + \left(\frac{p}{3}\right)^3 = \frac{1}{4\alpha^2\beta^2} > 0 \tag{1.29}$$

(assuming $4\alpha^2/(27\beta) \ll 1$, which can be easily checked). Using Equation (1.15), the solution is simply given by

$$\varphi_d = \frac{2k_b T}{3e}\ln(\alpha\beta). \tag{1.30}$$

In electrokinetic properties, the zeta potential represents the electrical potential of the diffuse layer at the position of the hydrodynamic shear plane, which is defined as the position of zero relative velocity between the solid and liquid phases. The exact position of the zeta potential is unknown but likely pretty close to the mineral surface. If we assume that the zeta potential represents the potential on the OHP (see Figure 1.1 for the position of this plane), it follows from Equation (1.30) that we can write the zeta potential as (Revil et al., 1999a, b)

$$\zeta = b\log_{10}C_f + c, \tag{1.31}$$

where

$$b = \frac{k_b T}{3e}\ln 10, \tag{1.32}$$

$$c = \frac{2k_b T}{3e}\ln\left[\frac{(8 \times 10^3 \varepsilon_f k_b TN)^{1/2}}{2eK_{(-)}\Gamma_S^0}10^{-pH}\right]. \tag{1.33}$$

This equation shows how the zeta potential depends on the salinity C_f for simple supporting 1:1 electrolytes. Note that Pride and Morgan (1991, their Figure 4) came to Equation (1.31) on purely empirical grounds, fitting experimental data with such an equation and getting empirically the values of b and c. Typically, the seismo-electric community has been using Equation (1.31) only as an empirical equation while it can derived from physical grounds as demonstrated by Revil et al. (1999a). The previous model yields $b = 20$ mV per tenfold change in concentration (salinity) for a 1:1 electrolyte. A comparison between the prediction of Equation (1.31) and a broad dataset of experimental data is shown in Figure 1.3. The slope b of the experimentally determined zeta potential is actually closer to 24 mV per tenfold change in concentration, therefore fairly close to the predicted value.

Equation (1.31) is not valid at very high salinities (10^{-1} mol l^{-1} and above). Jaafar et al. (2009) presented measurements of the streaming potential coupling coefficient in sandstone core samples saturated with NaCl solutions at concentrations up to 5.5 mol l^{-1} (Figure 1.3). Using measurement of the streaming potential coupling coefficient, they were able to determine the zeta potential up to the saturated concentration limit in salinity. They found that the magnitude of the zeta potential also decreases with increasing salinity, as discussed previously and as predicted by Equation (1.31), but approaches a constant value at high salinity around −20 mV. This value is, so far, not captured by exiting models.

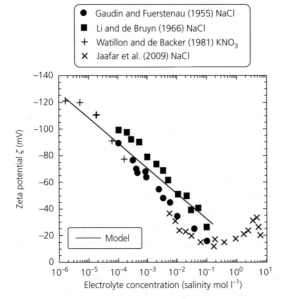

Figure 1.3 Zeta potential ζ on the surface of a silica grain. Comparison between the analytical model developed in the main text (plain line, Eqs. 1.31–1.33) and experimental data from the literature. These data are from Gaudin and Fuerstenau (1955), Li and De Bruyn (1966), Watilllon and De Backer (1970), and Jaafar et al. (2009). We use pH = 5.6 (pH of pure water in equilibrium with the atmosphere), $K_{(-)} = 10^{-7.4}$, and a density of surface active site at the surface of silica of $\Gamma_S^0 = 7$ sites nm^{-2}. Note the high salinity values are not captured by the model.

Table 1.1 Equilibrium constants for surface complexes at the surface of a silica sand.

Reactions	Equilibrium constants
$>SiOH + H^+ \leftrightarrow > SiOH^{2+}$	$10^{-2.2}$
$>SiOH \leftrightarrow > SiO^- + H^+$	$10^{-6.2}$
$>SiO^- + Na^+ \leftrightarrow >SiO^- Na^+$	$10^{-4.5}$
$>SiOH + Cu^{2+} \leftrightarrow >SiOCu^+ + H^+$	$10^{-3.4}$
$2 > SiOH + Cu^{2+} \leftrightarrow 2(>SiO)^- Cu^{2+} + 2H^+$	$10^{-8.8}$
$>SiOH + SO_4^{2-} + H^+ \leftrightarrow >SiSO_4^- + H_2O$	$10^{5.0}$

From Sverjensky (2005).

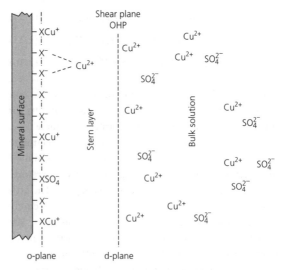

Figure 1.4 Sketch of the electrical double layer showing the speciation of copper and sulfate for a solution of copper sulfate in contact with a silica surface. Sorption of copper on the mineral surface (inner-sphere ligand) occurs as a monodentate complex (immobile), while sorption in the Stern layer (outer-sphere ligand) occurs as a (mobile) bidentate complex. This type of sorption has a strong effect on the value of the zeta potential and can, under given conditions, reverse the polarity of the zeta potential on the surface of silica. The o-plane denotes the mineral surface, and the d-plane denotes the outer Helmholtz plane (OHP) on which the zeta potential ζ is considered.

In addition, the analysis made earlier is correct only for silica in contact with simple supporting electrolytes such as NaCl or KCl with a weak sorption of the counterions. As mentioned briefly previously, the composition of the pore water can, however, strongly influence the value and even the sign of the zeta potential. In the case of strong sorptions, it is necessary to account for more intricate complexation reactions on the surface of silica like the one shown in Table 1.1 for copper. Figure 1.4 shows the speciation of copper on the mineral surface forming both monodentate and bidentate complexes. In the presence of such strong sorption phenomena, the zeta potential can reverse sign and drastically change in magnitude. This is especially true in the case of the sorption of cations of high valence (e.g., Al^{3+}) directly on the mineral surface. In such inner-sphere complex, the cation loses part of the hydration layer. The charge density of the counterions in the Stern layer can be high enough to overcome the charge density on the surface of the mineral. In this case, the charge of the diffuse layer and its associated zeta potential have a reversed polarity, at a given pH, with respect to what is normal for a simple supporting binary electrolyte like NaCl or KCl. Electrokinetic phenomena like the seismoelectric effect are very sensitive to these types of chemical changes because they are directly controlled by the properties of the electrical double layer and by the zeta potential.

1.1.1.2 The general case

A complete electrical double layer model for silica is now discussed avoiding most of the assumption used previously. The drawbacks of such approach, however, are that there are no analytical solutions of the system of equation and we have to use a numerical approach to determine the zeta potential and the surface charge for a given set of environmental conditions. We consider again silica grains in contact with a binary symmetric electrolyte like NaCl for the simplicity of the presentation and comparison with the experimental data. In the pH range 4–10, the surface mineral reactions at the silanol surface sites can be written as

$$> SiO^- + H^+ \Leftrightarrow > SiOH, \quad K_1, \quad (1.34)$$

$$> SiOH + H^+ \Leftrightarrow > SiOH_2^+, \quad K_2, \quad (1.35)$$

$$> SiO^- + Na^+ \Leftrightarrow > SiONa, \quad K_3. \quad (1.36)$$

The symbol ">" refers to the mineral framework, and K_1, K_2, K_3 are the associated equilibrium constants for the different reactions reported earlier (see Table 1.1). Additional reactions for a multicomponent electrolyte can be easily incorporated by adding reactions similar to Equation (1.36) or exchange reactions. Therefore, the present model is not limited to a binary salt. The protonation of surface siloxane groups $>SiO_2$ is extremely low, and these groups can be considered as inert. We neglect here the adsorption of anion Cl^- at the surface of the $>SiOH_2^+$ sites which occurs at pH < pH (pzc) ≈ 3, where pzc denotes the point of zero charge of silica:

$$pH(pzc) = \frac{1}{2}(\log K_1 + \log K_2). \quad (1.37)$$

Consequently, the value of K_2 is determined from the value of K_1 and pH (pzc) ≈ 3. The surface charge density Q_0 (in $C\,m^{-2}$) at the surface of the minerals can be expressed as follows:

$$Q_0 = e\left(\Gamma^0_{SiOH_2} - \Gamma^0_{SiO} - \Gamma^0_{SiONa}\right), \quad (1.38)$$

where Γ^0_i denotes the surface site density of species i (in sites m^{-2}). The surface charge density Q_β in the Stern layer is determined according to

$$Q_\beta = e\Gamma^0_{SiONa}. \quad (1.39)$$

The surface charge density in the diffuse layer is calculated using the classical Gouy–Chapman relationship in the case of a symmetric monovalent electrolyte:

$$Q_S = -\sqrt{8\varepsilon k_b T C_f} \sinh\left(\frac{e\varphi_d}{2k_b T}\right), \quad (1.40)$$

where C_f is the salinity in the free electrolyte (in $mol\,l^{-1}$), T is the temperature (in K), ε_f is the permittivity of the pore water ($\varepsilon_f = 81\,\varepsilon_0$, $\varepsilon_0 \sim 8.85 \times 10^{-12}\,F\,m^{-1}$), e represents the elementary charge (taken positive, $e = 1.6 \times 10^{-19}$ C), and k_b is the Boltzmann constant ($1.381 \times 10^{-23}\,J\,K^{-1}$). The electrical potential φ_d (in V) is the electrical potential at the OHP (see Figure 1.1). We make again the assumption that the electrical potential φ_d is equal to the zeta potential ζ placed at the shear plane. The shear plane is the hydrodynamic surface on which the relative velocity between the mineral grains and the pore water is null.

The continuity equation for the surface sites yields

$$\Gamma^0_1 = \Gamma^0_{SiO} + \Gamma^0_{SiOH} + \Gamma^0_{SiOH_2} + \Gamma^0_{SiONa}, \quad (1.41)$$

where Γ^0_1 (in sites m^{-2}) is the total surface site density of the mineral. We use the equilibrium constants associated with the half reactions to calculate the surface site densities Γ^0_i. Solving Equation (1.41) with the expressions of the equilibrium constants defined through Equations (1.34)–(1.36) yields

$$\Gamma^0_{SiO} = A\Gamma^0_1, \quad (1.42)$$

$$\Gamma^0_{SiOH} = A\Gamma^0_1 K_1 C^f_{H^+} \exp\left(-\frac{e\varphi_0}{k_b T}\right), \quad (1.43)$$

$$\Gamma^0_{SiOH_2^+} = A\Gamma^0_1 K_1 K_2 C^{f\,2}_{H^+} \exp\left(-\frac{2e\varphi_0}{k_b T}\right), \quad (1.44)$$

$$\Gamma^0_{SiONa} = A\Gamma^0_1 K_3 C^f_{Na^+} \exp\left(-\frac{e\varphi_\beta}{k_b T}\right), \quad (1.45)$$

$$A = 1 + K_1 C^f_{H^+} \exp\left(-\frac{e\varphi_0}{k_b T}\right) + K_1 K_2 C^{f\,2}_{H^+} \exp\left(-\frac{2e\varphi_0}{k_b T}\right)$$

$$+ K_3 C^f_{Na^+} \exp\left(-\frac{e\varphi_\beta}{k_b T}\right), \quad (1.46)$$

where φ_0 and φ_β are, respectively, the electrical potential at the o-plane corresponding to the mineral surface and the electrical potential at the β-plane corresponding to the plane of the Stern layer (see Figure 1.1). The electrical potentials φ_0, φ_β, and φ_d are related by

$$\varphi_0 - \varphi_\beta = \frac{Q_0}{C_1}, \tag{1.47}$$

$$\varphi_\beta - \varphi_d = -\frac{Q_S}{C_2}, \tag{1.48}$$

where C_1 and C_2 (in F m^{-2}) are the (constant) integral capacities of the inner and outer parts of the Stern layer, respectively. The global electroneutrality equation for the mineral/water interface is

$$Q_0 + Q_\beta + Q_S = 0. \tag{1.49}$$

We calculate the φ_d potential—thanks to Equations (1.38)–(1.49)—using an iterative method to solve the system of equations. We use $\Gamma_1^0 = 5$ sites m^{-2} and $C_2 = 0.2$ F m^{-2}. We use the values of K_1, K_3, and C_1 reported in Figure 1.5 to calculate the surface charge density Q_0 at the surface of silica mineral and the potential φ_d. The predictions of this double layer model are compared to the literature data (zeta potential and surface charge) in Figure 1.5. With the same model parameters, the surface charge of the mineral and the zeta potential can be described by this model as a function of the pH and salinity. Such type of model can also be used to predict the effect of specific sorption of cations like Cu^{2+} on the zeta potential/surface charge density of the silica surface.

As shown previously, the counterions are both located in the Stern and in the diffuse layer. The fraction of counterions located in the Stern layer is defined by

$$f = \frac{\Gamma_{SiONa}^0}{\Gamma_{SiONa}^0 + \Gamma_{Na}^D}, \tag{1.50}$$

where the surface charge density of the counterions in the diffuse layer is given by

$$\Gamma_{Na}^D \equiv \int_0^\infty \left[C_{Na^+}^D(\chi) - C_{Na^+}^f \right] d\chi = C_{Na^+}^f \int_0^\infty \left\{ \exp\left[-\frac{e\varphi(\chi)}{k_b T} \right] - 1 \right\} d\chi \tag{1.51}$$

and the electrical potential in the diffuse layer φ is given by

$$\varphi(\chi) = \frac{4 k_b T}{e} \tanh^{-1}\left[\tanh\left(\frac{e\varphi_d}{4 k_b T} \right) \exp\left(-\frac{\chi}{\chi_d} \right) \right]. \tag{1.52}$$

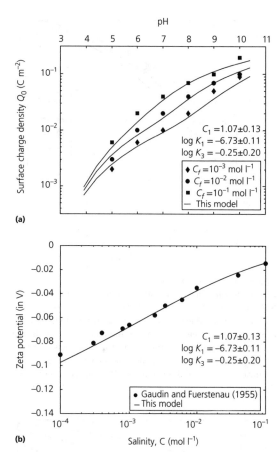

(a)

(b)

Figure 1.5 Comparison between the predictions of the triple layer model described in the main text at the end of Section 1.1 and experimental data in the case of silica. **a)** Comparison between the prediction of the model and surface charge density measurements obtained by potentiometric titrations at three different salinities (NaCl) and in the pH range 5–10 (Data from Kitamura et al., 1999). **b)** Comparison between the model prediction and measurements of the zeta potential at different salinities and pH = 6.5 (Data from Gaudin & Fuerstenau, 1955). The same model parameters are used for the two simulations.

In Equation (1.52), χ denotes the distance defined locally normal from the interface between the pore water and the solid grain, χ_d is the Debye screening length (in m), and Γ_{Na}^D is the equivalent surface density of the counterions in the diffuse layer. Figures 1.6 and 1.7 show that the fraction of counterions located in the Stern layer, f, depends strongly on the salinity and pH of the pore water solution. For example, at pH = 9 and at low salinities ($\leq 10^{-3}$ mol l^{-1}), most of the counterions are located in

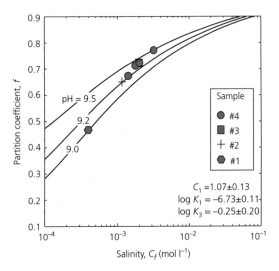

Figure 1.6 Partition coefficient versus the salinity of the free electrolyte with the TLM parameters indicated on Figure 1.2 for NaCl (pH = 9, 9.2, 9.5). The symbols correspond to the partition coefficient determined from the complex conductivity data for the seven experiments described in the main text. The data are determined from spectral induced polarization measurements (see Leroy et al., 2008). They show an increase of the partition coefficient with the salinity and the pH in fair agreement with the model.

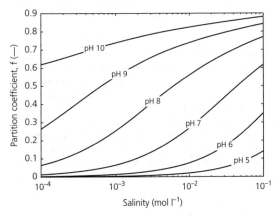

Figure 1.7 Determination of the partition coefficient though a triple layer model for silica for different values of the pH and salinity of NaCl solutions.

the diffuse layer, while at high salinity ($>10^{-3}$ mol l^{-1}), the counterions are mostly located in the Stern layer.

1.1.2 The case of clays

Clays are ubiquitous in nature, and as such, their influence on electrical properties in general and the seismoelectric

properties in particular is very important. A second reason to be interested by clays comes from their very small particle size (typically smaller than 5 μm) and the charged nature of their crystalline planes (Figure 1.8). The small size of the clay particles implies that they carry a huge charge per unit pore volume of porous rocks. There are at least two families of clay minerals depending on whether the space between the clay crystals is open or closed: on the one hand, kaolinite, chlorite, and illite have no open interlayer porosity, while on the other hand, smectite has an interlayer porosity strongly influencing its swelling properties. Figure 1.8 shows that the surface charge density of the clay particles has two distinct origins: one is located essentially on the basal planes that is mostly due to isomorphic substitutions in the crystalline framework and is pH independent (this charge is dominant for smectite). The second charge density is mostly located on the edge of the crystals due to amphoteric (pH-dependent) active sites.

The small particle size of clay minerals implies in turn a high specific surface area and a high cation exchange capacity (CEC) by comparison with other minerals. The specific surface area S_{sp} (in m^2 kg^{-1}) corresponds to the amount of surface area divided by the mass of grains. The CEC (in C kg^{-1}) corresponds to the amount of charge that can be titrated on the mineral surface divided by the mass of mineral. The ratio of the CEC by the specific surface area corresponds to the effective charge density on the mineral surface:

$$Q_0^S = \frac{CEC}{S_{sp}}. \tag{1.53}$$

As shown in Figure 1.9, the charge per unit surface is pretty constant for all clay minerals and comprised between 1 and 3 elementary charges per nm^2 at near-neutral pH values. Because part of the charge on the surface of the clay minerals is pH dependent, Maes et al. (1979) proposed for $3.9 \leq pH \leq 5.9$ and for montmorillonite (a special type of smectite) the following pH-dependent relationship for the CEC: CEC (in meq g^{-1}) = 79.9 + 5.04 pH for monovalent cations and CEC (in meq g^{-1}) = 96.1 + 3.93 pH for divalent cations.

A theory for the electrical double layer of clay minerals is now introduced. This theory can be used to predict the amount of charge on the mineral surface and in the Stern layer or more directly the zeta potential. It can also be used to predict a highly important parameter used in

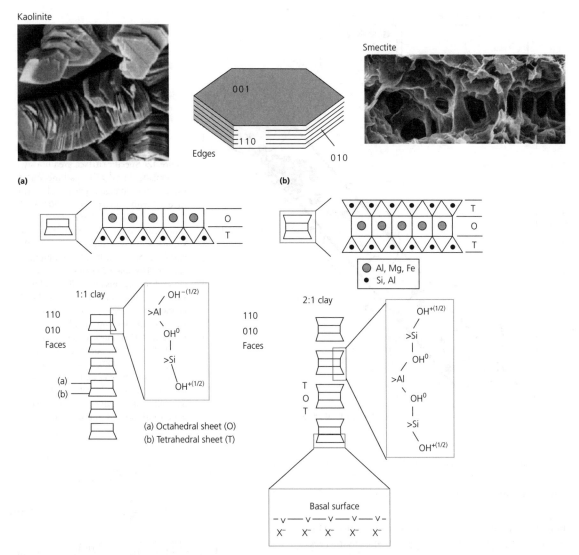

Figure 1.8 Active surface sites at the edge of **a)** 1:1 clays (kaolinite) and **b)** 2:1 clays (smectite or illite). In the case of kaolinite, the surface sites are mainly located on the edge of the mineral ({110} and {010} planes). In the case of smectite and illite and in the pH range near neutrality (5–9), the surface sites are mainly located on the basal plane ({001} plane), and they are due to isomorphic substitutions inside the crystalline framework (Modified from Leroy & Revil, 2004). Note also the difference in the morphology of the clay particles. T and O represent tetrahedral and octahedral sheets, respectively.

the determination of the complex conductivity of these minerals. This parameter is called the partition coefficient f (dimensionless). It describes the amount of counterions in the Stern layer by comparison with the total amount of counterions in the Stern and diffuse layers together.

We consider first a kaolinite crystal in contact with a binary symmetric electrolyte like NaCl. We restrict our analysis to the pH range 4–10, which is the pH range useful for most practical applications in geophysics. In this pH range and in the case of kaolinite, the surface mineral reactions at the aluminol, silanol, and >Al–O–Si< surface sites can be written as

$$> \text{AlOH}_2^{1/2+} \Leftrightarrow\, > \text{AlOH}^{1/2-} + \text{H}^+, \quad K_1, \qquad (1.54)$$

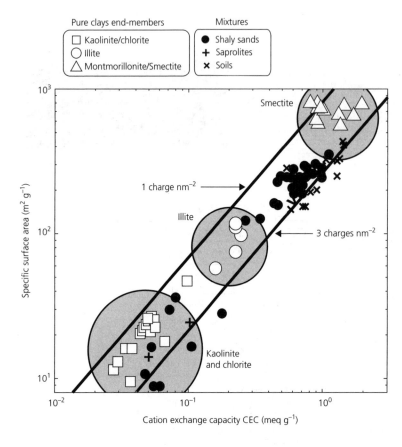

Figure 1.9 Specific surface area of clay minerals S_s (in $m^2\ g^{-1}$) as a function of the cation exchange capacity (CEC) (in $meq\ g^{-1}$ with $1\ meq\ g^{-1} = 96{,}320\ C\ kg^{-1}$ in SI units) for various clay minerals. The ratio between the CEC and the specific surface area gives the equivalent total surface charge density of the mineral surface. The shaded circles correspond to generalized regions for kaolinite, illite, and smectite. The two lines correspond to 1–3 elementary charges per unit surface area. Data for the clay end members are from Patchett (1975), Lipsicas (1984), Zundel and Siffert (1985), Lockhart (1980), Sinitsyn et al. (2000), Avena and De Pauli (1998), Shainberg et al. (1988), Su et al. (2000), and Ma and Eggleton (1999). Saprolite data: Revil et al. (2013). Soil data: Chittoori and Puppala (2011).

$$> SiOH^{1/2+} \Leftrightarrow > SiO^{1/2-} + H^+,\ \ K_2, \qquad (1.55)$$

$$> Al-ONa-Si < \ \Leftrightarrow \ > Al-O^- -Si < + Na^+,\ \ K_3, \quad (1.56)$$

where K_1, K_2, and K_3 are the equilibrium constants of reactions (1.54)–(1.56) and the sign ">" refers to the crystalline framework. The surface site >Al–O–Si< carries a net (-1) negative charge (Avena & DePauli, 1998). We assume that the surface complexation reactions occur on the {010} and {110} planes of kaolinite.

The availability of the surface sites introduced by the chemical reactions described previously at the surface of the {010} and {110} planes can be described by the conservation equations for the three types of sites (aluminol, silanol, and >Al–O–Si< surface sites). Solving these equations, we obtain the concentrations of the different surface sites:

$$\Gamma_{AlOH}^0 = \frac{\Gamma_1^0}{A}, \qquad (1.57)$$

$$\Gamma_{AlOH_2}^0 = \frac{\Gamma_1^0}{A}\left(\frac{C_{H^+}^f}{K_1}\right)\exp\left(-\frac{e\varphi_0}{k_b T}\right), \qquad (1.58)$$

$$\Gamma_{SiO}^0 = \frac{\Gamma_2^0}{B}, \qquad (1.59)$$

$$\Gamma_{SiOH}^0 = \frac{\Gamma_2^0}{B}\left(\frac{C_{H^+}^f}{K_2}\right)\exp\left(-\frac{e\varphi_0}{k_b T}\right), \qquad (1.60)$$

$$\Gamma_{AlOSi}^0 = \frac{\Gamma_3^0}{C}, \qquad (1.61)$$

$$\Gamma_{AlONaSi}^0 = \frac{\Gamma_3^0}{C}\left(\frac{C_{Na^+}^f}{K_3}\right)\exp\left(-\frac{e\varphi_\beta}{k_b T}\right), \qquad (1.62)$$

where A, B, and C are given by

$$A = 1 + \frac{C_{H^+}^f}{K_1}\exp\left(-\frac{e\varphi_0}{k_b T}\right), \qquad (1.63)$$

$$B = 1 + \frac{C_{H^+}^f}{K_2}\exp\left(-\frac{e\varphi_0}{k_b T}\right), \qquad (1.64)$$

$$C = 1 + \frac{C_{Na^+}^f}{K_3}\exp\left(-\frac{e\varphi_\beta}{k_b T}\right), \qquad (1.65)$$

where e is the elementary charge (in C), T is the temperature (in degree K), k_b is the Boltzmann constant, Γ_i^0 is the surface site density of site i, and $\Gamma_1^0, \Gamma_2^0, \Gamma_3^0$ (in sites per nm^2) are the total surface site densities of the three type of sites introduced earlier (aluminol, silanol, and >Al–O–Si< groups, respectively). The parameters C_i^f where i $= Na^+, H^+$ are the ionic concentrations (in mol l^{-1}), and φ_0 and φ_β are the electrical potentials at the mineral surface (o-plane) and at the β-plane, respectively (Figure 1.1). The resulting mineral surface charge density, Q_0, and the surface charge density in the Stern layer, Q_β (in C m^{-2}), are found by summing the surface site densities of charged surface groups (see Leroy & Revil, 2004).

In the case of smectite and illite, the surface site densities are located mainly on the basal plane {001} (Tournassat et al., 2004). We use the TLM developed by Leroy et al. (2007) to determine the distribution of the counterions at the mineral/water interface of 2:1 clay minerals. In the pH range 6–8, the influence of the hydroxyl surface sites upon the distribution of the counterions at the mineral/water interface can be neglected because the charge density induced by edge sites is small relative to that due to permanent excess of negative charge associated with the isomorphic substitutions inside the crystalline network of the smectite (Tournassat et al., 2004). We therefore consider only these sites in the model denoted as the "X-sites" (see Figure 1.8). The adsorption of sodium is described by

$$>XNa \Leftrightarrow \;>X^- + Na^+, \quad K_4, \qquad (1.66)$$

$$\Gamma_{XNa}^0 = \Gamma_X^0\left(\frac{C_{Na^+}^f}{K_4}\right)\exp\left(-\frac{e\varphi_\beta}{k_b T}\right). \qquad (1.67)$$

The mineral surface charge density Q_0 (in C m^{-2}) of smectite associated with these sites is considered equal to the ratio between the CEC of smectite (1 meq g^{-1}) and its specific surface area (800 m^2 g^{-1}), which gives a value equal to 0.75 charge nm^{-2} (for illite, a similar analysis yields 1.25 charges nm^{-2}). These values allow the calculation of the surface site densities Γ_X^0 and Γ_{XNa}^0 knowing the expressions of the mineral surface charge density Q_0 (in C m^{-2}) as a function of the surface site densities (see Leroy et al., 2007).

There are three distinct microscopic electric potentials in the inner part of the electrical layer. We note φ_0 as the mean potential on the surface of the mineral (Figure 1.1). The potential φ_β is located at the β-plane, and φ_d is the potential at the OHP (Figure 1.1). These potentials are related to each other by a classical capacitance model (Hunter, 1981):

$$\varphi_0 - \varphi_\beta = \frac{Q_0}{C_1}, \qquad (1.68)$$

$$\varphi_\beta - \varphi_d = -\frac{Q_S}{C_2}, \qquad (1.69)$$

where C_1 and C_2 (in F m^{-2}) are the (constant) integral capacities of the inner and outer parts of the Stern layer, respectively (Table 1.2). The parameter Q_S represents the surface charge density in the diffuse layer. The global electroneutrality equation for the mineral/water interface is

$$Q_0 + Q_\beta + Q_S = 0. \qquad (1.70)$$

We calculate the potential φ_d by using Equations (1.68)–(1.70) and the procedure reported by Leroy and Revil (2004) and Leroy et al. (2007) (the surface charge densities are expressed as a function of the corresponding surface site densities). We use the values of the equilibrium constants K_i and of the capacities C_1 and C_2 reported in Table 1.2. The system of equations was solved inside two MATLAB routines, one for kaolinite and one for illite and smectite. The counterions are both located in the Stern and in the diffuse layer. For all clay minerals, the fraction of counterions located in the Stern layer is defined by Equations (1.50)–(1.52) like for the silica surface.

Table 1.2 Optimized double layer parameters for the three main types of clay minerals (at 25°C).

Parameters	Kaolinite	Illite	Smectite
K_1 (at 25°C)	10^{-10*}	—	—
K_2 (at 25°C)	$8 \times 10^{-6*}$	—	—
K_3 (at 25°C)	$5 \times 10^{-2*}$	—	—
K_4 (at 25°C)	—	$0.8^\dagger$	$0.8^\dagger$
C_1 (F m^{-2})	1.58^*	$1^\dagger$	$1^\dagger$
C_2 (F m^{-2})	0.2^*	$0.2^\dagger$	$0.2^\dagger$

*From Leroy and Revil (2004).
†From Leroy et al. (2007).

As shown by Leroy and Revil (2004) and Leroy et al. (2007), the previous set of equations can be solved numerically using the parameters given in Table 1.2 as input parameters. The parameters of Table 1.2 have been optimized from a number of experimental data, especially zeta potential resulting from electrokinetic measurements and surface conductivity data (see Leroy & Revil, 2004; Leroy et al., 2007), and remain unchanged in the present work. The output parameters of the numerical TLM are the surface site densities in the Stern and diffuse layers and therefore the partition coefficient f. Some TLM computations of the fractions of counterions in the Stern layer show that f is typically in the range 0.80–0.99, indicating that clay minerals have a much larger fraction of counterions in the Stern layer by comparison with glass beads at the same salinities. The values of the partition coefficient determined from the present model are also consistent with values determined by other methods, for instance, using radioactive tracers (Jougnot et al., 2009) and osmotic pressure (Gonçalvès et al., 2007; Jougnot et al., 2009). This shows that the present electrochemical model is consistent because it can explain a wide diversity of properties.

1.1.3 Implications

As discussed in Sections 1.1.1 and 1.1.2, all minerals of a porous material in contact with water are coated by the electrical double layer shown in Figure 1.1. The surface of the mineral is charged (due to isomorphic substitutions in the crystalline network or surface ionization of active sites such as hydroxyl >OH sites). The surface charge is balanced by charges located in the Stern layer and in the diffuse layer. There are three fundamental implications associated with the existence of this electrical double layer at the surface of silicates and clays:

1 Pore water is never neutral. There is an excess of charge in the pore water that can be written as

$$\bar{Q}_V = (1-f)\rho_g\left(\frac{1-\phi}{\phi}\right)\text{CEC}, \qquad (1.71)$$

where f denotes the fraction of counterions in the Stern layer (attached to the grains) and therefore $(1-f)$ denotes the fraction of charge contained in the diffuse layer, ρ_g denotes the mass density of the grains (kg m^{-3}), ϕ denotes the porosity, and CEC denotes the cation exchange capacity of the material

(in C kg^{-1}). We will see later that the flow of the pore water relative to the grain framework drags an effective charge density $\hat{Q}_V^0$. We expect that for permeable porous media, we have

$$\hat{Q}_V^0 \ll \bar{Q}_V. \qquad (1.72)$$

In sandstones, the diffuse layer is relatively thin with respect to the size of the pores and especially for the pores that controlled the flow of the pore water. In other words, most of the water is neutral with the exception of the pore water surrounding the surface of the grains. In addition, only a small fraction of the diffuse layer is carried along the pore water flow. We will see that the charge density $\hat{Q}_V^0$ should be understood as an effective charge density that is controlled by the flow properties (especially the permeability) and that has little to do with the CEC itself. It should be clear therefore that the CEC cannot be determined from the effective charge density $\hat{Q}_V^0$, which will be properly defined later.

2 There is an excess of electrical conductivity in the vicinity of the pore water–mineral interface responsible for the so-called surface conductivity (see Figure 1.1). This surface conductivity exists for any minerals in contact with water including clean sands. That said, the magnitude of surface conductivity is much stronger in the presence of clay minerals due to their very high surface area (surface area of the pore water–mineral interface for a given pore volume).

3 The double layer is responsible for the (nondielectric) low-frequency polarization of the porous material. This polarization is coming from the polarization of the electrical double layer in the presence of an electrical field applied to the porous material. In the case of seismoelectric effects, it implies a phase lag between the pore fluid pressure and the electric field, but this phase lag is expected to be small (typically <10 mrad except at very low salinities where the magnitude of the phase can reach 30 mrad) and most of the time is neglected (for instance, by Pride, 1994).

The first consequence is fundamental to understanding the nature of electrical currents associated with the flow of pore water relative to the mineral framework (termed streaming currents); therefore, the occurrence of electrokinetic (macroscopic) electrical fields is due to the flow of pore water relative to the mineral framework associated with seismic waves. The second consequence is crucial to the understanding of electrical conductivity in porous materials. Electrical conductivity of porous media has two contributions, one associated with conduction in the bulk pore water and one associated with the electrical double layer (surface conductivity). Contrary to what is erroneously assumed in a growing number of scientific papers in hydrogeophysics, the formation factor will not be defined as the ratio of the conductivity of the pore water by the conductivity of the porous material. We will show that surface conductivity is crucial in obtaining an intrinsic formation factor characterizing the topology of the pore space of porous materials. The third consequence is important to the understanding of induced polarization, which translates into the frequency dependence of the electrical conductivity. Because of this low-frequency polarization, the conductivity appears, generally speaking, as a second-order symmetric tensor with components that are frequency dependent and complex. The real (or inphase) components are associated with electromigration, while the imaginary (quadrature) components are associated with polarization (i.e., the reversible storage of electrical charges in the porous material). We will show in the following that the model of Pride (1994) does not account correctly for the frequency dependence of electrical conductivity and is incomplete in its description of the electrical double layer (no speciation and no description of the Stern layer).

1.2 The streaming current density

We evaluate in this section the first consequence associated with the existence of the electrical double layer coating the surface of the mineral grains in a porous material. We have established in Section 1.1.1 that there is an excess charge density in pore water. We have defined the macroscopic charge density (charge per unit pore volume, in $C\,m^{-3}$) that is dragged by the flow of pore water as $\hat{Q}_V^0$ (the reason for the superscript 0 will be explored in Chapter 3). We showed that pore water, in proximity to the mineral grain surface, is characterized by a local charge density $\rho(\mathbf{x})$ (in $C\,m^{-3}$) at position $\mathbf{x}$ due to the presence of the electrical diffuse layer ($\rho(\mathbf{x}) = 0$ in the bulk pore water that is electroneutral). We note $\mathbf{v}_m(\mathbf{x})$ as the local instantaneous velocity of the pore water relative to the solid (in $m\,s^{-1}$). The macroscopic charge density $\hat{Q}_V^0$ is defined by

$$\hat{Q}_V^0 \langle \mathbf{v}_m(\mathbf{x}) \rangle = \langle \rho(\mathbf{x}) \mathbf{v}_m(\mathbf{x}) \rangle \qquad (1.73)$$

where the brackets denote a pore volume averaging,

$$\langle \cdot \rangle = \frac{1}{V_p} \int_{V_p} (\cdot) d\tau, \qquad (1.74)$$

and $d\tau$ denotes an elementary volume around point $M(\mathbf{x})$, and $\langle \mathbf{v}_m(\mathbf{x}) \rangle$ denotes the mean velocity averaged over the pore space. Equation (1.73) is valid whatever the size of the diffuse layer with respect to the size of the pores. In the case of a thin double layer (the thickness of the diffuse layer is much smaller than the thickness of the pores), the charge density $\hat{Q}_V^0$ is substantially smaller than the (total) charge density associated with the diffuse layer $\bar{Q}_V$, which explains why $\hat{Q}_V^0$ cannot be used to estimate the CEC of the minerals. In other words, there is no direct relationship between the effective charge density $\hat{Q}_V^0$ and the CEC of the minerals.

As shown in Figure 1.10, the drag of the excess of charge of the pore water (more precisely the drag of a fraction of the diffuse layer) is responsible for a macroscopic streaming source current density $\mathbf{J}_S$ at the scale of a representative elementary volume of the porous material. This macroscopic source density is related to the microscopic (pore scale) current density $\mathbf{j}_S$ associated with the local advective transfer of electrical charges by

$$\mathbf{J}_S = \phi \langle \mathbf{j}_S \rangle, \qquad (1.75)$$

$$\mathbf{J}_S = \phi \langle \rho \mathbf{v}_m \rangle, \qquad (1.76)$$

$$\mathbf{J}_S = \hat{Q}_V^0 \phi \langle \mathbf{v}_m \rangle, \qquad (1.77)$$

$$\mathbf{J}_S = \hat{Q}_V^0 \dot{\mathbf{w}} \qquad (1.78)$$

where $\dot{\mathbf{w}} = \phi \langle \mathbf{v}_m \rangle$ (Dupuit–Forchheimer equation) denotes the macroscopic Darcy velocity (in $m\,s^{-1}$; see Darcy, 1856). This Darcy velocity is not a true pore water velocity. It represents the flux of water through a cross section of the porous material (volume of water passing per surface area and per surface time across a cross section of the porous material). In Figure 1.10, we show the effect of the charge distribution and the flow regime on the source current density. At low frequencies, the flow is dominated by viscous effects, and the regime is called the viscous laminar flow regime. At high frequencies, the flow is controlled by the inertial term of the Navier–Stokes equation, and the flow regime is called

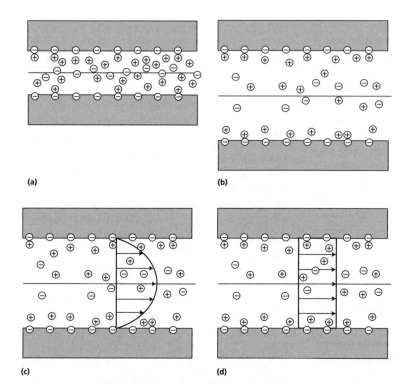

(a)

(b)

(c)

(d)

Figure 1.10 Sketch of the charge distribution and flow regime in a single pore. The grey areas correspond to the solid phase. There are four end members to consider depending on the pore size with respect to the double layer thickness and depending on the frequency. **a)** Thick double layer (the counterions of the diffuse layer are uniformly distributed in the pore space). **b)** Thin double layer (the thickness of the diffuse layer is much smaller than the size of the pores). **c)** Viscous laminar flow regime occurring at low frequencies. **d)** Inertial laminar flow regime occurring at high frequencies (Modified from Revil & Mahardika, 2013).

the inertial flow regime. This regime occurs for Reynolds numbers higher than 1 but smaller than the critical Reynolds number corresponding to turbulent flow (typically 200–300). For a broad range of porous media, the effective charge density, $\hat{Q}_V^0$, can be related directly to the permeability, k_0, as shown in Figure 1.11. This relationship is very useful to compute or invert the seismoelectric data since it offers a key relationship between the parameter that is controlling the seismoelectric coupling (as shown later) and the key hydraulic parameter of porous media, namely, the permeability.

The macroscopic source current density can be expressed directly as a function of the pore pressure gradient using Darcy's law. This law can be seen as a constitutive equation for the flow of the pore water at the scale of a representative elementary volume or can be seen as a macroscopic momentum conservation equation for the pore fluid. It is given by (Darcy, 1856)

$$\dot{\mathbf{w}} = -\frac{k_0}{\eta_f}\nabla p, \qquad (1.79)$$

where η_f denotes the dynamic viscosity of the pore water (in Pa s), k_0 (in m^2) denotes the low-frequency

permeability of the porous material, and p denotes the pore fluid (mechanical) pressure. Therefore, the streaming current density can be given by

$$\mathbf{J}_S = -\frac{\hat{Q}_V^0 k_0}{\eta_f}\nabla p. \qquad (1.80)$$

A popular alternative that can be derived by volume averaging the Nernst–Planck equation is the macroscopic Helmholtz–Smoluchowski equation. For the streaming current density, it takes the following form (e.g., Pride, 1994):

$$\mathbf{J}_S = \frac{\varepsilon_f \zeta}{\eta_f F}\nabla p, \qquad (1.81)$$

where F is called the electrical formation factor (dimensionless) and corresponds to a parameter that is properly defined in the modeling of the electrical conductivity of porous media (see Section 1.3). Equation (1.81) assumes a thin electrical double layer with respect to the size of the pores, while Equation (1.80) does not require such an assumption. A comparison between the two equations shows that the salinity dependence of $\hat{Q}_V^0$ should be the

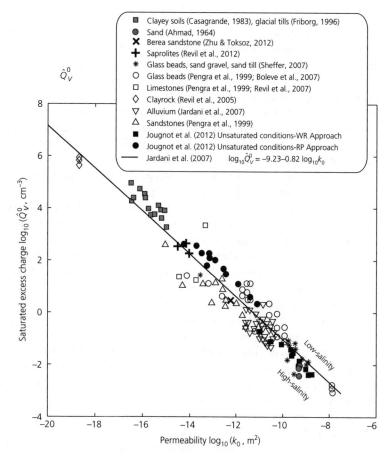

Figure 1.11 Quasistatic charge density $\hat{Q}_V^0$ (excess pore charge moveable by the quasistatic pore water flow) versus the quasistatic permeability k_0 for a broad collection of core samples and porous materials. This charge density is derived directly from laboratory measurements of the streaming potential coupling coefficient. Data from Ahmad (1969), Bolève et al. (2007), Casagrande (1983), Friborg (1996), Jougnot et al. (2012), Jardani et al. (2007), Pendra et al. (1999), Revil et al. (2005, 2007), Sheffer (2007), Revil et al. (2012), and Zhu and Toksöz (2013). The effective charge density $\hat{Q}_V^0$ cannot be used to predict the cation exchange capacity of the porous material. We also show the smaller effect of salinity.

same as the salinity dependence of the zeta potential, ζ. The polarity of $\hat{Q}_V^0$ is opposite to the polarity of ζ, and any change affecting the zeta potential would modify the effective charge density $\hat{Q}_V^0$ in the same way. A comparison between Equations (1.81) and (1.80) implies that at first approximation we have the following equivalence between the parameters: $\hat{Q}_V^0 k_0 \Leftrightarrow \varepsilon_f \zeta / F$.

1.3 The complex conductivity

In this section, we examine the second and third consequences associated with the electrical double layer, namely, the existence of surface conductivity and the existence of low-frequency polarization associated with the quadrature electrical conductivity. At low frequencies (below few kHz), porous media and colloids are not only conductive, but they store, reversibly, electrical

charges (Marshall & Madden, 1959; Titov et al., 2004; Leroy et al., 2008; Grosse, 2009). The total current density $\mathbf{J}$ can be decomposed into a contribution associated with the electromigration of the charge carriers plus a contribution associated with the "true" polarization of the material:

$$\mathbf{J} = \sum_{i=1}^{N} q_i \mathbf{J}_i + \frac{\partial \mathbf{D}}{\partial t}, \tag{1.82}$$

where $\mathbf{J}_i$ denotes the flux density of species i (the number of species passing per unit surface area and per unit time) and $\mathbf{D}$ is the displacement field associated with dielectric polarization of the porous material. In nonequilibrium thermodynamics, the flux densities $\mathbf{J}_i$ are coupled to other transport mechanisms in the porous media. These ionic fluxes are directly controlled by the gradient of the electrochemical potentials, introduced in Section 1.1, and the flow of pore water. These factors generate the source current density, $\mathbf{J}_S$, of electrokinetic origin. These couplings were first investigated by Marshall and Madden (1959) and imply the existence of low-frequency polarization mechanisms in the porous material. It is not our goal to develop a complete theory of polarization in this book, but rather to provide a practical view of the problem that can be used to analyze seismoelectric effects.

One of the most effective mechanisms of polarization is the coupling of the flux densities with the electrochemical potential gradients as discussed by Marshall and Madden (1959). The polarization implies a phase lag between the current and the electrical field and defines the frequency dependence of the conductivity of the material. Despite the fact that the seismoelectric theory contains an electroosmotic polarization effect (which is the one used by Pride (1994)), it has been known since Marshall and Madden (1959) that this mechanism cannot explain the low-frequency dependence of the conductivity of the material. While this assumption is clearly stated in Pride (1994), it seems to have been lost in translation in all the following works. In those works, the model of Pride is used to explain the low-frequency polarization of porous rocks, and as such, those authors have considered the mathematical expression of Pride (1994) as valid to describe the complex conductivity of porous materials. This is unfortunately not correct since the model of Pride does not account for low-frequency polarization mechanisms known to control the quadrature conductivity.

Continuing from the preceding text, the total current density entering, for instance, Ampère's law is

$$\mathbf{J} = \sigma^* \mathbf{E} + \mathbf{J}_S + \frac{\partial \mathbf{D}}{\partial t}, \tag{1.83}$$

where the first term on the right side of Equation (1.83) corresponds to a frequency-dependent electrical conductivity, σ^*, characterized by a real (inphase) component σ' and a quadrature (out-of-phase) component σ'':

$$\sigma^* = \sigma' + i\sigma'', \tag{1.84}$$

where i denotes the pure imaginary number $\left(i = \sqrt{-1}\right)$. The second term of Equation (1.83) corresponds to the source current density of electrokinetic origin, and the third term corresponds to the displacement current density. Note that in clayey materials, while σ' and σ'' both depend on frequency, this dependence is weak as shown

and discussed in detail by Vinegar and Waxman (1984) and more recently by Revil (2012, 2013a, b). This dependency will be therefore neglected in the following. The quadrature conductivity of clean sands and sandstones shows a clear frequency peak, but the magnitude of the quadrature conductivity is usually low. The only case of a strong and highly frequency-dependent induced polarization effect is the case of disseminated ores (e.g., sulfides like pyrite and oxides like magnetite). In this case, there is the possibility (still unexplored) to use the seismoelectric method to detect and image ore bodies.

1.3.1 Effective conductivity

The displacement field is related to the electrical field by $\mathbf{D} = \varepsilon \mathbf{E}$ where ε denotes the permittivity or dielectric constant (in F m^{-1}) of the material. We consider a harmonic external electrical field:

$$\mathbf{E} = \mathbf{E}_0 \exp(-i\omega t), \tag{1.85}$$

where f is the frequency in Hz, $\omega = 2\pi f$ denotes the angular frequency (pulsation in rad s^{-1}), and $\mathbf{E}_0$ represents the amplitude of the alternating electrical field. Equation (1.83) can be written as

$$\mathbf{J} = (\sigma^* - i\omega\varepsilon)\mathbf{E} + \mathbf{J}_S. \tag{1.86}$$

This total current density can be written as an apparent Ohm's law:

$$\mathbf{J} = \sigma_{\text{eff}}{}^* \mathbf{E} + \mathbf{J}_S, \tag{1.87}$$

where $\sigma_{\text{eff}}{}^* = \sigma_{\text{eff}} - i\omega\varepsilon_{\text{eff}}$ is the effective complex conductivity and σ_{eff} and ε_{eff} are real positive frequency-dependent scalars (at least in isotropic media) defined by

$$\sigma_{\text{eff}} = \sigma', \tag{1.88}$$

$$\varepsilon_{\text{eff}} = \varepsilon' - \sigma''/\omega. \tag{1.89}$$

Equations (1.88) and (1.89) are a direct consequence of Ampère's law in which the conductivity is considered complex (ion drift is coupled to diffusion), the permittivity is real, and the Maxwell–Wagner polarization and the polarization of the water molecules at few GHz are neglected. The effective properties measured in the laboratory or in the field contain both dielectric and conduction components. It is clear from Equation (1.89) that the effective permittivity is expected to be very strong at low frequencies

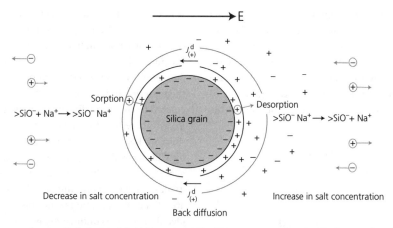

Figure 1.12 The presence of an applied electrical field **E** creates a dipole moment associated with the transfer of the counterions in both the Stern and the diffuse layers around a silica grain. This dipole moment points in the direction that is opposite to the applied field. The charge attached to the mineral framework remains fixed. The movement of the counterions in the Stern layer is mainly tangential along the surface of the grain. However, sorption and desorption of the counterions are in principle possible. Back diffusion of the counterions can occur both in the Stern and diffuse layers, and diffusion of the salt occurs in the pore space. In both cases, the diffusion of the counterions occurs over a distance that is equal to the diameter of the grain.

due to the quadrature conductivity-related term σ''/ω. A discussion of the frequency dependence of σ'' upon the effective permittivity can be found in Revil (2013a, b).

The polarization of the electrical double layer (called the α-polarization in electrochemistry) plays a dominant role at low frequencies through the apparent permittivity of the material (see Figure 1.12). This is in contrast with ideas expressed in the geophysical literature since Poley et al. (1978). In the prior geophysical literature, low-frequency polarization is envisioned to be dominated by the Maxwell–Wagner polarization (also called "space charge" or "interfacial" polarization) due to the discontinuity of the displacement current at the interfaces between the different phases of a porous composite.

1.3.2 Saturated clayey media

Assuming that clayey materials exhibit a fractal or self-affine behavior through a broad range of scales (e.g., Hunt et al., 2012), the inphase and quadrature conductivities are expected to be weakly dependent on frequency as discussed in detail by Vinegar and Waxman (1984) and Revil (2012). This has been shown for a range of frequencies typically used in laboratory measurements (0.1 mHz to 0.1 MHz). Revil (2013a) recently developed a model to describe the complex conductivity of clayey materials using a volume-averaging approach. According to this model, the inphase conductivity σ' (S m^{-1}) is given

as a function of the pore water conductivity σ_w (in S m^{-1}) by the following expression:

$$\sigma' = \frac{1}{F}\left\{ \sigma_w + \left(\frac{F-1}{F\phi}\right)\rho_S\left[\beta_{(+)}(1-f) + \beta_{(+)}^S f\right]\text{CEC} \right\}, \quad (1.90)$$

where F is the so-called formation factor, ϕ is the porosity, f denotes the fraction of counterions in the Stern layer, ρ_S denotes the mass density of the solid phase (typically 2650 kg m^{-3}), $\beta_{(+)}$ corresponds to the mobility of the counterions in the diffuse layer, and $\beta_{(+)}^S$ denotes the mobility of the counterions in the Stern layer (both in m^2 s^{-1} V^{-1}). The partition coefficient, f, is salinity dependent as discussed in Sections 1.1.1 and 1.1.2. For clay minerals (and for silica as well), the mobility of the counterions in the diffuse layer is equal to the mobility of the same counterions in the bulk pore water (e.g., $\beta_{(+)}(\text{Na}^+, 25°\text{C}) = 5.2 \times 10^{-8}$ m^2 s^{-1} V^{-1}). The mobility of the counterions in the Stern layer is substantially smaller and equal to $\beta_{(+)}^S(25°\text{C}, \text{Na}^+) = 1.5 \times 10^{-10}$ m^2 s^{-1} V^{-1} for clay minerals (Revil, 2012, 2013a, b), therefore about 350 times less mobile than in bulk solution. We can rewrite the inphase conductivity equation as

$$\sigma' \approx \frac{1}{F}\sigma_w + \left(\frac{1}{F\phi}\right)\rho_S\left[\beta_{(+)}(1-f) + \beta_{(+)}^S f\right]\text{CEC} \quad (1.91)$$

The surface conductivity corresponds to the last term of Equation (1.91). Equation (1.91) represents the full saturation version of a more general model. This model implies that the surface conductivity is controlled either by the grain diameter (or from the grain diameter probability distribution as discussed by Revil & Florsch, 2010; see Figure 1.13a) or by the CEC (Figure 1.13b). Surface conductivity could be also expressed as a function of the specific surface area. Indeed, the CEC and the specific

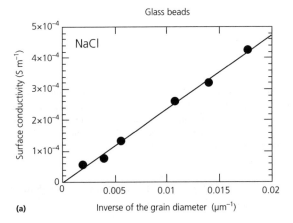

Glass beads

(a)

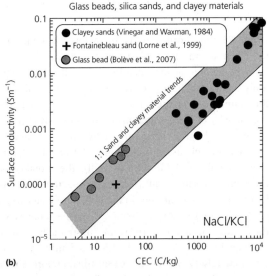

(b)

Figure 1.13 Surface conductivity. **a)** For glass beads and silica sands, the surface conductivity is controlled by the size of the grains (Data from Bolève et al., 2007). **b)** All the data for glass beads, silica sands, and shaly sands are on the same trend when plotted as a function of the (total) CEC. This is consistent with a surface conductivity model dominated by the contribution of the diffuse layer (Data from Vinegar & Waxman, 1984 (shaly sands, NaCl); Bolève et al., 2007 (glass beads, NaCl); and Lorne et al., 1999a, b (Fontainebleau sand KCl)).

surface area are related to each other by Equation (1.34): $Q_0^S = CEC/S_{sp}$ where Q_0^S, the surface charge density of the counterions, is about 0.32 C m^{-2} for clay minerals. For silica grains, there is a relationship between the mean grain diameter and the surface area or the equivalent CEC of the material. Indeed, the specific surface area S_{sp} was calculated from the median grain diameter, d, using $S_{sp} = 6/(\rho_s d)$ where $\rho_s = 2650$ kg m^{-3} denotes the density of the silica grains. This also yields an equivalent CEC given by CEC $= 6$ CEC $= 6Q_0^S/(\rho_s d)$ with $Q_0^S = 0.64$ C m^{-2}, and $\rho_s = 2650$ kg m^{-3}. In Figure 1.13b, the surface conductivity data of silica sands and glass beads and clayey media are all along a unique trend. This is consistent with the idea that surface conductivity is dominated by the diffuse layer. Indeed, the mobility of the counterions in the Stern layer is much smaller than the mobility of the counterions in the bulk pore water (see discussion in Revil, 2012, 2013a, b).

The quadrature conductivity expression obtained by Revil (2013a) is

$$\sigma'' \approx -\left(\frac{1}{F\phi}\right)\rho_s \beta_{(+)}^S f\, CEC. \qquad (1.92)$$

For the reasons explained previously for the surface conductivity, the quadrature conductivity can be expressed as a function of the specific surface area or as a function of the CEC. At saturation, a comparison between the equation for the quadrature conductivity and experimental data is shown in Figure 1.14 where we used the relationship between the CEC and the specific surface area given by Equation (1.53).

For clayey sands, taking $\beta_{(+)}^S(Na^+) = 1.5 \times 10^{-10}$ m^{2}s^{-1}V^{-1} at 25°C, $f = 0.90$, $Q_0^S = 0.32$ C m^{-2}, and $\rho_s = 2650$ kg m^{-3}, yields $\sigma'' \approx -cS_{sp}$ with $c = 7.6 \times 10^{-8}$ S kg m^{-3}. For the clean sands and sandstones, using $\beta_{(+)}(Na^+) = 5.2 \times 10^{-8}$ m^2 s^{-1} V^{-1}, $f = 0.50$, $Q_0^S = 0.64$ C m^{-2}, and $\rho_s = 2650$ kg m^{-3}, yields $c = 2.9 \times 10^{-5}$ S kg m^{-3}. The difference between the trend for the clean sands and sandstones and the trend for the clayey materials illustrates how different are the mobilities of the counterions in the Stern layer of silica and clays. In Figure 1.15, we plot directly the quadrature conductivity as a function of the CEC for shaly sands. In Figure 1.16, we plot the quadrature conductivity of sands as a function of the mean grain diameter. Using the transform given previously between the mean grain diameter and the CEC, we plot in Figure 1.17 the quadrature conductivity as a function

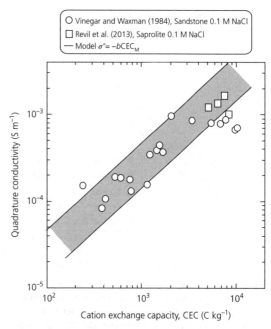

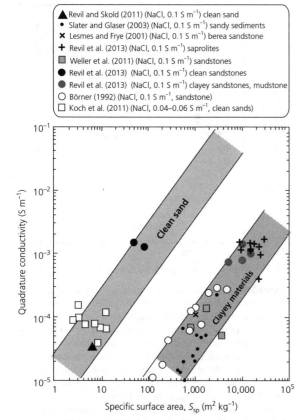

Figure 1.14 Influence of the specific surface area S_{Sp} upon the quadrature conductivity, which characterizes charge accumulation (polarization) at low frequencies. The trend determined for the clean sands and the clayey materials are from the model developed by Revil (2012) at 0.1 S m^{-1} NaCl. The measurements are reported at 10 Hz. Data from Revil and Skold (2011), Koch et al. (2011), Slater and Glaser (2003), Lesmes and Frye (2001), Revil et al. (2013), and Börner (1992).

Figure 1.15 Influence of the cation exchange capacity (CEC) upon the quadrature conductivity of clayey materials. The trend is determined for the clayey materials from the model developed by Revil (2012, 2013) at 0.1 mol l^{-1} NaCl (about 1 S m^{-1}). The measurements are from Vinegar and Waxman (1984) (shaly sands) and Revil et al. (2013) (saprolites). Note that the slope of this trend is salinity dependent.

of the CEC. The data are corrected for the dependence of the partition coefficient f with the salinity using the approach developed by Revil and Skold (2011). These data exhibit two distinct trends indicating that the mobility of the counterions in the Stern layer of silica is equal to the mobility of the same ions in the bulk pore water, while the mobility of the counterions at the surface of clays is much smaller than in the bulk pore water. For clayey materials, it is also clear that the surface conductivity can be directly related to the quadrature conductivity as discussed by Revil (2013a, b).

The following dimensionless number can be defined as $R \equiv -\sigma''/\sigma_S \geq 0$, which corresponds therefore to the ratio

of quadrature, σ'', to surface conductivity, σ_s. With this definition, the complex conductivity of a partially saturated porous siliciclastic sediment can be written as

$$\sigma^* = \frac{1}{F}\sigma_w[1 + \text{Du}(1 - i\text{R})], \qquad (1.93)$$

$$\text{Du} = \frac{F\sigma_S}{\sigma_w}. \qquad (1.94)$$

As briefly discussed by Revil and Skold (2011) and Revil (2012, 2013a), the ratio R can be related to the partition coefficient f. In the present case, we obtain

$$R = \frac{\beta_{(+)}^S f}{\left[\beta_{(+)}(1-f) + \beta_{(+)}^S f\right]}, \qquad (1.95)$$

$$R(\text{clay}) \approx \frac{\beta_{(+)}^S f}{\beta_{(+)}(1-f)}, \qquad (1.96)$$

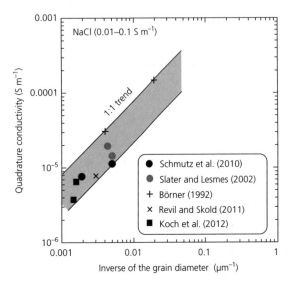

Figure 1.16 Influence of the mean grain diameter upon the quadrature conductivity of sands. Pore water conductivity in the range 0.01–0.1 S m⁻¹ NaCl. The measurements are from Schmutz et al. (2010), Slater and Lesmes (2002), Börner (1992), Revil and Skold (2011), and Koch et al. (2012). The quadrature conductivities in this figure are reported at the relaxation peak.

$$R(\text{sand}) \approx f. \qquad (1.97)$$

We can analyze the value of R for sands and clays. For sands, taking $\beta_{(+)}^S(\text{Na}^+, 25°C) = \beta_{(+)}(\text{Na}^+, 25°C) = 5.2 \times 10^{-8}\,\text{m}^2\,\text{s}^{-1}\,\text{V}^{-1}$, $f = 0.50$ (f depends actually on pH and salinity; see Figure 1.6), we have $R \approx 0.50$. In the case of clay minerals, taking $\beta_{(+)}^S(\text{Na}^+, 25°C) = 1.5 \times 10^{-10}\,\text{m}^2\,\text{s}^{-1}\,\text{V}^{-1}$ and $\beta_{(+)}(\text{Na}^+, 25°C) = 5.2 \times 10^{-8}\,\text{m}^2\,\text{s}^{-1}\,\text{V}^{-1}$, $f = 0.90$, yields $R = 0.0260$. In both cases, the results are consistent with the experimental results displayed in Figures 1.13 and 1.17.

1.4 Principles of the seismoelectric method

Now that the electrical double layer has been described and the direct consequences of the existence of this electrical double layer discussed, we need to introduce the key concepts behind the seismoelectric method.

1.4.1 Main ideas

The electroseismic (electric to seismic) and seismoelectric (seismic to electric) phenomena correspond to two

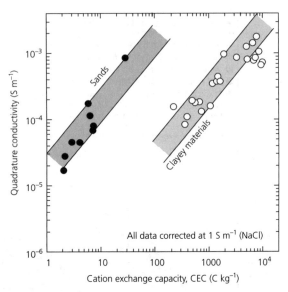

Figure 1.17 Trends for the quadrature conductivity versus CEC for sands and clayey materials. All the experimental data are corrected for salinity to bring them to a pore water conductivity of 1 S m⁻¹ (NaCl) using the salinity dependence of f the fraction of counterions in the Stern layer. The two different trends between the silica sands and the clayey materials are an indication that the mobility of the counterions in the Stern layer is much smaller for clay minerals than for silica. This plot shows how difficult it is to extract the petrophysical properties of formations from the quadrature conductivity alone. Indeed, formations with very different permeabilities and lithologies can have the same quadrature conductivity.

symmetric couplings existing between EM and seismic disturbances in a porous material (Frenkel, 1944; Pride, 1994). The electroseismic effects correspond to the generation of seismic waves when a porous material is submitted to a harmonic electrical field or electrical current. The seismoelectric effects correspond to the generation of electrical (possibly EM) disturbances when a porous material is submitted to the passage of seismic waves. The electroseismic and seismoelectric couplings are both controlled by the relative displacement between the charged solid phase (with the Stern layer attached to it) and the pore water (with its diffuse layer and consequently an excess of electrical charges per unit pore volume).

Figure 1.18 sketches the general idea underlying the seismoelectric theory. We consider porous media in which seismic waves propagate. The description of the propagation of the seismic waves depends on the

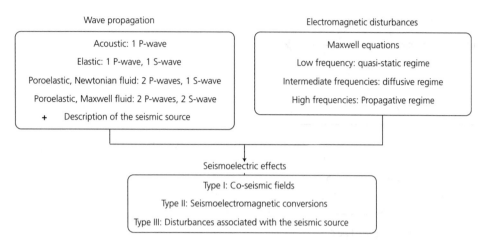

Wave propagation

Acoustic: 1 P-wave

Elastic: 1 P-wave, 1 S-wave

Poroelastic, Newtonian fluid: 2 P-waves, 1 S-wave

Poroelastic, Maxwell fluid: 2 P-waves, 2 S-wave

+   Description of the seismic source

Electromagnetic disturbances

Maxwell equations

Low frequency: quasi-static regime

Intermediate frequencies: diffusive regime

High frequencies: Propagative regime

Seismoelectric effects

Type I: Co-seismic fields

Type II: Seismoelectromagnetic conversions

Type III: Disturbances associated with the seismic source

Figure 1.18 General concept of seismoelectric disturbances. Seismoelectricity combined the propagation of seismic (compressional and rotational) waves in porous materials. These porous materials are considered to be composites with phases carrying a net charge density (the material as a whole is neutral). The mechanical equations are coupled to the Maxwell equations through a source current density and a source of momentum. Three types of seismoelectric disturbances can be observed: Type I corresponds to the electromagnetic fields traveling with the seismic wave itself (coseismic fields). Type II corresponds to the electromagnetic disturbances associated with the passage of a seismic wave through a macroscopic heterogeneity (seismoelectric conversion). Type III corresponds to the electromagnetic fields associated with a seismic source. S corresponds to shear wave, while P corresponds to compressional (pressure) wave (the letters P and S are also used in terms of arrival time: primary and secondary waves).

rheology of the material and its dispersive properties. In this book, we will consider various rheological behaviors leading to different types and numbers of waves. For instance, the acoustic theory applied to porous media can only be used to describe, in an approximate way, the propagation of compressional (*P* for pressure) waves in porous media. A refinement of this theory is to consider the elastic case. The elastic case implies that two types of seismic waves are generated: pressure (P-)waves and shear (S-)waves. That said, porous media are composite of mineral and fluids, so a more complicated theory exists to describe more accurately seismic wave propagation in porous media. This corresponds to the theory of poroelasticity.

A macroscopic linear poroelastic theory of wave propagation has been first proposed by Biot (1962a, b). Microscopic theories of poroelasticity have been also proposed by various authors, but they will not be reviewed in this book. The poroelastic theory proposed by Biot leads to two P-waves (a slow P-wave and a fast one) and one S-wave. Revil and Jardani (2010) generalized the Biot–Frenkel theory to the case where the fluid can sustain shear stresses. This theory will be developed in Chapter 2. It yields to two P-wave and two S-wave propagation models and provides a symmetric theory, while

the biotheory is asymmetric in its equations as the fluid does not sustain shear stresses. Whatever the theory used, the propagation of the seismic waves through a porous material is responsible for a movement of the water with respect to the solid phase. In the presence of an electrical double layer, this relative displacement is the source for an electrical current density. This current density acts a source term in the Maxwell equations generating EM disturbances. These EM disturbances are described by the Maxwell equations, which takes the general form of the coupled telegraph equations for the electrical and magnetic fields (these partial differential equations contain propagative and diffusive terms). If we consider low-frequency seismic waves, the time-dependent terms can be dropped from the telegraphist's equations, and we end up with Poisson equations for the electrical and magnetic fields. Three types of EM disturbances can be generated: Type I corresponds to the EM fields traveling with the seismic waves themselves. These are generally called the coseismic fields and can only be observed in the volume directly affected by the propagation of the seismic waves. Type II corresponds to EM effects associated with the passage of seismic waves through a macroscopic heterogeneity (seismoelectric conversion). They can be remotely observed, away from

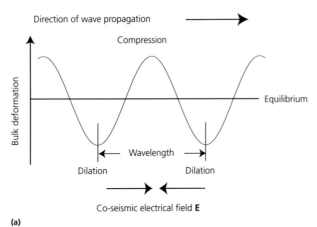

(a)

(b)   **(c)**

Figure 1.19 The coseismic electrical field and the coseismic (streaming) electrical current. **a)** The propagation of a compressional (pressure or P-)wave through a porous material generates areas of compression and dilation (expansion). **b)** In response to the change in the mechanical stresses, the pore water flows from the compressed regions to the dilated regions. **c)** The flow generates a streaming current density that is locally counterbalanced by the conduction current density creating, locally, an electrical field **E** of electrokinetic nature.

the interface. Finally, Type III corresponds to the EM fields associated with the seismic source itself.

We first describe Type I and II seismoelectric effects. When seismic waves propagate in a linear poroelastic porous material, two types of electrical disturbances are observed (Figures 1.19 and 1.20). The propagation of compressional P-waves generates an electrical source current associated with the displacement of the electrical diffuse layer in a Lagrangian framework attached to the solid phase (Figure 1.19). Inside a wavelength of a compressional wave, there are areas of dilation and compression (Figure 1.19a). These dilations and compressions of the solid skeleton are responsible for the flow of the pore water from the compressed regions to the regions where expansion occurs (Figure 1.19b). As the result of the flow of the pore water, there is the advective drag of a fraction of the excess of charge of the pore water. This advective drag is responsible for the streaming current density. Shear (S-)waves do not create coseismic electrical field.

These coseismic electrical signals travel at the same speed as the seismic waves (Pride, 1994). The amplitudes of the coseismic EM signals are controlled by the properties of the porous material (the formation factor) and by the properties of the pore fluid/solid interface (the zeta potential in the theory of Pride (1994), the excess charge per unit pore volume in the formulation used in this book).

In addition to the coseismic signals, another phenomenon occurs when a seismic wave moves through a sharp interface characterized by a change in the textural properties or a change in salinity or clay content or mineralogy (Figure 1.20). In this situation, a fraction of the mechanical energy is converted into EM energy, and a dipolar EM excitation is produced at the interface. The resulting EM disturbances diffuse very quickly away through the interface and can be recorded, nearly instantaneously, by electrodes or antennas located at the ground surface, in boreholes, or at the seafloor, for instance. In the seismoelectric method, we are mostly

Seismoelectric conversion

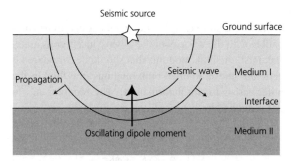

Figure 1.20 The seismoelectric conversion (called sometimes the interface response) results from the generation of an unbalanced source current density at an interface during the passage of a seismic wave. The divergence of the source current density at the interface is mathematically similar to an oscillating dipole moment generated at the interface in the first Fresnel zone. The star represents the seismic source.

interested in the information content associated with these conversions. However, the magnitude of these conversions decreases very quickly with the distance from the interface. In Chapters 2 and 3, we will provide a general theory of the coseismic field and seismoelectric conversion in saturated and partially saturated cases, respectively, and we provide in Section 1.4.2 a simple modeling approach based on the acoustic approximation.

A third effect corresponds to the EM fields generated directly by a seismic source. We will see that these EM fields (especially the electric component) can be combined with the radiated seismic fields and used to localize the seismic source and characterized the moment tensor of the source. These effects will be fully explored in Chapter 5.

1.4.2 Simple modeling with the acoustic approximation

1.4.2.1 The acoustic approximation in a fluid

We consider a fluid with its viscous effects considered to be negligible in the momentum conservation equation. The propagation of a seismic wave in this fluid can be described in terms of a pressure perturbation $p(\mathbf{r}, t)$ (true pressure minus the equilibrium pressure) or in terms of a fluid displacement $\mathbf{u}(\mathbf{r}, t)$. The volume strain θ is related to the displacement by

$$\theta = \frac{\Delta V}{V} = \nabla \cdot \mathbf{u}, \tag{1.98}$$

where $\Delta V/V$ represents a relative variation of volume during the passage of the seismic wave. The compressibility of the fluid is defined, in isothermal conditions, as

$$\beta_f = \frac{1}{K_f} = -\frac{1}{V}\left(\frac{\partial V}{\partial p}\right)_T, \tag{1.99}$$

and K_f denotes the bulk modulus (in Pa). Therefore, the pressure perturbation is related to the displacement by

$$p = -K_f\theta = -K_f\nabla \cdot \mathbf{u}. \tag{1.100}$$

Equation (1.100) corresponds to Hooke's law for a fluid and is valid for small deformation ($\theta \ll 1$). Equation (1.100) corresponds to a constitutive equation. To get the field equation for the pressure perturbation p, we need to combine Equation (1.100) with a conservation equation. Newton's law provides the required conservation equation for the momentum

$$-\nabla p = \rho_f \ddot{\mathbf{u}}, \tag{1.101}$$

where ρ_f denotes the density of the fluid (assumed to be constant) and $\ddot{\mathbf{u}}$ corresponds to the acceleration of the material. As we need to express the divergence of the fluid displacement in terms of fluid pressure, we want to take the divergence of Equation (1.101). This yields

$$-\nabla^2 p = \rho_f \frac{\partial^2}{\partial t^2}(\nabla \cdot \mathbf{u}), \tag{1.102}$$

$$\nabla^2 p = \frac{\rho_f}{K_f}\frac{\partial^2 p}{\partial t^2}, \tag{1.103}$$

$$\nabla^2 p - \frac{1}{c_f^2}\frac{\partial^2 p}{\partial t^2} = 0 \tag{1.104}$$

where

$$c_f = \left(\frac{K_f}{\rho_f}\right)^{1/2}. \tag{1.105}$$

Equation (1.104) corresponds to the wave equation for the fluid pressure with the velocity given by Equation (1.105). If we wish to determine the displacement of the fluid, we can, for instance, consider a wave with a sinusoidal time dependence given by

$$p(\mathbf{r}, t) = p_\omega(\mathbf{r})exp(-i\omega t), \tag{1.106}$$

where i denotes the pure imaginary number and ω is the angular frequency. Inserting Equation (1.106) into Equation (1.104) shows that the amplitudes obey the scalar Helmholtz equation:

$$\nabla^2 p_\omega(\mathbf{r}) + \frac{\omega^2}{c_f^2} p_\omega(\mathbf{r}) = 0. \qquad (1.107)$$

If we consider Newton's law, Equation (1.101), we obtain the following equation for the displacement of the fluid:

$$\mathbf{u}(\mathbf{r}, t) = -\frac{1}{\rho_f \omega^2} \nabla p_\omega(\mathbf{r}) exp(-i\omega t). \qquad (1.108)$$

Because the curl of a gradient is always equal to zero, we have the property

$$\nabla \times \mathbf{u}(\mathbf{r}, t) = 0. \qquad (1.109)$$

The displacement is irrotational, which means that the pressure wave is purely longitudinal and corresponds to a P-wave with a seismic velocity given by Equation (1.105).

1.4.2.2 Extension to porous media

Usually, the seismoelectric problem is formulated in terms of a coupling between the Maxwell equations and the Biot–Frenkel theory (e.g., Pride, 1994; Revil & Jardani, 2010), and the Biot–Frenkel theory will be discussed in Chapter 2. In the present section, we adapt the acoustic wave developed in the previous section to a porous body, and we simplify the seismoelectric theory to make it compatible with this acoustic approximation. We need an acoustic approximation solving now for the macroscopic pressure perturbation P of a porous material containing a fluid that cannot support shear stress (for instance, water). This formulation is obtained by adapting Equation (1.104) to a porous material:

$$\frac{\partial^2 P}{\partial t^2} - K\nabla \cdot \left(\frac{1}{\rho}\nabla P\right) = f(\mathbf{r}, t), \qquad (1.110)$$

where P denotes the confining pressure (in Pa), ρ is the mass density of the material (in kg m^{-3}), K is the bulk modulus of the porous material (in Pa), and $f(\mathbf{r}, t)$ denotes the source function (seismic source) at position $\mathbf{r}$ and time t. The acoustic pressure corresponds to the hydrostatic part of the stress tensor $\mathbf{T}$:

$$P = -\frac{1}{3}\mathrm{Trace}\,\mathbf{T}. \qquad (1.111)$$

Still, Equation (1.111) is not good enough to be compatible with the velocity of the P-wave in a porous material. First, the bulk modulus that should be considered is generally the undrained bulk modulus (the fluid in the pores resists to the deformation). The modulus K_u (in Pa) is defined by

$$K_u = \frac{K_f(K_s - K) + \phi K(K_s - K_f)}{K_f(1 - \phi - K/K_s) + \phi K_s}. \qquad (1.112)$$

In addition, the porous material, at the opposite of a viscous fluid, can sustain shear stresses. This means that Equation (1.110) needs to be replaced by

$$\frac{\partial^2 P}{\partial t^2} - \left(K_u + \frac{4}{3}G\right)\nabla \cdot \left(\frac{1}{\rho}\nabla P\right) = f(\mathbf{r}, t), \qquad (1.113)$$

where G is described as the shear modulus of the skeleton (frame) of the porous material (the reason for the term $K_u + (4/3)G$ can be obtained from elastic theory). Equation (1.113) can be used to solve the poroacoustic problem for the P-wave propagation in a porous material. Assuming that the viscous coupling between the pore water and the solid phase can be neglected, the velocity of the P-waves is approximated by (Gassmann, 1951)

$$c_p = \left(\frac{K_u + \frac{4}{3}G}{\rho}\right)^{\frac{1}{2}}. \qquad (1.114)$$

Now, we need to describe how the macroscopic perturbation P on the elastic skeleton affects the pore fluid pressure p, at least in an approximate way. In the undrained regime of poroelasticity, the pressure, P, is related to the so-called undrained pore fluid pressure p by (see Section 1.5.3)

$$p = BP, \qquad (1.115)$$

where $0 \leq B \leq 1$ is called the Skempton coefficient (see Section 1.5). It is given by

$$B = \frac{1 - K/K_u}{1 - K/K_s}, \qquad (1.116)$$

where K is the bulk modulus (in Pa), K_u is the undrained bulk modulus (in Pa), and K_s is the bulk modulus of the solid phase (in Pa). The passage of the wave generates a confining pressure fluctuation, P, on a representative elementary volume of the rock, more precisely on the elastic skeleton. This change in confining pressure generates in turn a change in the pore fluid pressure and therefore the flow of the pore water according to Darcy's law. Moreover, the flow of pore water relative to the solid phase generates a source current density given by

$$\mathbf{J}_S = \hat{Q}_V^0 \dot{\mathbf{w}} = -\frac{\hat{Q}_V^0 k_0}{\eta_f}\nabla p, \qquad (1.117)$$

where $\dot{\mathbf{w}}$ denotes the Darcy velocity discussed in Section 1.2. In getting Equation (1.117), we have neglected the inertial terms, and therefore, this equation considers only low-frequency disturbances (below 1 kHz).

1.4.3 Numerical example of the coseismic and seismoelectric conversions

We use now the acoustic approximation, implemented in the finite-element package COMSOL Multiphysics, to describe the effects of coseismic field and seismoelectric conversion. We start with the coseismic field associated with the traveling of a compressional pressure (P-)wave in a homogeneous material. Figures 1.21, 1.22, and 1.23 show three snapshots showing the propagation of such a wave between a seismic source and a geophone collocated with an electrode (the reference for the electrical potential at the electrode is taken to infinity). We see an electrical signal at the position of the electrode only when the seismic wave arrives at this location. Therefore, the coseismic electrical field is only localized in the spatial support of the seismic wave.

Figure 1.24 shows a snapshot of the electrical potential distribution when a seismic wave hits an interface between two media of different mechanical and electrical properties. In this example, the two layers are only characterized by a difference in electrical conductivity. The seismoelectric conversion at the interface is similar to the one that would be generated by a dipole distribution located along the interface in the first Fresnel zone of then seismic wave. The concept of Fresnel zone will be discussed further in Chapter 4.

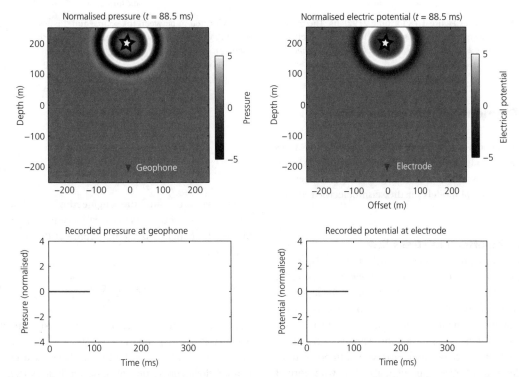

Figure 1.21 Synthetic example computed with the acoustic approximation to demonstrate the effect of the coseismic electrical field. A seismic source generates a pressure wave in a homogeneous material. We see here the normalized pressure (left panel) and the normalized electrical potential (right panel) at time 88.5 ms. The star represents the seismic source. (*See insert for color representation of the figure.*)

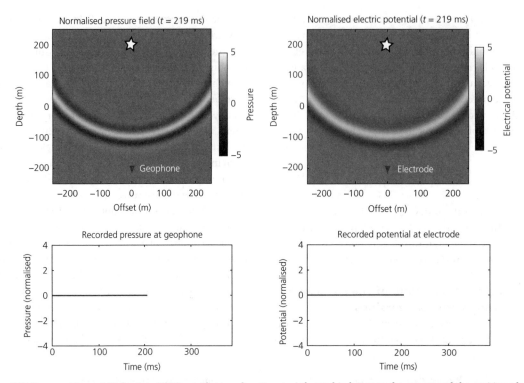

Figure 1.22 Same as Figure 1.18 for time 219.0 ms. The traveling P-wave is located in between the source and the position of the recording geophone located with an electrode (reference electrode to infinity). We see no disturbances at the position of the geophone and electrode. (*See insert for color representation of the figure.*)

1.5 Elements of poroelasticity

In this section, we provide some of the basic tools required to understand static and dynamic poroelasticity. These elements will be used further in Chapter 2 to establish some general models for the seismoelectric theory.

1.5.1 The effective stress law

We consider a porous material with a connected porosity and a solid frame (the skeleton) that behaves elastically. The material is homogeneous at the scale of the representative elementary volume and monomineralic. The stress tensor applied to the porous material is written as **T**. The component of the stress tensor corresponds to T_{ij}, which represents the force per unit surface area imposed from outside in the ith-direction along a surface with normal in the jth-direction (Figure 1.25). The strain (deformation) tensor is defined by

$$\varepsilon_{ij} = \frac{1}{2}\left(\frac{\partial u_i}{\partial x_j} + \frac{\partial u_j}{\partial x_i}\right),\qquad(1.118)$$

$$\varepsilon_{ij} = \frac{1}{2}\left(u_{i,j} + u_{j,i}\right),\qquad(1.119)$$

where we are using the engineering notations and where the vector **u** refers to the displacement of the solid phase. The (small) change in volume of the material is described as

$$\varepsilon_{kk} = \mathrm{Tr}\left(\varepsilon_{ij}\right) = \nabla\cdot\mathbf{u} = \frac{\delta V}{V},\qquad(1.120)$$

where Tr refers to the trace of the tensor (the sum of the diagonal elements) and $\delta V/V$ denotes a volume element increment. In Equation (1.120), we have used Einstein (summation) convention on repeated indices (Einstein, 1916). This convention implies summation over a set of repeated indices in an equation with the goal to

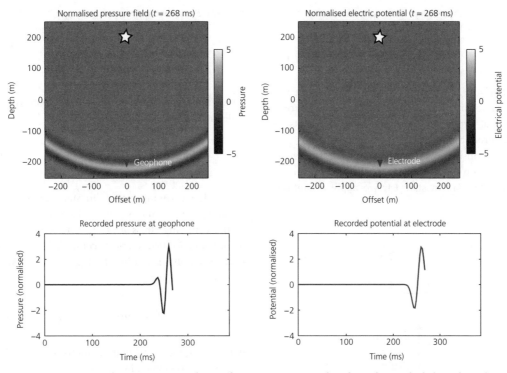

Figure 1.23 Same as Figure 1.18 for time 268.0 ms. The traveling (P-wave) seismic disturbance has reached ith an electrode (reference electrode to infinity). The geophone and the electrode record a pressure and an associated electrical disturbance. The electrical field corresponding to the electrical disturbance is called the coseismic field. (*See insert for color representation of the figure.*)

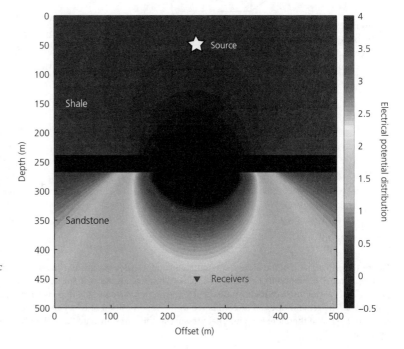

Figure 1.24 Snaphot of the seismoelectrical disturbance (seismoelectric conversion) generated at an interface between two media of different electrical conductivities (here a conductive shale and a less conductive sandstone) during the passage of a seismic P-wave. The star represents the seismic source. The radiation occurs in the first Fresnel zone of the seismic wave. (*See insert for color representation of the figure.*)

achieve notational brevity. The strain tensor is related to its components by

$$\boldsymbol{\varepsilon} = \varepsilon_{ij}\mathbf{x}_i \otimes \mathbf{x}_j, \qquad (1.121)$$

where $\mathbf{x}_i$ ($i = 1, 2, 3$) denote the basis vectors of the Cartesian framework of reference ($\mathbf{x}_i \cdot \mathbf{x}_j = \delta_{ij}$ where δ_{ij} denotes the Kronecker delta) and $\mathbf{a} \otimes \mathbf{b}$ represents the tensorial product between vectors $\mathbf{a}$ and $\mathbf{b}$.

We first consider two states, E1 and E2, that will be combined soon to determine the effective pressure in a porous material. In state E1, we apply a confining

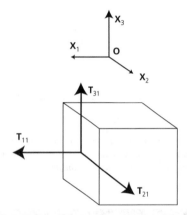

Figure 1.25 Definition of the stress tensor components on a cube of an elastic material. Each component defines a force on one of the face of the cube in a certain direction. These forces are imposed from the external world.

pressure P, and we consider no change in the fluid pressure $p = 0$. In this case, the bulk deformation is given by

$$\varepsilon_{kk}(P,0) \equiv \nabla \cdot \mathbf{u}(P,0) = -\frac{P}{K}, \qquad (1.122)$$

where the drained bulk modulus of the porous material is defined by

$$\frac{1}{K} = -\frac{1}{V}\left(\frac{\partial V}{\partial P}\right)_{p=0,T} \qquad (1.123)$$

and T denotes temperature. In state E2, we apply a confining stress P, and we imposed a fluid pressure equal to the confining stress $p = P$. In this second state, the deformation of the material is controlled by the stiffness of the solid phase (not by the stiffness of the skeleton). Therefore, we have

$$\varepsilon_{kk}(P=p,p) \equiv \nabla \cdot \mathbf{u}(P=p,p) = -\frac{P}{K_S}, \qquad (1.124)$$

where K_S, the drained bulk modulus of the solid phase (e.g., silica), is defined by

$$\frac{1}{K_S} = -\frac{1}{V}\left(\frac{\partial V}{\partial P}\right)_{p=P,T}. \qquad (1.125)$$

Now, we can describe the general bulk deformation of a porous material in state E as the linear superposition of the two states E1 and E2 as shown in Figure 1.26. This can be written as

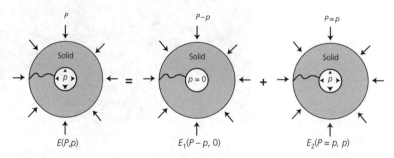

Figure 1.26 Application of the superposition principle used to determine the effective pressure law in a linear poroelastic material. The porous material is composed of a solid phase with an interconnected pore space, which is also connected to the external world. In the general case, we impose a confining pressure P to the porous material, and there is an internal pressure corresponding to the pore fluid pressure p. The general case can be considered as the superposition of two states E1 and E2. In the first state, we apply a confining pressure $P - p$, and therefore is no internal pressure in the pores. In state E2, we apply a confining pressure that is equal to the fluid pressure.

$$E(P,p) = E_1(P-p,0) + E_2(P=p,p). \qquad (1.126)$$

Equation (1.124) can be written in terms of bulk deformation as

$$\varepsilon_{kk}(P,p) = \varepsilon_{kk}(P-p,0) + \varepsilon_{kk}(P=p,p), \qquad (1.127)$$

$$\varepsilon_{kk}(P,p) = -\frac{1}{K}(P-p) - \frac{1}{K_S}p. \qquad (1.128)$$

The bulk deformation can be written in terms of an effective pressure P^* as

$$\varepsilon_{kk}(P,p) = -\frac{1}{K}P^*, \qquad (1.129)$$

$$P^* = P - \alpha p, \qquad (1.130)$$

$$\alpha = 1 - \frac{K}{K_S}. \qquad (1.131)$$

The coefficient α is called the effective stress coefficient or Biot coefficient. Equation (1.129) means that from the standpoint of the bulk deformation, having a confining pressure P and an internal pore fluid pressure p is equivalent of having a confining pressure P^* and no fluid pressure. We can proceed now to a complete analysis of the Biot coefficient α. We have $K_S \geq K$ and therefore $\alpha \geq 0$. If $K_S \gg K$ (for instance, at high porosity, the compressibility of the skeleton is high), we have $\alpha = 1$. In this last case, the effective pressure is equal to the differential pressure $P-p$. This situation is known as the Terzaghi effective stress principle (Terzaghi, 1943) and was developed initially for porous soils. In the case where the porosity is equal to zero, we have $K = K_S$, and therefore, $\alpha = 0$, which is consistent with the idea that at very low porosity (e.g., for crystalline rocks), the fluid pressure will play no role. A typical value of the Biot coefficient for a sandstone is 0.8.

1.5.2 Hooke's law in poroelastic media
In a linear poroelastic material, the constitutive Hooke's law is written as

$$T_{ij} = \left(K - \frac{2}{3}G\right)\varepsilon_{kk}\delta_{ij} + 2G\varepsilon_{ij} - \alpha p \delta_{ij}, \qquad (1.132)$$

where δ_{ij} is the Kronecker delta defined by

$$\delta_{ij} = \begin{cases} 0 & \text{if } i \neq j \\ 1 & \text{if } i = j \end{cases} \qquad (1.133)$$

and G denotes the shear modulus of the material, which is equal to the shear modulus of the skeleton (as long as the fluids do not sustain shear stresses). The two first terms of the right-hand side of Equation (1.132) correspond to the classical Hooke's law in elastic media, and the last term corresponds to the effect of the pore fluid pressure. We are going to demonstrate that this equation is consistent with the effective pressure law determined in Section 1.5.1.

Taking the nondiagonal components of Equation (1.132) yields

$$T_{ij}(i \neq j) = 2G\varepsilon_{ij}, \qquad (1.134)$$

which corresponds to pure shear. We look now for Hooke's law for the bulk deformation. We have

$$T_{kk} \equiv \text{Tr}(\mathbf{T}), \qquad (1.135)$$

$$T_{kk} \equiv 3(K\varepsilon_{kk} - \alpha p), \qquad (1.136)$$

where $\text{Tr}(\cdot)$ denotes the trace of the matrix corresponding to the tensor (sum of the diagonal elements). The confining pressure is related to the bulk stress by

$$P \equiv -\frac{1}{3}T_{kk} = -K\varepsilon_{kk} - \alpha p, \qquad (1.137)$$

and therefore, we can check the property

$$\varepsilon_{kk} = -\frac{1}{K}(P - \alpha p) = -\frac{1}{K}P^*, \qquad (1.138)$$

which is consistent with Equation (1.129).

1.5.3 Drained versus undrained regimes
In the drained regime, the fluid is free to move in and out of a representative elementary volume that is undergoing deformation, and doing so, it dissipates energy. The fluid pressure can be considered as constant, and the mass of fluid per unit volume of rock is variable. In the undrained regime, the fluid cannot (or has no time to) move in/out with respect to the representative elementary volume. In this case, the fluid resists to the deformation. The drained and the undrained bulk moduli are defined by

$$\frac{1}{K} = -\frac{1}{V}\left(\frac{\partial V}{\partial P}\right)_{p,T}, \tag{1.139}$$

$$\frac{1}{K_u} = -\frac{1}{V}\left(\frac{\partial V}{\partial P}\right)_{m,T}. \tag{1.140}$$

Obviously, it is harder to deform a rock if some fluid (for instance, water) is resisting the deformation. Consequently, we have the following property: $K_S \geq K_u \geq K$. The undrained regime is defined by replacing the porous material by an equivalent elastic material with a bulk modulus K_u (the shear modulus remains unchanged). In the undrained regime, Hooke's law is therefore given by

$$T_{ij} = \left(K_u - \frac{2}{3}G\right)\varepsilon_{kk}\delta_{ij} + 2G\varepsilon_{ij}. \tag{1.141}$$

We can look for the bulk deformation of the material in the undrained regime (the pure shear components are unchanged with respect to the drained case). This yields

$$T_{kk} = 3K_u\varepsilon_{kk}, \tag{1.142}$$

and therefore,

$$\varepsilon_{kk} = \frac{1}{3K_u}T_{kk} = -\frac{P}{K_u}. \tag{1.143}$$

What about the fluid pressure in the undrained case? We can still apply the complete form of Hooke's law by writing that the fluid pressure is equal to the undrained fluid pressure. The effective stress law yields

$$\varepsilon_{kk}(P,p_u) = -\frac{1}{K}(P - \alpha p_u). \tag{1.144}$$

Equating Equations (1.143) and (1.144) yields

$$\frac{P}{K_u} = \frac{1}{K}(P - \alpha p_u), \tag{1.145}$$

and therefore,

$$p_u = \frac{K_u - K}{\alpha K_u}P, \tag{1.146}$$

$$p_u = \frac{K - K_u}{\alpha}\varepsilon_{kk}. \tag{1.147}$$

Equation (1.146) can be used to define the Skempton coefficient B:

$$p_u = BP, \tag{1.148}$$

$$B \equiv \left(\frac{\partial p}{\partial P}\right)_{m,T}, \tag{1.149}$$

$$B = \frac{K_u - K}{\alpha K_u} = \frac{1 - K/K_u}{1 - K/K_S}. \tag{1.150}$$

As $K_S \geq K_u \geq K$, $0 \leq B \leq 1$, and $0 \leq p_u \leq P$. The Skempton coefficient can be used to determine the effect of a change of the confining pressure on a change on the fluid pressure in undrained conditions. Two extreme cases can be considered. When $K_u \to K$, we have $B \to 0$, for instance, when gas is present in the pore space of the porous material. In this case, an increase of the confining pressure has no effect on the fluid pressure because of the very high compressibility of the gas. The second case is when $K \to 0$, $K \ll (K_S, K_u)$. In this case, we have $(\alpha, B) \to 1$. This is the case of a very compressible soil with a very incompressible fluid like water.

To understand the implication of the drained and undrained regimes in terms of deformation, we can think about a tank filled by a water-saturated sand. We apply at $t = 0$ a mechanical load on the top of the sand (think about a building built very quickly on a porous soil). The deformation that follows can be considered using two steps: in the first step, the deformation is instantaneous and undrained with a bulk modulus K_u and a shear modulus G. In this step, we can compute the undrained fluid pressure everywhere, and the fluid has no time to move. In the second step, the fluid starts to flow in response to the generated (undrained) fluid pressure distribution. This flow is responsible for a delayed mechanical response: the soil creeps over time because of the flow of the pore water until an equilibrium situation is reached. Our poroacoustic analysis was proceeding along the same lines: we first determined the undrained fluid pressure associated with the stress or deformation associated with the propagation of the seismic waves (using an undrained bulk modulus; see Figure 1.19), and then, we let the fluid flows in response to the fluid pressure distribution. The flow of water is responsible for the streaming current density defined as the advective drag of the effective charge density $\hat{Q}_V^0$ contained in the pore water (Figure 1.19).

1.5.4 Wave modes in the pure undrained regime

Before we get into the realm of the dynamic Biot–Frenkel theory, it is instructive to review how the two wave modes (pressure or primary (P-)waves and shear or secondary (S-)waves) are obtained in the elastic case. In elastic media, Hooke's law can be written in two different equivalent forms using the elements of the tensors or the tensors themselves:

$$T_{ij} = \left(K_{\mathrm{u}} - \frac{2}{3}G \right) \varepsilon_{kk}\delta_{ij} + 2G\varepsilon_{ij}, \qquad (1.151)$$

$$\mathbf{T} = \lambda_{\mathrm{u}}\mathrm{Tr}(\boldsymbol{\varepsilon})\mathbf{I}_3 + 2G\boldsymbol{\varepsilon}, \qquad (1.152)$$

where the Lamé coefficient λ_{u} is given by

$$\lambda_{\mathrm{u}} \equiv K_{\mathrm{u}} - \frac{2}{3}G \qquad (1.153)$$

and $\mathbf{I}_3$ denotes the 3×3 identity matrix ($\mathbf{T}$ and $\boldsymbol{\varepsilon}$ denote the stress and deformation tensors). If we define the bulk deformation as

$$\theta \equiv \nabla \cdot \mathbf{u} = \varepsilon_{kk}, \qquad (1.154)$$

Hooke's law can be written as

$$\mathbf{T} = \lambda_u\theta\mathbf{I} + 2G\boldsymbol{\varepsilon}. \qquad (1.155)$$

In order to find the field equation for the displacement of the solid phase, we need to combine Hooke's law (which is a constitutive equation) with a continuity equation, actually the momentum conservation equation applied to the elastic material. This equation corresponds to Newton's law:

$$\nabla \cdot \mathbf{T} = \rho \frac{\partial^2 \mathbf{u}}{\partial t^2}, \qquad (1.156)$$

$$T_{ij,j} = \rho\ddot{u}_i, \qquad (1.157)$$

where ρ denotes the bulk density of the material.

The divergence of the stress tensor can be computed as follows:

$$T_{ij} = \lambda_u u_{k,k}\delta_{ij} + G\left(u_{i,j} + u_{j,i} \right), \qquad (1.158)$$

$$T_{ij,j} = \lambda_u u_{k,ki} + G\left(u_{i,jj} + u_{j,ij} \right), \qquad (1.159)$$

$$T_{ij,j} = (\lambda_u + G)u_{j,ij} + Gu_{i,jj}, \qquad (1.160)$$

$$T_{ij,j} = (\lambda_u + G)u_{i,jj} + Gu_{j,ji}. \qquad (1.161)$$

After straightforward algebraic manipulations using the properties of the Kronecker delta and the fact that we can change the order of the derivatives, we obtain

$$\left(\frac{\lambda_u + G}{\rho} \right)u_{j,ji} + \left(\frac{G}{\rho} \right)u_{i,jj} = \ddot{u}_i. \qquad (1.162)$$

Equation (1.162) can be written in vectorial notations as

$$\left(\frac{\lambda_u + G}{\rho} \right)\nabla(\nabla\cdot\mathbf{u}) + \left(\frac{G}{\rho} \right)\nabla^2\mathbf{u} = \ddot{\mathbf{u}}. \qquad (1.163)$$

Using the general property

$$\nabla^2\mathbf{u} = \nabla(\nabla\cdot\mathbf{u}) - \nabla \times \nabla \times \mathbf{u}, \qquad (1.164)$$

we obtain

$$\left(\frac{\lambda_u + 2G}{\rho} \right)\nabla(\nabla\cdot\mathbf{u}) - \left(\frac{G}{\rho} \right)\nabla \times \nabla \times \mathbf{u} = \ddot{\mathbf{u}}. \qquad (1.165)$$

Now, we define the following two parameters:

$$c_{\mathrm{p}}^2 = \frac{\lambda_{\mathrm{u}} + 2G}{\rho}, \qquad (1.166)$$

$$c_{\mathrm{s}}^2 = \frac{G}{\rho}. \qquad (1.167)$$

From Equations (1.165)–(1.167), we have

$$c_{\mathrm{p}}^2\nabla(\nabla\cdot\mathbf{u}) - c_{\mathrm{s}}^2\nabla \times \nabla \times \mathbf{u} = \ddot{\mathbf{u}}. \qquad (1.168)$$

Equation (1.168) corresponds to the wave equation of elasticity. To find the different wave modes, we need to use a Helmholtz decomposition of the displacement field:

$$\mathbf{u} = \nabla\varphi + \nabla \times \boldsymbol{\Psi}, \qquad (1.169)$$

$$\nabla\cdot\boldsymbol{\Psi} = 0, \qquad (1.170)$$

where φ is a scalar potential and $\boldsymbol{\Psi}$ is a vectorial potential. Equation (1.170) corresponds to the so-called coulomb gauge. The Helmholtz decomposition of a vector field consists in writing this field as the sum of the gradient of a scalar potential and the curl of a vector potential. Using Equations (1.169) and (1.170) in Equation (1.168) and using the properties

$$\nabla \cdot (\nabla \times \Psi) = 0, \tag{1.171}$$

$$\nabla \times (\nabla \varphi) = 0, \tag{1.172}$$

we end up with two equations:

$$c_p^2 \nabla^2 \varphi = \ddot{\varphi}, \tag{1.173}$$

$$c_S^2 \nabla^2 \Psi = \ddot{\Psi}. \tag{1.174}$$

The first mode corresponds to the compressional wave with velocity c_p, while the second mode corresponds to the shear mode with velocity c_S. The velocity ratio is such that we have

$$\frac{c_p}{c_S} = \sqrt{2 + \frac{\lambda_u}{G}} \geq 1. \tag{1.175}$$

Therefore, the P-wave propagates faster than the shear wave. We will show in the next chapters that there are more than two wave modes in poroelastic media. The different wave modes have different EM signatures in seismoelectric theory, and we will discuss in the next chapters these different signatures and their potential applications to probe the Earth using the seismoelectric method.

1.6 Short history

Now that we have discussed some of the key concepts required to understand the seismoelectric theory, we can provide a short introduction to the history of the seismoelectric method. The physics of the streaming potential takes its roots in the experimental work done initially by Quincke (1859) who discovered that the flow of water through a capillary generates a measurable difference of electrical potential. Helmholtz (1879) obtained theoretical expressions for the streaming current density for glass capillaries. The first observations of the seismoelectric effect itself were made by Thompson (1936). Thompson was a geophysicist working for the Humble Oil and Refining Company. His work on the so-called "seismic electric"

effect was inspired by Louis Statham in the same company (see Blau & Staham, 1936). They made indeed the first documented observations of electric potential changes between a pair of electrodes when seismic waves passed through these electrodes. Thompson (1936) was looking for the effect of seismic waves on the electrical resistivity of Earth materials. His goal was to detect seismic waves by measuring continuously the electrical resistivity between a set of electrodes. His paper starts as follows: "The seismic electric effect is the name which has been given to the variation of earth resistivity with elastic deformation." The concept of "seismic electric" effect was modeled in the same issue of *Geophysics* by Slotnick (1936) on the base of an equivalent electrical circuit. In the case of Thompson (1936), he measured clear fluctuations in the electrical potentials that could not be explained by a change in the value of the resistance (or apparent resistivity) between the electrodes. In these works, there were no concept that the propagation of the seismic waves could create a source current density or an electrical field on their own and no ideas related to electrokinetic effects and the associated electrical double layer theory.

Ivanov (1939) was the first to record seismoelectric effects in USRR completely passively (no current injection). His pioneering work was followed few years later by the seminal paper of Frenkel (1944). Frenkel developed the first electrokinetic theory of the coseismic electric field in water-saturated porous rocks to explain the observations made by Ivanov. Ivanov and Frenkel should be therefore considered as the pioneers of the seismoelectric method in geophysics. A rigorous treatment of wave propagation in water-saturated porous media was, however, only introduced more than a decade later by Biot (1962a, b). A huge number of field and theoretical works were done in USRR after WWII in using the seismoelectric effect to determine the thickness of the weathered zone to help in interpreting seismic data or for mineral exploration.

Decades after, Frenkel (1944) and Thompson and Gist (1991, 1993) presented a case study for the exploration of oil and gas reservoirs using seismoelectric converted electrical signals. They used adapted data processing and common midpoint (CMP) techniques to produce a seismoelectric image of the subsurface to depths on the order of a few hundred meters. They concluded that seismoelectric conversions could be detected from a depth of 300 m. Thompson and Gist (1993) suggested that these methods could be used for much greater depths (several kilometers in the case of the electroseismic method). The seismoelectric method has also been used for a variety of applications in near-surface geophysics (for instance, Migunov & Kokorev, 1977; Fourie, 2003; Kulessa et al., 2006). Mikhailov et al. (2000) described crosshole seismoelectric measurements in a small-scale laboratory experiment with vertical and inclined fractures located between the source and the receivers. They recorded not only the coseismic electric signals generated by the seismic wave arriving at the receivers but also the EM wave associated with the Stoneley wave excited in the fracture. They claimed that a tomography image with the travel times extracted from the seismoelectric measurements could be possibly constructed.

Several modeling attempts have been developed to comprehend both the seismoelectric and electroseismic effects. Neev and Yeatts (1989) developed a theory of these effects, but as discussed by Pride (1994), this theory was incomplete. The model developed in the seminal paper of Pride (1994) couples fully the Biot–Frenkel theory to the Maxwell equations via a source current density of electrokinetic origin. This model was obtained by volume averaging the local Navier–Stokes and Nernst–Planck equations as well as the Maxwell equations. It has opened the door to numerical modeling of both the coseismic and seismoelectric conversions using finite-difference or finite-element methods and was used to assess the usefulness of these methods for various applications (e.g., White, 2005; White & Zhou, 2006). The model introduced by Pride has been the fundamental tool used to model the electroseismic and seismoelectric responses of porous rocks in the last two decades (see Haartsen & Toksoz, 1996; Haartsen & Pride, 1997; Garambois & Dietrich, 2001, 2002, for some examples). This approach is however open to some criticisms as discussed later in this book in Chapter 2. Pride's model is unable to describe correctly the surface conductivity, the frequency dependence of the conductivity of the material, and the quadrature conductivity of porous rocks. It is not based on the electrical double layer theory (only the diffuse layer is accounted for), and therefore, some fundamental elements are missing in this theory. That said, Pride has been the first to provide the complete set of macroscopic equations, and his model can easily be corrected to account for the missing components (induced polarization and frequency dependence of the electrical conductivity).

Various authors have used the finite-difference technique to simulate the 2D seismoelectric response of a heterogeneous medium, taking into account all the poroelastic wave modes (fast and slow P-waves and shear (S-)wave) and their coseismic electrical signals plus the seismoelectric conversions (note that the isochoric shear wave does not produce any coseismic electrical field in perfectly homogeneous porous media). Pain et al. (2005) presented a 2D mixed finite-element algorithm to solve the poroelastic Biot equations including the electrokinetic coupling in order to study the sensitivity of the seismoelectric method to material properties, like porosity and permeability of geological formations surrounding a borehole.

Several works have focused on producing some full waveform modeling of the seismoelectric signals using Pride's theory (see, for instance, Grobbe et al., 2012; Grobbe & Slob, 2013). The code developed by Niels Grobbe, electroseismic and seismoelectric modeling (ESSEMOD), is able to model all existing seismoelectric source–receiver combinations (using Pride's theory) in layered media, considering fully coupled Maxwell equations. This code can be also used to model seismo-electric laboratory configurations of a sample in a water tank (i.e., fluid/porous medium/fluid transitions; see Smeulders et al., 2014). This code can also be used to generate all required fields for the theoretical interferometric seismoelectric Green's function retrieval. This allows, for instance, to improve the signal-to-noise ratio of the weak seismoelectric conversions (or interface response fields). By applying interferometric techniques (e.g., Schoemaker et al., 2012), stacking inherently takes place with possible signal-to-noise ratio improvements as well. Along a similar idea, Sava and Revil (2012) introduced recently a simplified poroacoustic formulation to describe the seismoelectric coupling in porous media, and they introduce a new method called seismoelectric beamforming. The idea is to focus seismic waves on a grid of specific points and to use the seismoelectric conversion to image heterogeneities in mechanical and electrical properties (see Chapter 6). The poroacoustic approximation allows handling the computation of the seismoelectric signals in very complex geometries very quickly, and the beamforming approach is used to enhance the seismoelectric conversion over the coseismic signals.

Most of the current efforts in seismoelectric theory are also directed to understand the seismoelectric conversions in unsaturated or in porous media saturated by two immiscible fluid phases. In most cases, the water-saturated case is extended to unsaturated flow assuming that the nonwetting fluid is air at atmospheric pressure and corresponds to a very compressible phase (Revil et al., 2014). Recently, Smeulders et al. (2014) have used Pride's model to perform laboratory experiments between water and water-saturated core samples or oil-saturated core samples. They found that the contrast between water and water-saturated porous glass samples is larger than the contrast between water and oil-saturated porous glass samples. The contrast between water and water-saturated Fontainebleau sandstone is observed to be larger than the contrast between oil and water-saturated Fontainebleau sandstone in agreement with the models of Revil and Mahardika (2013) and Revil et al. (2014). A complete theory of the seismoelectric conversions in porous media saturated by two immiscible fluids will be given in this book in Chapter 3.

In parallel to the history of the seismoelectric method in geophysics, there is a rich history in the development of the so-called electroacoustic spectroscopic methods to study colloidal suspensions, concentrated dispersion, emulsions, and microemulsions (Booth & Enderly, 1952; Marlow et al., 1983; Valdez, 1993). Debye (1933) predicted that the passage of an acoustic wave through an electrolyte would generate an electrical field, the so-called ion vibration potential (IVP). Indeed, as a sound wave passes through a solution, it is responsible for a charge separation due to differences in the effective masses and frictional coefficients of the solvated anions and cations. The resulting sum of these tiny dipoles leads to a macroscopic electrical field, which depends on the sound wave frequency. This effect was observed a decade later by Yaeger et al. (1949) and Derouet and Denizot (1951).

Hermans (1938) and Rutgers et al. (1958) were the first to report a colloid vibration potential (CVP), investigating therefore the electrical field associated with the passage of acoustic waves through a colloidal suspension. Colloidal suspensions represent a suspension of very small solid particles (less than few micrometers) in water. They can be understood, therefore, as a special case of very high-porosity porous media. Generally speaking, CVP and colloid vibration current (CVI) are two phenomena where acoustic waves are applied to a colloidal system and a resultant electric field or current is created by the vibration of the colloid electric double layers. Several hundred of experimental and theoretical works have been published in this field since the 1950s. For brevity, we mention here only a few key papers. Enderby (1951) and Booth and Enderby (1952) developed the first theory for CVP in the early 1950s. The first quantitative experiments were made in the 1960s by Zana and Yeager (1967a, b, c, 1982). Oja et al. (1985) observed an inverse electroacoustic effect called the electrosonic amplitude (ESA). ESA involves the generation of acoustic waves caused by the driving force of an applied electric field or electrical current and is therefore associated with electroosmotic effects (i.e., the movement of the water in response to an electrical field due to the excess of charge present in the pore water). It is however nothing else than electroseismic effects in the wording commonly used in geophysics.

The first commercially available electroacoustic instruments were developed by Pen Kem, Inc. (Marlow et al., 1990). There are now several commercially available instruments manufactured by Colloidal Dynamics, Dispersion Technology, and Matec. The electroacoustic spectroscopic methods are used to determine the particle size distribution as well as the zeta potential of the particles. Like the seismoelectric and electroseismic methods, there are two different electroacoustic methods depending on what field is used as a driving force. CVI is the phenomenon where acoustic waves are applied to a system and a resultant electric field or current is created by the vibration of the colloid electric double layers. Scales and Jones (1992) were the first to recognize the effect of polydispersity (particle size distribution) on the electroacoustic measurements. A comprehensive theory was developed by O'Brien (1991) and O'Brien et al. (1993) including for concentrated systems, that is, relatively low-porosity materials. A review of the method can be found in Hunter (1998) and Greenwood (2003).

1.7 Conclusions

We summarize the ideas developed in this chapter as follows: (1) the surface of minerals in contact with water is charged. This charge is compensated by a charge in the pore water, which is therefore not neutral in proximity to the mineral grain surface. The surface charge of the minerals is counterbalanced locally by ions that are sorbed (and therefore "attached" to the mineral surface), forming the Stern layer and some ions forming a diffuse layer that interacts with the mineral surface charge only through the coulombic interaction. (2) The flow of pore water with respect to the solid phase (formed by the assemblage of grains) generates a source current density due to the drag of the charge density, near the mineral grain surface, contained in the pore water. This streaming current depends on either the zeta potential, ζ, or an effective charge density $\hat{Q}_V^0$. We established a bridge between these two quantities. (3) The seismoelectric response due to the flow of water relative to the mineral framework is triggered by the propagation of seismic waves and, possibly as discussed later in Chapter 5, by the seismic source itself. We must therefore account for three types of EM disturbances: those associated with the seismic sources, the coseismoelectric effects associated with the propagation of the seismic waves, and the seismoelectric conversion (also called interface response) associated with the conversion of hydromechanical to EM energy at macroscopic interfaces in the investigated material. (4) The distribution of the resulting electrical field is controlled not only by the source current density (direction and magnitude) generated in the porous material but also by the conductivity distribution in the material. This conductivity is frequency dependent (to some extent) and is characterized by both an inphase component and a quadrature component. This implies that the electrical field is not necessarily in phase with the source current density. The frequency dependence of the electrical conductivity is termed induced polarization in geophysics and is not captured by the theory of Pride.

We have also discussed previously the main ingredients that are necessary to set the stage for the introduction of the seismoelectric method. Chapter 2 will be devoted to the development of a seismoelectric theory in fully water-saturated conditions. This requires discussion of the equations describing the propagation of seismic waves in porous media characterized by a linear elastic skeleton (dynamic poroelasticity) and the Maxwell equations, usually taken in their quasistatic forms.

References

Ahmad, M.U. (1969) A laboratory study of streaming potentials. *Geophysical Prospecting*, **12**(**1**), 49–64.

Avena, M.J. & De Pauli, C.P. (1998) Proton adsorption and electrokinetics of an Argentinean Montmorillonite. *Journal of Colloid and Interface Science*, **202**, 195–204, doi: 10.1006/jcis.1998.5402.

Biot, M.A. (1962a) Generalized theory of acoustic propagation in porous dissipative media. *Journal of the Acoustical Society of America*, **34**(**9**), 1254–1264.

Biot, M.A. (1962b) Mechanics of deformation and acoustic propagation in porous media. *Journal of Applied Physics*, **33**(**4**), 1482–1498.

Blau, L. & Statham, L. (1936) Method and apparatus for seismic electric prospecting. Technical Report, 2054067, U.S. Patent and Trademark Office (USPTO), Washington, DC.

Bolève, A., Crespy, A., Revil, A., Janod, F., & Mattiuzzo J.L. (2007) Streaming potentials of granular media: Influence of the Dukhin and Reynolds numbers. *Journal of Geophysical Research*, **112**, B08204, doi:10.1029/2006JB004673.

Booth, F. & Enderby, J. (1952) On electrical effects due to sound Waves in Colloidal Suspensions. *Proceedings of the American Physical Society*, **208A**, 32–42.

Börner, F.D. (1992) Complex conductivity measurements of reservoir properties. Proceedings of the Third European Core Analysis Symposium, Paris, 359–386.

Casagrande, L. (1983) Stabilization of soils by means of electro-osmosis: State of art. *Journal of Boston Society of Civil Engineers*, **69**(**2**), 255–302.

Chapman, D.L. (1913) A contribution to the theory of electrocapillarity. *Philosophical Magazine*, **25**(**6**), 475–481.

Chittoori, B. & Puppala, A.J. (2011) Quantitative estimation of clay mineralogy in fine-grained soils. *Journal of Geotechnical and Geoenvironmental Engineering*, **137**(**11**), 997–1008, doi: 10.1061/(ASCE)GT.1943-5606.0000521.

Cicerone, D.S., Regazzoni, A.E., & Blesa, M.A. (1992) Electrokinetic properties of the calcite/water interface in the presence of magnesium and organic matter. *Journal of Colloid and Interface Science*, **154**, 423–433.

Darcy, H. (1856) *Les fontaines publiques de la Ville de Dijon*, Dalmont, Paris.

Debye, P. (1933) A method for the determination of the mass of electrolyte ions. *Journal of Chemical Physics*, **1**, 13–16.

Derouet, B. & Denizot, F. (1951) Comptes Rendus de l'Académie des Sciences, *Paris*, **233**, 368.

Einstein, A. (1916) The foundations of the general theory of relativity. *Annalen der Physik*, **354**(**7**), 769–822, doi:10.1002/andp.19163540702.

Enderby, J.A. (1951) On electrical effects due to sound waves in colloidal suspensions. *Proceedings of the Royal Society, London*, **A207**, 329–342.

Fourie, F.D. (2003) Application of Electroseismic Techniques to Geohydrological Investigations in Karoo Rocks, PhD Thesis, University of the Free State, Bloemfontein, South Africa, 195 pp.

Frenkel, J. (1944) On the theory of seismic and seismoelectric phenomena in a moist soil, *Journal Physics (Soviet)*, **8**(**4**), 230–241.

Friborg, J. (1996) Experimental and theoretical investigations into the streaming potential phenomenon with special reference to applications in glaciated terrain. PhD thesis, Lulea University of Technology, Sweden.

Garambois, S. & Dietrich, M. (2001) Seismoelectric wave conversions in porous media: field measurements and transfer function analysis. *Geophysics*, **66**(5), 1417–1430.

Garambois, S. & Dietrich, M. (2002) Full waveform numerical simulations of seismoelectromagnetic wave conversions in fluid-saturated stratified porous media. *Journal of Geophysical Research*, **107**(B7), ESE5-1–ESE5-18.

Gassmann, F. (1951) Über die elastizität poröser medien. *Vierteljahrsschrift der Naturforschenden Gesellschaft Zuerich*, **96**, 1–23.

Gaudin, A. M. & Fuerstenau, D.W. (1955) Quartz flotation with anionic collectors *Transactions of the American Institute of Mining, Metallurgical, and Petroleum Engineers*, **202**, 66–72.

Gonçalvès, J., Rousseau-Gueutin, P., & Revil, A. (2007) Introducing interacting diffuse layers in TLM calculations: a reappraisal of the influence of the pore size on the swelling pressure and the osmotic efficiency of compacted bentonites. *Journal of Colloid and Interface Science*, **316**, 92–99.

Gouy, G.L. (1910) Sur la constitution de la charge électrique à la surface d'un électrolyte. *Journal de Physique Théorique et Appliquée*, **9**(4), 457–468.

Greenwood, R. (2003) Review of the measurement of zeta potentials in concentrated aqueous suspensions using electroacoustics. *Advances in Colloid and Interface Science*, **106**, 55–81, doi:10.1016/S0001-8686(03)00105-2.

Grobbe, N. & Slob, E. (2013) Validation of an electroseismic and seismoelectric modeling code, for layered earth models, by the explicit homogeneous space solutions. In: *SEG Technical Program Expanded Abstracts*, Society of Petroleum Geologists, Houston. pp. 1847–1851.

Grobbe, N., Thorbecke, J., & Slob, E. (2012) ESSEMOD— electroseismic and seismoelectric flux-normalized modeling for horizontally layered, radially symmetric configurations. *Geophysical Research Abstract*, **14**:10011.

Grosse, C. (2009) Generalization of a classic theory of the low frequency dielectric dispersion of colloidal suspensions to electrolyte solutions with different ion valences. *Journal of Physical Chemistry B*, **113**, 11201–11215.

Haartsen, M.W. & Pride, S.R. (1997) Electroseismic waves from point sources in layered media. *Journal of Geophysical Research*, **102**(B11), 24 745–24 769.

Haartsen, M. W. & Toksoz, M. N. (1996) Dynamic streaming currents from seismic point sources in homogeneous poroelastic media. *Geophysical Journal International*, **132**, 256– 274.

Hermans, J. (1938) Charged colloid particles in an ultrasonic field. *Philosophical Magazine*, **25**, 426.

Helmholtz, H. (1879) Study concerning electrical boundary boundary layers. *Weidemann Annal Physik Chemie*, **7**, 337–382, 3rd Ser.

Hiorth, A., Cathles, L.M., & Madland, M.V. (2010) The impact of pore water chemistry on carbonate surface charge and oil wettability. *Transport in Porous Media*, **85**(1), 1–21.

Hunt, A., Huisman, J.A. & Vereecken, H. (2012) On the origin of slow processes of charge transport in porous media. *Philosophical Magazine*, **92**(36), 4628–4648.

Hunter, R.J. (1981) *Zeta Potential in Colloid Science: Principles and Applications*, Academic Press, New York.

Hunter R.J. (1998) Review recent developments in the electroacoustic characterisation of colloidal suspensions and emulsions. *Colloids Surface A*, **141**, 37–65.

Ivanov, A. G. (1939) Effect of electrization of earth layers by elastic waves passing through them. *Proceedings of the USSR Academy of Sciences (Dokl. Akad. Nauk SSSR)*, **24**, 42–45.

Jaafar, M. Z., Vinogradov, J., & M. D. Jackson (2009) Measurement of streaming potential coupling coefficient in sandstones saturated with high salinity NaCl brine. *Geophysical Research Letters*, **36**, L21306, doi:10.1029/2009GL040549.

Jardani, A., Revil, A., Bolève, A., Dupont, J.P., Barrash, W., & Malama B (2007) Tomography of groundwater flow from self-potential (SP) data. *Geophysical Research Letters*, **34**, L24403.

Jougnot, D., Revil, A., & Leroy P. (2009) Diffusion of ionic tracers in the Callovo-Oxfordian clay-rock using the Donnan equilibrium model and the electrical formation factor. *Geochemica et Cosmochemica Acta*, **73**, 2712–2726.

Jougnot, D., Linde, N., Revil, A., & Doussan, C. (2012) Derivation of soil-specific streaming potential electrical parameters from hydrodynamic characteristics of partially saturated soils. *Vadoze Zone Journal*, **11**(1), 1–15, doi:10.2136/vzj2011.0086.

Kitamura, A., Fujiwara, K., Yamamoto, T., Nishikawa, S., & Moriyama, H. (1999) Analysis of adsorption behavior of cations onto quartz surface by electrical double-layer model, *Journal of Nuclear Science and Technology*, **36**, 1167–1175.

Koch, K., Kemna, A., Irving, J., & Holliger, K. (2011) Impact of changes in grain size and pore space on the hydraulic conductivity and spectral induced polarization response of sand. *Hydrological Earth System Science*, **15**, 1785–1794, doi:10.5194/hess-15-1785-2011

Koch, K., Revil, A. & Holliger, K. (2012) Relating the permeability of quartz sands to their grain size and spectral induced polarization characteristics. *Geophysical Journal International*, **190**, 230–242, doi: 10.1111/j.1365-246X.2012.05510.x.

Kulessa, B., Murray, T. & Rippin, D. (2006) Active seismoelectric exploration of glaciers. *Geophysical Research Letters*, **33**, L07503, doi:10.1029/2006GL025758.

Leroy, P. & Revil A. (2004) A triple layer model of the surface electrochemical properties of clay minerals. *Journal of Colloid and Interface Science*, **270**(2), 371–380.

Leroy, P., Revil, A., Altmann, S., & Tournassat, C. (2007) Modeling the composition of the pore water in a clay-rock geological formation (Callovo-Oxfordian, France). *Geochimica et Cosmochimica Acta*, **71**(5), 1087–1097, doi: 10.1016/j. gca.2006.11.009.

Leroy, P., Revil, A., Kemna, A., Cosenza, P., & Ghorbani, A. (2008) Spectral induced polarization of water-saturated packs of glass beads. *Journal of Colloid and Interface Science*, **321**(1), 103–117.

Lesmes, D.P. & Frye, K.M. (2001) Influence of pore fluid chemistry on the complex conductivity and induced polarization responses of Berea sandstone. *Journal of Geophysical Research*, **106**(**B3**), 4079–4090.

Li, H.C. & De Bruyn, P.L. (1966) Electrokinetic and adsorption studies on quartz. *Surface Science*, **5**, 203–220.

Lipsicas, M. (1984) Molecular and surface interactions in clay intercalates, in *Physics and Chemistry of Porous Media*, edited by D. L. Johnson and P. N. Sen, eds., pp. 191–202, American Institute of Physics, College Park.

Lockhart, N. C. (1980) Electrical properties and the surface characteristics and structure of clays, II, Kaolinite: a nonswelling clay. *Journal of Colloid and Interface Science*, **74**, 520–529.

Lorne, B., Perrier, F., & Avouac, J.-P. (1999a) Streaming potential measurements. 1. Properties of the electrical double layer from crushed rock samples. *Journal of Geophysical Research*, **104**(**B8**), 17857–17877.

Lorne, B., Perrier, F., & Avouac, J.-P. (1999b) Streaming potential measurements. 2. Relationship between electrical and hydraulic flow patterns from rocks samples during deformations. *Journal of Geophysical Research*, **104**(**B8**), 17,879–17,896.

Ma, C. & Eggleton, R. A. (1999) Cation exchange capacity of kaolinite. *Clays and Clay Minerals*, **47**, 174–180.

Maes, A., Stul, M.S. & Cremers, A. (1979) Layer charge-cation-exchange capacity relationships in montmorillonite, *Clays and Clay Minerals*, **27**, 387–392.

Marlow, B.J., Fairhurst, D., & Pendse, H.P. (1983) Colloid vibration potential and the electrokinetic characterization of concentrated colloids. *Langmuir*, **4**(3), 611–626.

Marlow, B.J., Oja, T., & Goetz, P.J. (1990) Colloid Analyzer, U.S. Patent 4,907,453.

Marshall, D. J. & Madden, T. R. (1959) Induced polarization, a study of its causes. *Geophysics*, **24**, 790–816.

Migunov, N. & Kokorev, A. (1977) Dynamic properties of the seismoelectric effect of water saturated rocks. *Izvestiya, Earth Physics*, **13**(6), 443–446.

Mikhailov, O.V., Queen, J. & Toksoz, M.N. (2000) Using borehole electroseismic measurements to detect and characterize fractured (permeable) zones. *Geophysics*, **65**(4), 1098–1112, doi:10.1190/1.1444803.

Neev, J. & Yeatts, F.R. (1989) Electrokinetic effects in fluid-saturated poro-elastic media, *Physical Review B*, **40**(13), 9135–9141.

O'Brien, R.W. (1991) Determination of Particle Size and Electric Charge, U.S. Patent 5,059,909, October 22, 1991.

O'Brien, R.W., Rowlands, W.N., & Hunter R.J. (1993) Determining charge and size with the acoustosizer, in: National Institute of Standards and Technology Special Publication 856, tric point. USA Department of Commerce, Washington, DC, pp. 1–22.

Oja, T., Petersen, G., & Cannon, D. (1985) Measurement of electric-kinetic properties of a solution. U.S. Patent 4,497,208.

Pain, C.C., Saunders, J.H., Worthington, M.H. et al. (2005) A mixed finite-element method for solving the poroelastic Biot equation with electrokinetic coupling. *Geophysical Journal of International*, **160**, 592–608.

Patchett, J. G. (1975) An investigation of shale conductivity, Society of Professional Well Logging Analysis 16th Logging Symposium, Paper U, Houston, TX, 41 pp.

Pengra, D.B., Li, S.X., & Wong P.-Z. (1999) Determination of rock properties by low-frequency AC electrokinetics. *Journal of Geophysical Research*, **104**(**B12**), 29485–29508.

Poley, J.P., Nooteboom, J.J. & de Waal, P.J. (1978) Use of V.H.R. dielectric measurements for borehole formation analysis. *The Log Analyst*, **19**(3), 8–30.

Pride, S. (1994) Governing equations for the coupled electromagnetics and acoustics of porous media. *Physical Review B*, **50**, 15678–15696.

Pride, S. & Morgan, F.D. (1991) Electrokinetic dissipation induced by seismic waves. *Geophysics*, **56**(7), 914–925.

Quincke, G. (1859) Concerning a new type of electrical current. *Annalen der Physics und Chemie (Poggendorff's Annal., Ser. 2)*, **107**, 1–47.

Revil, A., Schwaeger, H., Cathles, L.M., & Manhardt, P. (1999a) Streaming potential in porous media. 2. Theory and application to geothermal systems. *Journal of Geophysical Research*, **104**(**B9**), 20033–20048.

Revil, A., Pezard, P.A., & Glover, P.W.J. (1999b) Streaming potential in porous media. 1. Theory of the zeta-potential. *Journal of Geophysical Research*, **104**(**B9**), 20021–20031.

Revil, A., Leroy, P., & Titov, K. (2005) *Characterization of transport properties of argillaceous sediments. Application to the Callovo-Oxfordian Argillite*. *Journal of Geophysical Research*, **110**, B06202, doi: 10.1029/2004JB003442.

Revil, A., Linde, N., Cerepi, A., Jougnot, D., Matthäi, S., & Finsterle, S. (2007) Electrokinetic coupling in unsaturated porous media, *Journal of Colloid and Interface Science*, **313**(1), 315–327, doi:10.1016/j.jcis.2007.03.037.

Revil, A. & Florsch, N. (2010) Determination of permeability from spectral induced polarization data in granular media, *Geophysical Journal International*, **181**, 1480–1498, doi: 10.1111/j.1365-246X.2010.04573.

Revil A. & Jardani, A. (2010) Seismoelectric response of heavy oil reservoirs. Theory and numerical modelling. *Geophysical Journal International*, **180**, 781–797, doi: 10.1111/j.1365-246X.2009.04439.x.

Revil, A. & Skold, M. (2011) Salinity dependence of spectral induced polarization in sands and sandstones. *Geophysical Journal International*, **187**, 813–824, doi: 10.1111/j.1365-246X.2011.05181.x.

Revil, A., Karaoulis, M., Johnson, T., & Kemna, A. (2012) Review: Some low-frequency electrical methods for subsurface characterization and monitoring in hydrogeology, *Hydrogeology Journal*, **20**(4), 617–658, doi:10.1007/s10040-011-0819-x, 2012.

Revil, A. (2012) Spectral induced polarization of shaly sands: Influence of the electrical double layer. *Water Resources Reseacrh*, **48**, W02517, doi:10.1029/2011WR011260.

Revil, A. & H. Mahardika (2013) Coupled hydromechanical and electromagnetic disturbances in unsaturated clayey materials. *Water Resources Research*, **49**, doi:10.1002/wrcr.20092.

Revil, A., Skold, M., Hubbard, S.S., Wu, Y., Watson, D., & Karaoulis, M. (2013) Petrophysical properties of saprolites from the Oak Ridge Integrated Field Research Challenge site, Tennessee. *Geophysics*, **78(1)**, D21–D40, doi: 10.1190/geo2012-0176.1.

Revil, A. (2013a) Effective conductivity and permittivity of unsaturated porous materials in the frequency range 1 mHz–1GHz. *Water Resources Research*, **49**, doi:10.1029/2012WR012700.

Revil, A. (2013b) On charge accumulations in heterogeneous porous materials under the influence of an electrical field. *Geophysics*, **78(4)**, D271–D291, doi: 10.1190/GEO2012-0503.1.

Revil A., Barnier, G., Karaoulis, M., & Sava, P. (2014) Seismoelectric coupling in unsaturated porous media: theory, petrophysics, and saturation front localization using an electroacoustic approach. *Geophysical Journal International*, **196(2)**: 867–884, doi: 10.1093/gji/ggt440.

Rutgers, A.J. & Rigole, W. (1958) Ultrasonic vibration potentials in colloid solutions, in solutions of electrolytes and pure liquids, *Transactions of the Faraday Society*, **54**, 139–143.

Sava, P. & Revil, A. (2012) Virtual electrode current injection using seismic focusing and seismoelectric conversion. *Geophysical Journal International*, **191(3)**, 1205–1209, doi: 10.1111/j.1365-246X.2012.05700.x.

Scales P.J. & Jones, E. (1992) Effect of particle size distribution on the accuracy of electroacoustic mobilities. *Langmuir*, **8(2)**, 385–389.

Schmutz, M., Revil, A., Vaudelet, P., Batzle, M., Femenia Vinao, P., & Werkema, D.D. (2010) Influence of oil saturation upon spectral induced polarization of oil bearing sands. *Geophysical Journal International*, **183**, 211–224, doi:10.1111/j.1365-246X.2010.04751.x.

Schoemaker, F.C., Grobbe, N., Schakel, M. D., de Ridder, S. A. L. and Slob, E. C. & Smeulders, D. M. J. (2012) Experimental validation of the electrokinetic theory and development of seismoelectric interferometry by cross-correlation. *International Journal of Geophysics*, **514242**, 23 pp., doi:10.1155/2012/514242.

Shainberg, I., Alperovitch, N. & Keren, R. (1988) Effect of magnesium on the hydraulic conductivity of Na-smectite-sand mixtures. *Clays and Clay Minerals*, **36**, 432–438.

Sheffer, M.R. (2007) Forward modeling and inversion of streaming potential for the interpretation of hydraulic conditions from self-potential data. PhD thesis, University of British Columbia.

Sinitsyn, V. A., S. U. Aja, D. A. Kulik, & S. A. Wood (2000) Acid-base surface chemistry and sorption of some lanthanides on K^+-saturated Marblehead illite. I. Results of an experimental investigation. *Geochimica et Cosmochimica Acta*, **64**, 185–194.

Slater, L.D. & Glaser, D.R. (2003) Controls on induced polarization in sandy unconsolidated sediments and application to aquifer characterization. *Geophysics*, **68(5)**, 1547–1558, doi: 10.1190/1.1620628.

Slater, L. & Lesmes, D. P. (2002) Electrical-hydraulic relationships observed for unconsolidated sediments. *Water Resources Research*, **38(10)**, 1213, doi: 10.1029/2001WR001075.

Slotnick, M.M. (1936) A simplified circuit of the seismic electric method and its steady-state solution. *Geophysics*, **1**, 336–339.

Smeulders, D.M.J., Grobbe, N., Heller, H.K.J., & Schakel, M.D. (2014) Seismoelectric conversion for the detection of porous medium interfaces between wetting and nonwetting fluids. *Vadoze Zone Journal*, **5**, 7 pp, doi:10.2136/vzj2013.06.0106.

Stern, O. (1924) Zur Theorie der elektrolytischen Doppelschicht (The theory of the electrolytic double shift). *Zeitschrift Fur Elektrochemie Und Angewandte Physikalische Chemie*, **30**, 508–516.

Strand, S., Høgnesen, E.J. & Austad, T. (2006) Wettability alteration of carbonates-effects of potential determining ions (Ca and SO4) and temperature. *Colloids and Surfaces A: Physiochemical and Engineering Aspects*, **275**, 1–10.

Su, Q., Feng, Q. & Shang, Z. (2000) Electrical impedance variation with water saturation in rock, *Geophysics*, **65**, 68–75.

Sverjensky, D. A. (2005) Prediction of the speciation of alkaline earths adsorbed on mineral surfaces in salt solutions. *Geochimica et Cosmochimica Acta*, **69**, 225–257.

Tadros, Th.F. & Lyklema J. (1969) The electrical double layer on silica in the presence of bivalent counter-ions. *Journal of Electroanalytical Chemistry, Interfacial and Electrochemistry*, **22**, 1–7.

Terzaghi, K. (1943) *Theoretical Soil Mechanics*. John Wiley & Sons, New York.

Thompson, R.R. (1936) The seismic electric effect. *Geophysics*, **1**, 327–335.

Thompson, A. H. & Gist, G. (1991) Electroseismic prospecting, Society of Exploration Geophysics 61st Annual International Meeting, Expanded Abstracts, Houston, TX, pp. 425–427.

Thompson, A. H. & Gist, G. (1993) Geophysical applications of electrokinetic conversion, *Leading Edge*, **12**, 1169–1173.

Titov, K., Kemna, A., Tarasov, A., & Vereecken, H. (2004) Induced polarization of unsaturated sands determined through time-domain measurements. *Vadose Zone Journal*, **3**, 1160–1168.

Tournassat, C., Ferrage, E., Poinsignon, C., & Charlet L. (2004), The titration of clay minerals. Part II. Structural-based model and implications for clay reactivity. *Journal of Colloid and Interface Science*, **273**, 234–246.

Valdez J.L. (1993) Electroacoustics for characterisation of particulates and suspensions: Proceedings of Workshop Held at the National Institute of Standards and Technology, February 3–4, 1993, Gaithersburg, MD, in: S.G. Malghan (Ed.), US Department of Commerce, Washington, DC, 1993, pp. 111–128, NIST Special publication 856.

van Bekkum, H., Flanigen, E.M., & Jansen J.C. (2001) *Introduction to Zeolite Science and Practice, Studies in Surface Science and Catalysis*, vol. 58, Elsevier, Amsterdam.

Vinegar, H.J. & Waxman, M.H. (1984) Induced polarization of shaly sands. *Geophysics*, **49**, 1267–1287.

von Smoluchowski, M. (1906) Zur kinetischen Theorie der Brownschen Molekularbewegung und der Suspensionen (in German). *Annalen der Physik*, **326(14)**, 756–780.

Wang, M. & Revil, A. (2010) Electrochemical charge of silica surface at high ionic strength in narrow channels. *Journal of Colloid and Interface Science*, **343**, 381–386, 2010.

Watilllon, A. & De Backer, R. (1970) Potentiel d'écoulement, courant d'coulement et conductance de surface à l'interface eau-verre. *Journal of Electroanalytical Chemistry*, **25**, 181–196.

White, B.S. (2005) Asymptotic Theory of Electroseismic Prospecting. *SIAM Journal of Applied Mathematics*, **65(4)**, 1443–1462.

White, B.S. & Zhou, M. (2006) Electroseismic prospecting in layered media. *SIAM Journal of Applied Mathematics*, **67(1)**, 69–98.

Yeager, E., Bugosh, J., Hovorka, F., & McCarthy, J. (1949) The application of ultrasonics to the study of electrolyte solutions. II. The detection of Debye effect. *Journal of chemical Physics*, **17(4)**, 411–415.

Zana, R. & Yeager, E. (1967a) Quantitative studies of ultrasonic vibration potentials in polyelectrolyte solutions. *The Journal of Physical Chemistry*, **71(11)**, 3502–3520.

Zana, R. & Yeager, E. (1967b) Ultrasonic vibration potentials in tetraalkylammonium halide solutions. *The Journal of Physical Chemistry*, **71(13)**, 4241–4244.

Zana, R. & Yeager, E., (1967c) Ultrasonic vibration potentials and their use in the determination of ionic partial molal volumes, *The Journal of Physical Chemistry*, **71(13)**, 521–535.

Zana, R. & Yeager, E. (1982) Ultrasonic vibration potentials. *Modern Aspects of Electrochemistry*, **14**, 3–60.

Zhu, Z. & Toksöz, M.N. (2013) Experimental measurements of the streaming potential and seismoelectric conversion in Berea sandstone. *Geophysical Prospecting*, **61(3)**, 688–700, doi: 10.1111/j.1365-2478.2012.01110.

Zundel, J.P. & Siffert, B. (1985) Mécanisme de rétention de l'octylbenzene sulfonate de sodium sur les minéraux argileux. Solid-Liquid Interactions in Porous Media, pp. 447–462, Technip, Paris.

CHAPTER 2

Seismoelectric theory in saturated porous media

We present in this chapter a complete theory for the generation of seismoelectric effects in the quasistatic limit of the Maxwell equations and for various types of rheological constitutive laws for the porous material and the pore fluid. We start with a description of the poroelastic wave propagation in a poroelastic material filled by a viscoelastic fluid that can sustain shear stresses (extended Biot–Frenkel theory). This represents the general case of wave propagation that will be discussed in this chapter. Then we present the equations describing the propagation of the seismic waves in a poroelastic material saturated by a Newtonian fluid (classical Biot–Frenkel theory) as a special case of the more general theory. In this second case, we will describe the properties of the most important sensitivity coefficient entering into the coupled equations, the so-called streaming potential coupling coefficient.

2.1 Poroelastic medium filled with a viscoelastic fluid

Traditionally, we consider the wave propagation inside the framework of the classical Biot–Frenkel theory. In this theory, the skeleton of the porous material is considered to be linear elastic while the pore fluid is Newtonian. The present section provides a constitutive model of the seismoelectric response of a charged porous medium filled with a (Maxwell) viscoelastic solvent like a heavy wet oil. We will show that this approach is more general

than the classical Biot–Frenkel theory and provides interesting mechanisms for some resonance of the fluid inside the pore space of the material. At low frequencies, it has been observed that electrokinetic phenomena in oil-wet porous media are similar to those existing in water-saturated porous materials (Yasufuku et al., 1977; Alkafeef et al., 2001; Alkafeef & Smits, 2005). Various authors have shown that the components of oil that are responsible for wettability are also polar components (Buckley & Liu, 1998; Alkafeef et al., 2006) and therefore they are good solvents (Delgado et al., 2007). We will show in this chapter that the extended Biot–Frenkel symmetry for such a Maxwell fluid yields symmetric constitutive relationships and the existence of two (compressional) P-waves and two shear (S-)waves.

2.1.1 Properties of the two phases

We start this section by describing the constitutive model of the stress–strain relationships for a viscoelastic fluid such as a wet oil. A linear Maxwell fluid model consists of a linear dashpot in series with a linear spring. In this situation, the fluid behaves like a Newtonian viscous fluid at low frequency (or for long time scales) and like a solid at high frequencies (or short time scales). The transition frequency is discussed in the following text. We denote by T_{ij} (in Pa) and e_{ij} (dimensionless) the components of the stress and deformation tensors $\mathbf{T}_f$ and $\mathbf{e}_f$ of this fluid (indicated by the subscript f), respectively. Both $\mathbf{T}_f$ and $\mathbf{e}_f$ are symmetric second-order tensors. The stress

The Seismoelectric Method: Theory and Applications, First Edition. André Revil, Abderrahim Jardani, Paul Sava and Allan Haas.
© 2015 John Wiley & Sons, Ltd. Published 2015 by John Wiley & Sons, Ltd.

components of the elastic and viscous contributions of the Maxwell fluid are given by

$$\left(T_{ij}\right)^{e} = \lambda_{f}(e_{kk})^{e}\delta_{ij} + 2G_{f}\left(e_{ij}\right)^{e}, \tag{2.1}$$

$$\left(T_{ij}\right)^{v} = 2\eta_{f}\left(\dot{e}_{ij}\right)^{v} - p_{f}\delta_{ij}, \tag{2.2}$$

where the superscripts "e" and "v" stand for "elastic" and "viscous" contributions to the stress tensor and to the deformation tensor, the dot above the symbol denotes the first time derivative, λ_{f} and G_{f} (both in Pa) are the Lamé and shear moduli of the fluid, η_{f} is the dynamic (shear) viscosity, and p is the local fluid pressure. We use the (Einstein) convention that repeated indices represent summation.

The bulk modulus of the fluid K_{f} (in Pa) is defined by the relationship $-p_{f}\mathbf{I} = K_{f}(\nabla \cdot \mathbf{u}_{f})\mathbf{I}$ where $\mathbf{I}$ denotes the identity tensor and $\mathbf{u}_{f}$ (in m) the displacement of the fluid phase. The bulk modulus of the fluid is related to the Lamé constants by $K_{f} = \lambda_{f} + (2/3)G_{f}$. For a Maxwell fluid, the components of the stress tensor of the fluid, $\mathbf{T}_{f}$, are given by $T_{ij} = \left(T_{ij}\right)^{e} = \left(T_{ij}\right)^{v}$ and $\dot{e}_{ij} = \left(\dot{e}_{ij}\right)^{e} + \left(\dot{e}_{ij}\right)^{v}$. Using Equations (2.1) and (2.2) with $\dot{e}_{ij} = \left(\dot{e}_{ij}\right)^{e} + \left(\dot{e}_{ij}\right)^{v}$ yields

$$\dot{T}_{ij} + \frac{G_{f}}{\eta_{f}}\left(T_{ij} + p_{f}\delta_{ij}\right) = 2G_{f}\dot{e}_{ij} + \lambda_{f}(\dot{e}_{kk})^{e}\delta_{ij}. \tag{2.3}$$

Assuming harmonic variations of the stress (i.e., the stress oscillates in time as $e^{-i\omega t}$, i denotes the pure imaginary number as in Chapter 1), we can use the previous equation in the frequency domain using a Fourier transform (we keep the same notations of the time and frequency domain variables). The characteristic frequency is defined as $\omega_{m} = G_{f}/\eta_{f}$ and the associated relaxation time is $\tau_{m} = \eta_{f}/G_{f}$. Typically, for the Mexican crude oil investigated by Dante et al. (2007), the critical frequency depends on the temperature and is in the range 10–100 Hz in the temperature range 20–40°C. For heavy oils in tar sands, for example, the critical frequency of resonance can occur at much lower frequencies. Usually, heavy oils are heated to decrease their dynamic viscosity, and as a result, the resonance frequency can shift into the seismic frequency band (see Behura et al., 2007), depending on the resulting change in dynamic viscosity.

In the frequency domain, Equation (2.3) is written as

$$\begin{aligned} T_{ij} = &-p_{f}^{*}\delta_{ij} + 2\eta_{f}^{*}\left(-i\omega e_{ij}\right)^{v} \\ &- \frac{i\omega\tau_{m}}{1 - i\omega\tau_{m}}\left[\lambda_{f}(e_{kk})^{e}\delta_{ij} + 2G_{f}\left(e_{ij}\right)^{e}\right], \end{aligned} \tag{2.4}$$

$$p_{f}^{*} = \frac{p_{f}}{1 - i\omega\tau_{m}}, \tag{2.5}$$

$$\eta_{f}^{*} = \frac{\eta_{f}}{1 - i\omega\tau_{m}}. \tag{2.6}$$

The quantities p_{f}^{*} and η_{f}^{*} denote effective fluid pressure (in Pa) and dynamic viscosity (in Pa s), respectively. At low frequencies $\omega \ll \omega_{m}$, we have $p_{f}^{*} = p_{f}$ and $\eta_{f}^{*} = \eta_{f}$, the third term on the right-hand side of Equation (2.4) is negligible, and the stress tensor is given by (e.g., De Groot and Mazur, 1984)

$$\mathbf{T}_{f} = K_{f}(\nabla \cdot \mathbf{u}_{f})\mathbf{I} + 2\eta_{f}\left(\dot{\mathbf{d}}_{f}\right)^{v}, \tag{2.7}$$

where the $\left(\dot{\mathbf{d}}_{f}\right)^{v}$ corresponds to the viscous contribution to the fluid strain deviator,

$$\left(\dot{\mathbf{d}}_{f}\right)^{v} = \frac{1}{2}\left(\nabla \mathbf{v}_{f} + (\nabla \mathbf{v}_{f})^{T}\right) - \frac{1}{3}(\nabla \cdot \mathbf{v}_{f})\mathbf{I}. \tag{2.8}$$

The vector $\mathbf{d}_{f}$ (dimensionless) denotes the fluid strain deviator, $\mathbf{v}_{f}$ (m s^{-1}) corresponds to the velocity of the fluid, and the superscript T means transpose. The deformation tensor is related to the deviator by

$$\mathbf{e}_{f} = \left(\frac{1}{3}\right)(\nabla \cdot \mathbf{u}_{f})\mathbf{I} + \mathbf{d}_{f}. \tag{2.9}$$

At low frequencies, the fluid behaves therefore as a Newtonian fluid. At high frequencies $\omega >> \omega_{m}$, the stress tensor is given from Equation (2.4) by

$$\mathbf{T}_{f} = K_{f}(\nabla \cdot \mathbf{u}_{f})\mathbf{I} + 2G_{f}\mathbf{d}_{f}. \tag{2.10}$$

Therefore, at high frequencies, the pore fluid behaves as an elastic material.

A singular perfect Maxwell fluid has a relaxation time distribution described by a Dirac (delta) function. However, the real fluids we consider here are complex mixtures of various different fluids, and as such, these mixtures do not behave like a single, perfect Maxwell fluid. Instead, they behave more like a generalized Maxwell fluid with a variety of relaxation times. From this, we can create a more realistic and generalized model using a distribution of relaxation times. For instance, a Cole–Cole distribution of relaxation times can be used as a generalized model (e.g., Revil et al., 2006; Behura et al., 2007, for the Uvalde heavy oil) to demonstrate our analytical approach. The normalized probability distribution function for the relaxation time of a Cole–Cole distribution is given by (Cole & Cole, 1941)

$$f(\tau) = \frac{1}{2\pi} \frac{\sin[\pi(c-1)]}{\cosh[c\ln(\tau/\tau_m)] - \cos[\pi(c-1)]}, \quad (2.11)$$

where $0 \le c \le 1$ is called the Cole–Cole exponent ($c = 1$ corresponds to an ideal Maxwell fluid). The Cole–Cole distribution is symmetric with respect to the characteristic time $\tau = \tau_m$, which corresponds to the peak of the distribution. For $0.5 \le c \le 1$, the Cole–Cole distribution of relaxation times looks like a lognormal distribution. The tail of the Cole–Cole distribution is however increasingly broad as c decreases. A comparison between the Maxwell and Cole–Cole behavior of the resonances of a generalized Maxwell fluid is shown in Figure 2.1. In this figure, we compare experimental data from Castrejon-Pita et al. (2003) for a Maxwell fluid (a mixture of cetylpyridinium chloride and sodium salicylate in a capillary) with a model in which the Cole–Cole distribution of relaxation times is used. An extended Maxwell fluid with a Cole–Cole distribution of relaxation times is able to represent the hydrodynamic behavior of the fluid, while a simple Maxwell fluid (with a single relaxation time) cannot perform that task.

The behavior of the Maxwell fluid can be considered from the standpoint of the behavior of an equivalent Newtonian viscous fluid with frequency-dependent bulk modulus and dynamic shear viscosity:

$$\mathbf{T}_f = K_f^*(\omega)(\nabla \cdot \mathbf{u}_f)\mathbf{I} + 2\eta_f^*(\omega)\left(\dot{\mathbf{d}}_f\right)^v$$
$$- \frac{(i\omega\tau_m)^c}{1-(i\omega\tau_m)^c}[K_f(\nabla \cdot \mathbf{u}_f)\mathbf{I} + 2G_f(\mathbf{d}_f)^e], \quad (2.12)$$

$$K_f^*(\omega) = \frac{K_f}{1-(i\omega/\omega_m)^c}, \quad (2.13)$$

$$K_f^*(\omega)(\nabla \cdot \mathbf{u}_f) = -p_f^*, \quad (2.14)$$

$$p_f^*(\omega) = \frac{p_f}{1-(i\omega/\omega_m)^c}, \quad (2.15)$$

$$\eta_f^*(\omega) = \frac{\eta_f}{1-(i\omega/\omega_m)^c}. \quad (2.16)$$

Using the Boussinesq approximation, the flow of a fluid inside a deformable porous material is governed locally by the Navier–Stokes equation, which acts as a momentum conservation equation:

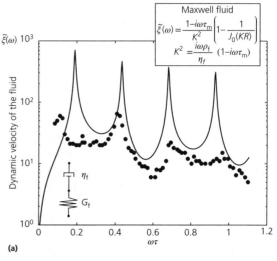

(a)

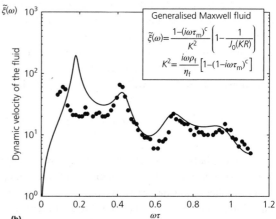

(b)

Figure 2.1 Hydrodynamic response of a viscoelastic fluid in a linear capillary of radius $R = 2.5$ cm subject to harmonic fluid pressure variations. The fluid is a mixture of cetylpyridinium chloride and sodium salicylate (CPyCl/NaSal 60:100) (data from Castrejon-Pita et al., 2003). The experimental data are scaled with respect to the viscous relaxation time $\tau = R^2\rho_f/\eta_f$ ($\eta_f = 60$ Pa s, $\rho_f = 1050$ kg m^{-3}). The parameter $\tilde{\xi}$ denotes the normalized conductance (relative to the DC value) and J_0 is the Bessel function. **a)** Comparison between the experimental data and the response predicted by a linear Maxwell fluid ($\tau_m = 1.9$ s). **b)** Comparison between the experimental data and the response predicted by a generalized Maxwell fluid using a Cole–Cole distribution of relaxation times ($\tau_m = 2.9$ s, Cole–Cole exponent $c = 0.9$, parameters optimized with a Newton–Raphson minimization method using the L$_2$ norm). J_0 denotes the Bessel function of zero order.

$$\rho_f \frac{\partial \mathbf{v}_f}{\partial t} = -\nabla p_f + \nabla \cdot \boldsymbol{\pi} + \mathbf{F}_f, \quad (2.17)$$

$$\nabla \cdot \mathbf{v}_{\mathrm{f}} = 0, \qquad (2.18)$$

respectively, where ρ_{f} denotes the local mass density of the fluid, $\mathbf{F}_{\mathrm{f}}$ is the body force, and

$$\boldsymbol{\pi} = \mathbf{T}_{\mathrm{f}} + p_{\mathrm{f}}^{*}\mathbf{I} + \frac{i\omega\tau_{\mathrm{m}}}{1 - i\omega\tau_{\mathrm{m}}}\left[K_{\mathrm{f}}(\nabla \cdot \mathbf{u}_{\mathrm{f}})\mathbf{I} + 2G_{\mathrm{f}}(\mathbf{d}_{\mathrm{f}})^{e} \right] \qquad (2.19)$$

represents the viscous (deviatoric) stress tensor of the fluid. For a purely viscous Newtonian fluid, we have $\nabla \cdot \boldsymbol{\pi} = \eta_{\mathrm{f}}\nabla^{2}\mathbf{v}_{\mathrm{f}}$. Assuming a linearized Maxwell model and using Equation (2.4), the viscous stress tensor $\boldsymbol{\pi}$ obeys

$$\dot{\boldsymbol{\pi}} + \left(\frac{G_{\mathrm{f}}}{\eta_{\mathrm{f}}} \right)\boldsymbol{\pi} = 2G_{\mathrm{f}}\dot{\mathbf{e}}_{\mathrm{f}}, \qquad (2.20)$$

$$\tau_{\mathrm{m}}\dot{\boldsymbol{\pi}} + \boldsymbol{\pi} = \eta_{\mathrm{f}}\nabla\mathbf{v}_{\mathrm{f}}. \qquad (2.21)$$

In the frequency domain, Equation (2.21) yields

$$(1 - i\omega\tau_{\mathrm{m}})\boldsymbol{\pi} = \eta_{\mathrm{f}}\nabla\mathbf{v}_{\mathrm{f}}, \qquad (2.22)$$

$$\nabla \cdot \boldsymbol{\pi} = \frac{\eta_{\mathrm{f}}}{1 - i\omega\tau_{\mathrm{m}}}\nabla^{2}\mathbf{v}_{\mathrm{f}}. \qquad (2.23)$$

Inserting Equation (2.23) inside the Navier–Stokes equation, Equation (2.17), yields

$$-i\omega\rho_{\mathrm{f}}\mathbf{v}_{\mathrm{f}} = -\nabla p_{\mathrm{f}} + \eta_{\mathrm{f}}^{*}\nabla^{2}\mathbf{v}_{\mathrm{f}} + \mathbf{F}_{\mathrm{f}}. \qquad (2.24)$$

Equation (2.24) has the same form than the Navier–Stokes equation of a viscous Newtonian fluid. This equation can be used by replacing the classical viscosity η_{f} by an effective (time- or frequency-dependent) viscosity η_{f}^{*}. The result of the analysis in this section is that a Cole–Cole distribution of relaxation times can be used to describe a generalized Maxwell fluid (see Figure 2.1b).

We consider that the solid phase is formed by a mono-mineralic isotropic solid material assumed to be perfectly elastic. The local elastic equation of motion for the solid phase is

$$\nabla \cdot \mathbf{T}_{\mathrm{s}} + \mathbf{F}_{\mathrm{s}} = \rho_{\mathrm{s}}\frac{\partial^{2}\mathbf{u}_{\mathrm{s}}}{\partial t^{2}}, \qquad (2.25)$$

where $\mathbf{u}_{\mathrm{s}}$ (in m) denotes the displacement of the solid phase, ρ_{s} (in kg m^{-3}) is the bulk density of the solid phase, and $\mathbf{F}_{s}$ is the body force applied to the grains (Pa m^{-1} or N m^{-3}). The microscopic solid stress tensor is given by

$$\mathbf{T}_{\mathrm{s}} = K_{\mathrm{s}}(\nabla \cdot \mathbf{u}_{\mathrm{s}})\mathbf{I} + 2G_{\mathrm{s}}\mathbf{d}_{\mathrm{s}}, \qquad (2.26)$$

$$\mathbf{d}_{\mathrm{s}} = \frac{1}{2}\left[\nabla\mathbf{u}_{\mathrm{s}} + \nabla\mathbf{u}_{\mathrm{s}}^{\mathrm{T}} \right] - \frac{1}{3}(\nabla \cdot \mathbf{u}_{\mathrm{s}})\mathbf{I}, \qquad (2.27)$$

where $\mathbf{d}_{\mathrm{s}}$ is the solid strain deviator (dimensionless), K_{s} (in Pa) is the bulk modulus of the solid phase assumed to be isotropic, and G_{s} is the shear modulus (in Pa).

2.1.2 Properties of the porous material

Next, we derive the coupled constitutive equations between the Darcy velocity and the total current density for a linear poroelastic material saturated by a polar (wet) oil (Figure 2.2). The non-mechanical properties used in this section are provided in Table 2.1 while the mechanical properties are summarized in Table 2.2. We look for an average force balance equation on the fluid in relative motion with respect to the solid phase. We introduce the relative flow velocity $\mathbf{v} = \mathbf{v}_{\mathrm{f}} - \bar{\mathbf{v}}_{\mathrm{s}}$ (nm s^{-1}) where $\bar{\mathbf{v}}_{\mathrm{s}} = \dot{\bar{\mathbf{u}}}_{\mathrm{s}}$ is

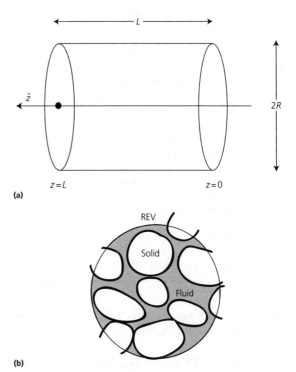

(a)

(b)

Figure 2.2 Sketch of the porous material. **a)** The representative elementary volume (REV) of the porous material is an averaging disk of radius R and length L. **b)** The REV corresponds to a porous body with an elastic skeleton filled with a viscoelastic fluid characterized by an extended Maxwell rheology.

Table 2.1 Nomenclature of the nonmechanical material properties.

Symbol	Meaning	Unit
F	Formation factor	Dimensionless
f	Fraction of counterions in the Stern layer	Dimensionless
m	Cementation exponent	Dimensionless
n	Saturation exponent	Dimensionless
λ	Brooks and Corey exponent	Dimensionless
r	Coupling coefficient saturation exponent	Dimensionless
q	Characteristic time saturation exponent	Dimensionless
k_s	Permeability at saturation	m^2
k_0	Low-frequency permeability	m^2
k_r	Relative permeability	Dimensionless
k^*	Complex permeability	m^2
K_h	Hydraulic conductivity	$m\,s^{-1}$
K_s	Hydraulic conductivity at saturation	$m\,s^{-1}$
C_s	Coupling coefficient at saturation	$V\,m^{-1}$
C_0	Low-frequency coupling coefficient	$V\,m^{-1}$
C_r	Relative coupling coefficient	Dimensionless
C_p^*	Complex streaming potential coupling coefficient	$V\,m^{-1}$
C_{os}^*	Electroosmotic coupling coefficient	$C\,m^{-3}$
ϕ	Porosity	Dimensionless
σ	Electrical conductivity	$S\,m^{-1}$
σ'	In-phase electrical conductivity	$S\,m^{-1}$
σ''	Quadrature conductivity	$S\,m^{-1}$
σ_{eff}^*	Complex effective electrical conductivity	$S\,m^{-1}$
σ_{eff}^*	(Real) effective electrical conductivity	$S\,m^{-1}$
σ_s	Surface conductivity of the solid phase	$S\,m^{-1}$
σ^*	Complex electrical conductivity	$S\,m^{-1}$
ε_{eff}	(Real) effective dielectric constant	$F\,m^{-1}$
ε	Dielectric constant	$F\,m^{-1}$
$L^*(\omega)$	Streaming current coupling coefficient	$A\,m^{-2}$
$\hat{Q}_V^0$	Moveable charge density at low frequency	$C\,m^{-3}$
$\hat{Q}_V^\infty$	Moveable charge density at high frequency	$C\,m^{-3}$
Q_V	Total charge density from the CEC	$C\,m^{-3}$
$\bar{Q}_V$	Charge density of the diffuse layer	$C\,m^{-3}$
μ	Magnetic permeability	$H\,m^{-1}$

Table 2.2 Nomenclature of the mechanical material properties.

Symbol	Meaning	Unit
C	Biot modulus	Pa
C_s	Biot modulus at saturation	Pa
M	Biot modulus	Pa
M_s	Biot modulus at saturation	Pa
α	Biot coefficient at saturation	Dimensionless
α_w	Biot coefficient at partial saturation	Dimensionless
$G = G_{fr}$	Shear modulus of the solid frame	Pa
$K = K_{fr}$	Bulk modulus of the solid frame	Pa
λ	Lamé coefficient	Pa
λ_u	Undrained Lamé coefficient	Pa
K_a	Bulk modulus of the air	Pa
K_o	Bulk modulus of the oil	Pa
K_w	Bulk modulus of the water	Pa
K_s	Bulk modulus of the solid phase	Pa
K_f	Bulk modulus of the fluid phase	Pa
K_u	Undrained bulk modulus	Pa

the generalized Maxwell fluid (see discussion in Section 2.1.1). The body force applied to the pore fluid corresponds to the body force due to the gravity field plus the electrostatic force associated with the existence of the local net electrical charge density ρ ($C\,m^{-3}$)

$$\mathbf{F}_f = \rho_f \mathbf{g} + \rho \mathbf{E} e^{-i\omega t}, \qquad (2.30)$$

where ρ_f (in $kg\,m^{-3}$) is the mass density of the pore water, $\mathbf{g}$ (in $m\,s^{-2}$) the acceleration of the gravity, and $\mathbf{E}$ is the electric field (in $V\,m^{-1}$). The average force $\bar{\mathbf{F}}_f$ denotes a source of momentum in the momentum conservation equation (see Section 2.1.4). Next, we average Equation (2.28) over the fluid phase which yields

$$-i\omega \rho_f \bar{\mathbf{v}} = -\nabla \bar{p}_f + i\omega \rho_f \bar{\mathbf{v}}_s + \eta_f^* \nabla^2 \bar{\mathbf{v}}_f + \bar{\mathbf{F}}_f. \qquad (2.31)$$

In the low frequency viscous laminar flow regime, the no-slip boundary condition at the surface of the pores is written as $\bar{\mathbf{v}} = 0$.

As discussed in Chapter 1, the local charge density, ρ (in $C\,m^{-3}$), is associated with the electrical diffuse layer part of the electrical double layer (see Figure 1.2) and is therefore dependent on the position inside the pore space. This local charge density is related to the concentrations of the ionic species that are affected by the coulombic field created by the effective surface charge density of the mineral surface. As discussed in Chapter 1,

the phase average velocity of the solid phase. In the frequency domain, using Equation (2.24), we have

$$-i\omega \rho_f \mathbf{v} = -\nabla p_f + i\omega \rho_f \bar{\mathbf{v}}_s + \eta_f^* \nabla^2 \mathbf{v}_f + \mathbf{F}_f, \qquad (2.28)$$

$$\eta_f^*(\omega) = \frac{\eta_f}{1 - (i\omega \tau_m)^c}, \qquad (2.29)$$

where c is the Cole–Cole distribution parameter corresponding to the distribution of the relaxation times for

once averaged, ρ can be replaced by the effective charge density:

$$\hat{Q}_V^0 \ll \bar{Q}_V = \sum_{i=1}^{N} q_i \bar{C}_i, \qquad (2.32)$$

where $\bar{C}_i$ is the phase average of the concentration of species i in the pore space and q_i the charge (in C) of the ionic species i ($q_i = 0$ for neutral species) and $\bar{Q}_V$ denotes the total charge density per unit pore volume of the diffuse layer. The effective charge density has been introduced in Chapter 1 and its derivation will not be repeated here.

The boundary value problem for the fluid flow is expressed on the surface of an averaging cylinder (see Pride, 1994) by

$$\eta_f^* \nabla^2 \mathbf{v} + i\omega \rho_f \mathbf{v} = \nabla p_f - \mathbf{F}_f, \qquad (2.33)$$

$$\nabla \cdot \mathbf{v} = 0, \qquad (2.34)$$

$$\mathbf{v} = 0 \text{ on } S, \qquad (2.35)$$

$$p_f = \begin{cases} \hat{\mathbf{z}} \cdot \left(\nabla \bar{p}_f - i\omega \rho_f \dot{\bar{\mathbf{u}}}_s \right), \ z = L \\ 0, \ z = 0 \end{cases} \qquad (2.36)$$

where S represents the solid–pore fluid interface and $\hat{\mathbf{z}}$ is the unit vector normal to the disk face of the representative elementary volume. This boundary value problem is exactly the same as the one given by Pride (1994) for a poroelastic porous body saturated by a Newtonian fluid, except that η_f should be replaced by η_f^*. We can therefore easily generalize the results obtained by Pride (1994) to a poroelastic material saturated by a viscoelastic solvent. This yields the following modified Darcy equation:

$$-i\omega \bar{\mathbf{w}} = \frac{k(\omega)}{\eta_f} \left[-\nabla \bar{p}_f + \rho_f \omega^2 \bar{\mathbf{u}}_s + \bar{\mathbf{F}}_f \right] - C(\omega)\sigma \mathbf{E}, \qquad (2.37)$$

where $\bar{\mathbf{w}} = \varphi(\bar{\mathbf{u}}_f - \bar{\mathbf{u}}_s)$ is the phase average filtration displacement of the fluid phase relative to the mineral framework, ϕ is the interconnected porosity, $k(\omega)$ is the dynamic permeability, $C(\omega)$ is the dynamic coupling coefficient described in the following text, and σ is the electrical conductivity of the porous material. Note that the time derivative of the averaged filtration displacement $\dot{\bar{\mathbf{w}}}(\omega) = \phi \bar{\mathbf{v}}(\omega)$ corresponds to the Darcy velocity, that is, the flux density of water (i.e., the volume of water

passing per unit surface area of a cross section of the porous material per unit time).

Expression for the electrical conductivity σ can be obtained by upscaling the local Nernst–Planck equation. We assume that the wet oil carries a net charge per unit volume to compensate the fixed charge density at the surface of the minerals. As discussed in Chapter 1, the total current density $\mathbf{J}$ (in A m^{-2}) of the porous material is given by $\mathbf{J} = \sigma \mathbf{E} + \hat{Q}_V^0 \dot{\bar{\mathbf{w}}}$. In this equation, the last term corresponds to the source current density given by the product between an effective charge density and the Darcy velocity. This equation differs from the more conventional form that uses the product of a cross-coupling coefficient with the gradient of fluid pressure and an inertial coupling term. The present approach has fewer parameters than the classical approach because the charge density, $\hat{Q}_V^0$, can be determined from the low-frequency permeability, k_0, alone (see Figure 1.9).

At low frequency, the electrical field, $\mathbf{E}$, is related to the electrical potential, ψ, by $\mathbf{E} = -\nabla \psi$, which satisfies $\nabla \times \mathbf{E} = 0$. The low-frequency coupling coefficient is given by

$$C_0 = \lim_{\omega \to 0} C(\omega) = \left(\frac{\partial \psi}{\partial p_f^0} \right)_{\mathbf{J}=0} = -\frac{\hat{Q}_V^0 k_0}{\eta_f \sigma}. \qquad (2.38)$$

The generalized Darcy's and Ohm's laws appear therefore as cross-coupled constitutive equations

$$\begin{bmatrix} -i\omega \bar{\mathbf{w}} \\ \mathbf{J} \end{bmatrix} = \begin{bmatrix} \dfrac{k(\omega)}{\eta_f} & -C(\omega)\sigma \\ -C(\omega)\sigma & \sigma \left(1 + \dfrac{c(\omega)^2 \eta_f \sigma}{k(\omega)} \right) \end{bmatrix} \begin{bmatrix} -\nabla \bar{p}_f + \rho_f \omega^2 \bar{\mathbf{u}}_s + \bar{\mathbf{F}}_f \\ \mathbf{E} \end{bmatrix}, \qquad (2.39)$$

where $\bar{\mathbf{F}}_f$ is the applied body force acting on the fluid phase.

The normalized dynamic permeability is given by

$$\tilde{k}(\omega) \equiv \frac{k(\omega)}{k_0} = \frac{1 - (i\omega\tau_m)^c}{1 - i(\omega/\omega_c)(1 - (i\omega\tau_m)^c)}, \qquad (2.40)$$

where $k_0 = \Lambda^2/(8F)$ (Johnson, 1986) is the permeability at low frequencies $\omega \ll \text{Min}(\omega_c, \omega_m)$, $\omega_m = 1/\tau_m$, and $\omega_c = \eta_f/(k_0 \rho_f F)$. Both F and Λ are two textural parameters defined by (Johnson, 1986)

$$\frac{1}{F} = \frac{1}{V} \int_{V_p} \mathbf{e}^2 dV_p, \tag{2.41}$$

$$\Lambda = 2 \frac{\int_{V_p} \mathbf{e}^2 dV_p}{\int_S \mathbf{e}^2 dS}, \tag{2.42}$$

where dV_p denotes an integration over the pore space and dS denotes an integration over the pore–solid interface. The normalized electrical field $\mathbf{e} = \nabla\Gamma$ is obtained by solving the following canonical Laplace problem for an average disk in the direction of the macroscopic electrical field (see Pride, 1994): Γ has units of length and satisfies the Laplace equation $\nabla^2\Gamma = 0$ throughout the pore space with boundary conditions $\Gamma = 0$ on $z = 0$ and $\Gamma = L$ on $z = L$ and $\hat{\mathbf{n}} \cdot \nabla\Gamma = 0$ on both the internal interface between the solid and fluid phases ($\hat{\mathbf{n}}$ is the unit vector at the solid–fluid interface pointing inside the solid) and on the external surface of the averaging disk ($\hat{\mathbf{n}}$ is the outward unit vector). This approach assumed implicitly the continuity of the solid phase at the scale of a representative elementary volume of the porous rock. Note that Pride did not use the formation factor in his model but instead he defined the bulk tortuosity of the pore space, which is $\alpha_\infty = F\phi$.

The dynamic streaming potential coupling coefficient is given by

$$C(\omega) = -\frac{\hat{Q}_V^0 k(\omega)}{\eta_f \sigma}. \tag{2.43}$$

The normalized coupling coefficient is given by

$$\tilde{c}(\omega) \equiv \frac{C(\omega)}{C_0} = \tilde{k}(\omega). \tag{2.44}$$

Therefore, in our approach, the normalized dynamic coupling coefficient is equal to the normalized dynamic permeability, the normalization being made with respect to the low-frequency values of these parameters k_0 and C_0.

Using the thin double layer assumption (small size of the diffuse layer with respect to the size of the pores), the dynamic relative coupling coefficient of a porous material saturated by a viscous Newtonian fluid is given by (Pride, 1994)

$$\tilde{c}(\omega) = \frac{1}{\sqrt{1 - 2i\omega/\omega_c}}, \tag{2.45}$$

and $c_0 = \varepsilon_f\zeta/(\eta_f\sigma_f)$ (Helmholtz–Smoluchowski equation; see von Smoluchowski, 1906) where ζ is the so-called zeta potential (the inner potential of the electrical diffuse layer; see Chapter 1).

Using the dynamic viscosity defined by Equation (2.16) and using Equation (2.45), the dynamic relative coupling coefficient of a porous material saturated by a generalized Maxwell fluid in the thin double layer assumption is given by

$$\tilde{c}(\omega) = \frac{1 - (i\omega\tau_m)^c}{\sqrt{1 - 2i(\omega/\omega_c)(1 - (i\omega\tau_m)^c)}}. \tag{2.46}$$

In Equation (2.46), the effect of the viscosity appears twice: in the numerator because of the dependence of C_0 with the viscosity and in the last term in the denominator because of the dependence of the critical frequency ω_c with the viscosity. It is interesting to see that Equation (2.46) has the same asymptotic limit at low frequencies as Equations (2.40) and (2.44) corresponding to a thick electrical double layer (using the Donnan assumption). This limit is given by $1 + i(\omega/\omega_c)$ for a Newtonian fluid.

The excitation of the pore fluid in the resonance frequency band of the viscoelastic fluid yields an amplification of the value of the streaming potential coupling coefficient by several orders of magnitude with respect to the value of the streaming potential coupling coefficient at low frequencies. We expect that the determination of $\tilde{c}(\omega)$ in a wide range of frequencies (e.g., 0.01–10 kHz in the laboratory) could help to characterize the properties of the oil contained in the pores through the determination of both τ_m, the peak of the relaxation time, and α, which characterizes the broadness of the distribution of relaxation times associated with the composition of the oil in terms of polymers.

With the basis established earlier in this section, we can determine, within an order of magnitude, the DC value of the streaming potential coupling coefficient of a sandstone filled with a polar oil. For the Berea sandstone (permeability $\sim 10^{-13}$ m^2, porosity ~ 0.18), an analysis of the data displayed by Alkafeef et al. (2006) for the streaming current density versus the fluid pressure gradient yields $\hat{Q}_V^0 = 10^{-4}$ C m^{-3}. Alkafeef et al. (2006) used a crude oil containing asphaltene, which is a strongly polar polymer responsible for the high dielectric constant for this fluid. The analysis by Alkafeef and coworkers implied a charge density that is 6 orders of magnitude smaller than for a sandstone of equal permeability filled by an electrolyte. The shear viscosity of the crude oil in the Alkafeef et al. (2006) samples was equal to 5×10^{-3} Pa s at 25°C. Using $\sigma = 2 \times 10^{-10}$ S m^{-1}, we compute a streaming potential coupling coefficient on the order of $C_0 = -4 \times 10^{-5}$ V Pa^{-1}. So the low value of the charge density is more than compensated for by the effect of the high value of the electrical resistivity of the fluid. Indeed, -4×10^{-5} V Pa^{-1} is a very high value by comparison with the value of the streaming potential coupling coefficient of a typical soil or reservoir rock saturated by a brine (typically in the range from 10^{-6} to 10^{-8} V Pa^{-1}: see Revil et al., 2003). This implies that even well below the relaxation frequency of the resonance of the solvent, the streaming potential coupling coefficient of a sandstone filled with such an oil can reach a very large value.

2.1.3 The mechanical equations

The bulk stress tensor is defined by

$$\bar{\mathbf{T}} = (1-\varphi)\bar{\mathbf{T}}_s + \varphi\bar{\mathbf{T}}_f \qquad (2.47)$$

where $\bar{\mathbf{T}}_s$ and $\bar{\mathbf{T}}_f$ are the phase average stress tensors in the solid phase and viscoelastic fluid, respectively. The confining pressure and the deviatoric stress of the porous body are defined by

$$P \equiv -\frac{tr(\bar{\mathbf{T}})}{3} = (1-\phi)\bar{p}_s + \phi\bar{p}_f, \qquad (2.48)$$

$$\bar{\mathbf{T}}^D \equiv \bar{\mathbf{T}} + P\mathbf{I}, \qquad (2.49)$$

where $\bar{p}_s$ is the mean pressure in the solid phase. The deviatoric stress is then given by

$$\bar{\mathbf{T}}^D = (1-\phi)\bar{\mathbf{T}}_s^D + \phi\bar{\boldsymbol{\pi}}, \qquad (2.50)$$

where $\bar{\mathbf{T}}_s^D = \bar{\mathbf{T}}_s + \bar{p}_s\mathbf{I}$ is the mean deviatoric stress of the solid phase and $\bar{\boldsymbol{\pi}}$ is the mean deviatoric stress in the pore fluid phase, which is described in Section 2.1.3.1.

2.1.3.1 Strain–stress relationships

In this section, we follow the approach of Biot and Frenkel, and we extend the classical dynamic poroelastic theory to the case of a poroelastic skeleton filled with a viscoelastic fluid. According to the rheological behavior of the pore fluid described in Section 2.1, its effective shear modulus is given in the frequency domain by

$$G_f^*(\omega) = \frac{-(i\omega/\omega_m)^c}{1-(i\omega/\omega_m)^c} G_f. \qquad (2.51)$$

The fluid does not bear any shear stresses at low frequencies (for $\omega \ll \omega_m$) and behaves as an elastic solid sustaining net shear stresses at high frequencies (for $\omega \gg \omega_m$). We define the two phase average deviators by

$$2\bar{\mathbf{d}}_s \equiv \nabla\bar{\mathbf{u}}_s + \nabla\bar{\mathbf{u}}_s^T - \frac{2}{3}(\nabla\cdot\bar{\mathbf{u}}_s)\bar{\mathbf{I}}, \qquad (2.52)$$

$$2\bar{\mathbf{d}}_f \equiv \nabla\bar{\mathbf{u}}_f + \nabla\bar{\mathbf{u}}_f^T - \frac{2}{3}(\nabla\cdot\bar{\mathbf{u}}_f)\bar{\mathbf{I}}, \qquad (2.53)$$

respectively. We introduce a relative deviator as

$$\bar{\mathbf{d}}_w \equiv \varphi(\bar{\mathbf{d}}_f - \bar{\mathbf{d}}_s), \qquad (2.54)$$

$$2\bar{\mathbf{d}}_w = \nabla[\varphi(\bar{\mathbf{u}}_f - \bar{\mathbf{u}}_s)] + \nabla[\varphi(\bar{\mathbf{u}}_f - \bar{\mathbf{u}}_s)]^T - \frac{2}{3}(\nabla\cdot[\varphi(\bar{\mathbf{u}}_f - \bar{\mathbf{u}}_s)])\bar{\mathbf{I}}, \qquad (2.55)$$

$$2\bar{\mathbf{d}}_w = \nabla\bar{\mathbf{w}} + \nabla\bar{\mathbf{w}}^T - \frac{2}{3}(\nabla\cdot\bar{\mathbf{w}})\bar{\mathbf{I}}, \qquad (2.56)$$

$$2\dot{\bar{\mathbf{d}}}_w = \nabla\dot{\bar{\mathbf{w}}} + \nabla\dot{\bar{\mathbf{w}}}^T - \frac{2}{3}(\nabla\cdot\dot{\bar{\mathbf{w}}})\bar{\mathbf{I}}. \qquad (2.57)$$

The total stress tensor and the fluid phase average stress tensors are given by

$$\bar{\mathbf{T}} = -P\mathbf{I} + \bar{\mathbf{T}}^D, \qquad (2.58)$$

$$\bar{\mathbf{T}}_f = -\bar{p}_f\mathbf{I} + \bar{\boldsymbol{\pi}}, \qquad (2.59)$$

where P is the confining pressure, $\bar{\mathbf{T}}^D$ is the macroscopic deviatoric stress, and $\bar{\boldsymbol{\pi}}$ is the phase average deviatoric stress of the pore fluid.

In linear poroelasticity, following Biot, we consider two distinct thought experiments to obtain the phase average tensors in linear isotropic porous materials. The first experiment corresponds to the case of an ideal drained experiment in which a confining pressure $P = (1-\phi)\bar{p}_s$ ($\bar{p}_s$ is the pressure in the solid phase) and deviatoric stress $\bar{\mathbf{T}}^D$ are applied to a porous sample with no change in the fluid pressure (e.g., considering an empty or drained porous material). In this case, the bulk deformation of the solid frame, the variation of the porosity, and the deviators are

$$\nabla \cdot \bar{\mathbf{u}}_s(P,0) = -\frac{P}{K_{fr}}, \tag{2.60}$$

$$\Delta \phi(P,0) = -\left(1 - \frac{K_{fr}}{K_s} - \phi\right)\frac{P}{K_{fr}}, \tag{2.61}$$

$$2\bar{\mathbf{d}}_s\left(\bar{\mathbf{T}}^D,0\right) = \frac{\mathbf{T}^D}{G_{fr}}, \tag{2.62}$$

$$2\phi(\bar{\mathbf{d}}_f)^e\left(\bar{\mathbf{T}}^D,0\right) = \left(1 - \frac{G_{fr}}{G_s} - \phi\right)\frac{1}{G_{fr}}\mathbf{T}^D, \tag{2.63}$$

where G_{fr} and K_{fr} are the shear and bulk moduli of the dry porous frame (in other words the shear and bulk moduli of the skeleton of the material, both expressed in Pa).

In the second thought experiment, we apply a fluid pressure $\bar{p}_f$ everywhere throughout the pore space and, simultaneously, a confining pressure $P = \bar{p}_f$ to the external surface of the sample. At the same time, we apply the mean deviatoric stress $\bar{\pi}$ to the pore space and, simultaneously, a deviatoric stress $\bar{\mathbf{T}}^D = \bar{\pi}$ to the external surface of the sample. In this case, the bulk deformation of the solid frame, the variation of the porosity, and the variation of the deviators are given by

$$\nabla \cdot \bar{\mathbf{u}}_s(\bar{p}_f,\bar{p}_f) = -\frac{\bar{p}_f}{K_s}, \tag{2.64}$$

$$\Delta \phi(\bar{p}_f,\bar{p}_f) = 0, \tag{2.65}$$

$$2\bar{\mathbf{d}}_s(\bar{\pi},\bar{\pi}) = \frac{\bar{\pi}}{G_s}, \tag{2.66}$$

$$2\varphi(\bar{\mathbf{d}}_f)^e(\bar{\pi},\bar{\pi}) = 0. \tag{2.67}$$

Using the superposition principle, we add together the results of the two thought experiments discussed. For the general case, the bulk deformation of the solid frame, the variation of the porosity, and the deviator are

$$\nabla \cdot \bar{\mathbf{u}}_s(P,\bar{p}_f) = \nabla \cdot \bar{\mathbf{u}}_s(P-\bar{p}_f,0) + \nabla \cdot \bar{\mathbf{u}}_s(\bar{p}_f,p_f), \tag{2.68}$$

$$\Delta \phi(P,\bar{p}_f) = \Delta \phi(P-\bar{p}_f,0) + \Delta \phi(\bar{p}_f,\bar{p}_f), \tag{2.69}$$

$$2\bar{\mathbf{d}}_s\left(\bar{\mathbf{T}}^D,\bar{\pi}\right) = 2\bar{\mathbf{d}}_s(\mathbf{T}^D-\pi,0) + 2\bar{\mathbf{d}}_s(\bar{\pi},\bar{\pi}), \tag{2.70}$$

$$2\varphi(\bar{\mathbf{d}}_f)^e\left(\bar{\mathbf{T}}^D,\bar{\pi}\right) = 2\bar{\mathbf{d}}_f(\mathbf{T}^D,0) + 2\bar{\mathbf{d}}_f(\bar{\pi},\bar{\pi}). \tag{2.71}$$

This yields

$$\nabla \cdot \bar{\mathbf{u}}_s(P,\bar{p}_f) = -\frac{(P-\bar{p}_f)}{K_{fr}} - \frac{\bar{p}_f}{K_s}, \tag{2.72}$$

$$\Delta \phi(P,\bar{p}_f) = -\left(1 - \frac{K_{fr}}{K_s} - \phi\right)\frac{P-\bar{p}_f}{K_{fr}}, \tag{2.73}$$

$$2\bar{\mathbf{d}}_s(\mathbf{T}^D,\bar{\pi}) = \frac{1}{G_{fr}}(\mathbf{T}^D-\bar{\pi}) + \frac{\bar{\pi}}{G_s}, \tag{2.74}$$

$$2\phi(\bar{\mathbf{d}}_f)^e\left(\bar{\mathbf{T}}^D,\bar{\pi}\right) = \left(1 - \frac{G_{fr}}{G_s} - \phi\right)\frac{1}{G_{fr}}\left(\bar{\mathbf{T}}^D-\bar{\pi}\right), \tag{2.75}$$

and therefore

$$\nabla \cdot \bar{\mathbf{u}}_s(P,\bar{p}_f) = -\frac{P}{K_{fr}} + \alpha\frac{\bar{p}_f}{K_{fr}}, \tag{2.76}$$

$$\Delta \phi(P,\bar{p}_f) = -(\alpha-\phi)\frac{P-\bar{p}_f}{K_{fr}}, \tag{2.77}$$

$$2\bar{\mathbf{d}}_s\left(\bar{\mathbf{T}}^D,\bar{\pi}\right) = \frac{1}{G_{fr}}\bar{\mathbf{T}}^D - \alpha_G\frac{\bar{\pi}}{G_{fr}}, \tag{2.78}$$

$$2\phi(\bar{\mathbf{d}}_f)^e\left(\bar{\mathbf{T}}^D,\bar{\pi}\right) = (\alpha_G-\phi)\frac{1}{G_{fr}}\left(\bar{\mathbf{T}}^D-\bar{\pi}\right), \tag{2.79}$$

where $\alpha = 1 - K_{fr}/K_s$ is the classical Biot coefficient ($\phi \le \alpha \le 1$) and $\alpha_G = 1 - G_{fr}/G_s$ is a second Biot shear coefficient introduced by Revil and Jardani (2010).

The equation for the porosity variation $\Delta \phi(P,\bar{p}_f)$ can be written as a function of the increment of the fluid content $\nabla \cdot \bar{\mathbf{w}}(P,\bar{p}_f)$ using the continuity equation for the mass of the pore fluid. This equation is given by

$$\Delta \phi + \frac{\phi}{K_f}\bar{p}_f + \nabla \cdot \bar{\mathbf{w}} + \phi \nabla \cdot \bar{\mathbf{u}}_s = 0, \tag{2.80}$$

$$\nabla \cdot \bar{\mathbf{w}} = (\alpha-\phi)\frac{P-\bar{p}_f}{K_{fr}} - \frac{\phi}{K_f}\bar{p}_f + \phi\frac{P}{K_{fr}} - \phi\alpha\frac{\bar{p}_f}{K_{fr}}, \tag{2.81}$$

$$\nabla \cdot \bar{\mathbf{w}} = \alpha\frac{P}{K_{fr}} - \frac{\bar{p}_f}{K_{fr}}\left[\alpha + \phi\left(\frac{K_{fr}}{K_f} - \frac{K_{fr}}{K_s}\right)\right]. \tag{2.82}$$

A similar approach can be undertaken for the deviators. Starting with the definition of the differential deviator $\dot{\bar{\mathbf{d}}}_w \equiv \varphi\left(\dot{\bar{\mathbf{d}}}_f - \dot{\bar{\mathbf{d}}}_s\right)$ and using

$$\dot{\bar{\mathbf{d}}}_f = \left(\dot{\bar{\mathbf{d}}}_f\right)^v + \left(\dot{\bar{\mathbf{d}}}_f\right)^e, \tag{2.83}$$

$$\dot{\bar{\mathbf{d}}}_f = \frac{1}{2\eta_f^*}\bar{\boldsymbol{\pi}} + \left(\dot{\bar{\mathbf{d}}}_f\right)^e, \tag{2.84}$$

where the superscripts v and e stand for the viscous and elastic contributions, respectively, we obtain

$$-\frac{\phi}{2\eta_f^*}\bar{\boldsymbol{\pi}} + \dot{\bar{\mathbf{d}}}_w + \phi\left[\dot{\bar{\mathbf{d}}}_s - \left(\dot{\bar{\mathbf{d}}}_f\right)^e\right] = 0, \tag{2.85}$$

$$-\frac{\phi}{2\eta_f}(1 - i\omega\tau_m)\bar{\boldsymbol{\pi}} - i\omega\left[\dot{\bar{\mathbf{d}}}_w + \phi\left(\dot{\bar{\mathbf{d}}}_s - \left(\dot{\bar{\mathbf{d}}}_f\right)^e\right)\right] = 0, \tag{2.86}$$

$$2\phi\left(\dot{\bar{\mathbf{d}}}_f\right)^e + \frac{\phi}{G_f^*}\bar{\boldsymbol{\pi}} + 2\dot{\bar{\mathbf{d}}}_w + 2\phi\dot{\bar{\mathbf{d}}}_s = 0, \tag{2.87}$$

where we have used $\tau_m = \eta_f/G_f$ and the Debye approximation for the distribution of the time constant. Using Equation (2.87), we can write the elastic contribution of the mean deviator in the fluid phase $2\phi\left(\dot{\bar{\mathbf{d}}}_f\right)^e\left(\bar{\mathbf{T}}^D, \bar{\boldsymbol{\pi}}\right)$ as a function of the differential deviator $\bar{\mathbf{d}}_w\left(\bar{\mathbf{T}}^D, \bar{\boldsymbol{\pi}}\right)$:

$$2\bar{\mathbf{d}}_w = -\alpha_G\frac{\bar{\mathbf{T}}^D}{G_{fr}} + \frac{\bar{\boldsymbol{\pi}}}{G_{fr}}\left[\alpha_G + \phi\left(\frac{G_{fr}}{G_f^*} - \frac{G_{fr}}{G_s}\right)\right]. \tag{2.88}$$

This yields for the nondeviatoric and the deviatoric contributions to the deformation

$$\begin{bmatrix} \nabla \cdot \bar{\mathbf{u}}_s \\ \nabla \cdot \bar{\mathbf{w}} \end{bmatrix} = -\frac{1}{K_{fr}}\begin{bmatrix} 1 & -\alpha \\ -\alpha & \phi\left(\dfrac{K_{fr}}{K_f} - \dfrac{K_{fr}}{K_s}\right) + \alpha \end{bmatrix} \cdot \begin{bmatrix} P \\ \bar{p}_f \end{bmatrix}, \tag{2.89}$$

$$2\begin{bmatrix} \bar{\mathbf{d}}_s \\ \bar{\mathbf{d}}_w \end{bmatrix} = \frac{1}{G_{fr}}\begin{bmatrix} 1 & -\alpha_G \\ -\alpha_G & \phi\left(\dfrac{G_{fr}}{G_f^*} - \dfrac{G_{fr}}{G_s}\right) + \alpha_G \end{bmatrix} \cdot \begin{bmatrix} \bar{\mathbf{T}}^D \\ \bar{\boldsymbol{\pi}} \end{bmatrix}, \tag{2.90}$$

respectively. These results generalize the Frenkel–Biot theory to linear poroelastic bodies filled with a linear generalized Maxwell fluid described by a Cole–Cole model. These equations can be also inverted to give

$$-\begin{bmatrix} P \\ \bar{p}_f \end{bmatrix} = \begin{bmatrix} K_U & C \\ C & C/\alpha \end{bmatrix} \cdot \begin{bmatrix} \nabla \cdot \bar{\mathbf{u}}_s \\ \nabla \cdot \bar{\mathbf{w}} \end{bmatrix}, \tag{2.91}$$

$$\begin{bmatrix} \bar{\mathbf{T}}^D \\ \bar{\boldsymbol{\pi}} \end{bmatrix} = \begin{bmatrix} G_U & C_G \\ C_G & C_G/\alpha_G \end{bmatrix} \cdot \begin{bmatrix} 2\bar{\mathbf{d}}_s \\ 2\bar{\mathbf{d}}_w \end{bmatrix}, \tag{2.92}$$

where the newly introduced material properties are defined as a function of the properties of the constituents by

$$G_U = \frac{(G_s - G_{fr})^2}{G_s\left[1 + \phi\left(G_s/G_f^* - 1\right)\right] - G_{fr}} + G_{fr}, \tag{2.93}$$

$$C_G = \frac{G_s(G_s - G_{fr})}{G_s\left[1 + \phi\left(G_s/G_f^* - 1\right)\right] - G_{fr}}, \tag{2.94}$$

$$M_G = \frac{G_s^2}{G_s\left[1 + \phi\left(G_s/G_f^* - 1\right)\right] - G_{fr}}, \tag{2.95}$$

and $M_G = C_G/\alpha_G$. We check that in the limit where $G_f^* = 0$, we obtain $G_U = G_{fr}$, $C_G = 0$, and $M_G = 0$, and we recover in this limit the classical equations of poroelasticity. The equations can also be used to generalize the Gassmann substitution formula (see Gassmann, 1951) for the undrained bulk and shear moduli

$$K_U = \frac{K_f(K_s - K_{fr}) + \phi K_{fr}(K_s - K_f)}{K_f(1 - \phi - K_{fr}/K_s) + \phi K_s}, \tag{2.96}$$

$$G_U(\omega) = \frac{G_f^*(\omega)(G_s - G_{fr}) + \phi G_{fr}\left(G_s - G_f^*(\omega)\right)}{G_f^*(\omega)(1 - \phi - G_{fr}/G_s) + \phi G_s}, \tag{2.97}$$

where $G_f^*(\omega)$ is given by Equation (2.51).

The two other Biot moduli are classically defined by

$$C = \frac{K_f(K_s - K_{fr})}{K_f(1 - \phi - K_{fr}/K_s) + \phi K_s}, \tag{2.98}$$

$$M = \frac{C}{\alpha} = \frac{K_f K_s}{K_f(1 - \phi - K_{fr}/K_s) + \phi K_s}. \tag{2.99}$$

Note that at low frequencies, the shear modulus reduced to the shear modulus of the skeleton

$$\lim_{\omega \to 0} G(\omega) = G_{fr}, \tag{2.100}$$

and therefore, we recover the case of a Newtonian fluid for Equations (2.96) and (2.97).

It is also possible to write Equation (2.91) as a function of the Skempton coefficient B (see discussion in Chapter 1) and the undrained bulk modulus K_U:

$$\begin{bmatrix} P \\ p_f \end{bmatrix} = -K_U \begin{bmatrix} 1 & B \\ B & B/\alpha \end{bmatrix} \begin{bmatrix} \nabla\cdot\bar{\mathbf{u}}_s \\ \nabla\cdot\bar{\mathbf{w}} \end{bmatrix}, \tag{2.101}$$

$$K_U = -\frac{P}{\nabla\cdot\bar{\mathbf{u}}_s}\Big|_{\nabla\cdot\mathbf{w}=0} = \frac{K}{1-B\alpha}, \tag{2.102}$$

$$B = \frac{p_f}{P}\Big|_{\nabla\cdot\mathbf{w}=0}, \tag{2.103}$$

$$\alpha = \frac{1}{B}\left(1 - \frac{K}{K_U}\right), \tag{2.104}$$

and $C = K_U B$ and α is the bulk Biot coefficient defined previously.

2.1.3.2 The field equations

We start with the Darcy equation, Equation (2.39), written in the following form and neglecting the electroosmotic term:

$$\frac{\eta_f}{k(\omega)}\dot{\mathbf{w}} = -\nabla\bar{p}_f - \rho_f\ddot{\mathbf{u}}_s + \mathbf{F}_f. \tag{2.105}$$

The dynamic permeability is written as (see Eq. 2.40)

$$\frac{1}{k(\omega)} \equiv \frac{1 - i(\omega/\omega_c)(1-(i\omega\tau_m)^c)}{k_0(1-(i\omega\tau_m)^c)}. \tag{2.106}$$

Using the fact that $\ddot{\mathbf{w}} = -i\omega\dot{\mathbf{w}}$, we can easily rewrite Equation (2.105) as

$$b(t)\otimes\dot{\mathbf{w}} + \tilde{\rho}_f\ddot{\mathbf{w}} = -\left(\nabla\bar{p}_f + \rho_f\ddot{\mathbf{u}}_s - \mathbf{F}_f\right), \tag{2.107}$$

where the effective fluid density is given by $\tilde{\rho}_f = \rho_f F$, where F is the formation factor defined by Chapter 1 (see Eq. 1.90). This is the same result than for a viscous Newtonian fluid. Indeed, for a viscous Newtonian fluid, the frequency-dependent permeability is given by

$$k(\omega) = \frac{k_0}{1 - i\omega/\omega_c}, \tag{2.108}$$

where $\omega_c = \eta_f/(k_0\rho_f F)$ where F is the formation factor. Equation (2.108) is a simplified version of Equation (236) of Pride (1994). Inserting Equation (2.108) into Equation (2.105) results in Equation (2.107) with $\tilde{\rho}_f = \rho_f F$.

The second point studied in this section is to go from Equation (2.107) and $\rho\ddot{\mathbf{u}}_s + \rho_f\ddot{\mathbf{w}} - \nabla\cdot\bar{\mathbf{T}} = \mathbf{F}$ (see Section

2.1.3.4 for a complete derivation) to the following equation (which will be solved numerically in the following text):

$$\nabla\cdot\bar{\bar{\mathbf{Y}}} = \bar{F}, \tag{2.109}$$

where

$$\bar{\bar{\mathbf{Y}}} = \begin{bmatrix} T_{xx} & T_{xz} \\ T_{zx} & T_{zz} \\ -\bar{p}_f & 0 \\ 0 & -\bar{p}_f \end{bmatrix}, \tag{2.110}$$

and

$$\bar{F} = \begin{bmatrix} -F_x + \rho\ddot{u}_x + \rho_f\ddot{w}_x \\ -F_z + \rho_f\ddot{u}_z + \rho_f\ddot{w}_z \\ \rho_f\ddot{u}_x + \tilde{\rho}_f\ddot{w}_x \\ \rho_f\ddot{u}_z + \tilde{\rho}_f\ddot{w}_z \end{bmatrix}. \tag{2.111}$$

In the frequency domain, the stress tensor is written as

$$\bar{\mathbf{T}} = \left(K_U - \frac{2}{3}G_U\right)(\nabla\cdot\bar{\mathbf{u}}_s)\mathbf{I} + \left(C - \frac{2}{3}C_G\right)(\nabla\cdot\bar{\mathbf{w}})\mathbf{I} \\ + G_U\left[\nabla\bar{\mathbf{u}}_s + \nabla\bar{\mathbf{u}}_s^T\right] + C_G\left[\nabla\bar{\mathbf{w}} + \nabla\bar{\mathbf{w}}^T\right]. \tag{2.112}$$

For a two-dimensional (2D) problem, the components of the stress tensor are

$$\bar{\mathbf{T}} \equiv \begin{bmatrix} T_{xx} & T_{xz} \\ T_{zx} & T_{zz} \end{bmatrix} \tag{2.113}$$

which are given by

$$T_{xx} = \left(K_U - \frac{2}{3}G_U\right)\left(\frac{\partial u_x}{\partial x} + \frac{\partial u_z}{\partial z}\right) + \left(C - \frac{2}{3}C_G\right)\left(\frac{\partial w_x}{\partial x} + \frac{\partial w_z}{\partial z}\right) \\ + 2G_U\left(\frac{\partial u_x}{\partial x}\right) + 2C_G\left(\frac{\partial w_x}{\partial x}\right), \tag{2.114}$$

$$T_{zz} = \left(K_U - \frac{2}{3}G_U\right)\left(\frac{\partial u_x}{\partial x} + \frac{\partial u_z}{\partial z}\right) + \left(C - \frac{2}{3}C_G\right)\left(\frac{\partial w_x}{\partial x} + \frac{\partial w_z}{\partial z}\right) \\ + 2G_U\left(\frac{\partial u_z}{\partial z}\right) + 2C_G\left(\frac{\partial w_z}{\partial z}\right), \tag{2.115}$$

$$T_{zx} = T_{xz} = G_U\left(\frac{\partial u_x}{\partial z} + \frac{\partial u_z}{\partial x}\right) + C_G\left(\frac{\partial w_x}{\partial z} + \frac{\partial w_z}{\partial x}\right). \tag{2.116}$$

With these notations, Equation (2.107) and $\rho\ddot{\mathbf{u}}_s + \rho_f\ddot{\mathbf{w}} - \nabla\cdot\bar{\mathbf{T}} = \mathbf{F}$ yield directly Equation (2.109). In the special case for which the fluid does not bear any shear stress (Newtonian case), we have $G_U = G_{fr}$ and $C_G = 0$ and we recover the classical poroelastic equations.

2.1.3.3 Note regarding the material properties

We look for an expression of $b(t) = \eta_f^*(t)/k_0$, where $b(t) = \mathrm{FT}^{-1}(\eta^*(\omega)/k_0)$, and

$$b(t) = \mathrm{FT}^{-1}\left(b(\omega) \equiv \frac{\eta_f}{k_0} \frac{1}{1-(i\omega\tau_m)^c}\right), \qquad (2.117)$$

where FT^{-1} stands for the inverse Fourier transform of a given frequency-dependent function. Let us start with the simpler case where the distribution of the relaxation time obeys a Dirac distribution (Debye relaxation, $c = 1$). In this case, we have

$$\mathrm{FT}^{-1}\left(\frac{\eta_f}{k_0}\frac{1}{1-i\omega\tau_m}\right) = \frac{\eta_f}{k_0}\left[1-\exp\left(-\frac{t}{\tau_m}\right)\right]. \qquad (2.118)$$

The more general case $c \leq 1$ can be found in Revil et al. (2006). It yields

$$b(t) = \frac{\eta_f}{k_0}\left[1 - \sum_{n=0}^{\infty}\frac{(-1)^n(t/\tau_m)^{nc}}{\Gamma(1+nc)}\right], \qquad (2.119)$$

where $\Gamma(\ )$ is the gamma function defined by

$$\Gamma(x) \equiv \int_0^{\infty} u^{x-1}e^{-u}du. \qquad (2.120)$$

We can easily check that with the case where $c = 1$, and Equation (2.119) becomes equal to (2.118) (Revil et al., 2006). We can apply the same approach to the properties of the fluid discussed in Section 2.1.1. For example, we can determine the inverse Fourier transform of the frequency-dependent shear modulus of the fluid defined by Eq. (2.51):

$$G_f^*(\omega) = -\frac{(i\omega/\omega_m)^c}{1-(i\omega/\omega_m)^c}G_f, \qquad (2.121)$$

and $G_f^*(t) = \mathrm{FT}^{-1}G_f^*(\omega)$. If the Cole–Cole exponent is equal to 1, we have $G_f^*(t) = G_f\exp(-t/\tau_m)$. The more general case $c \leq 1$ can be found easily as

$$G_f^*(t) = G_f\sum_{n=0}^{\infty}\frac{(-1)^n(t/\tau_m)^{nc}}{\Gamma(1+nc)}. \qquad (2.122)$$

2.1.3.4 Force balance equations

Performing a volume average of the force balance equations in each phase, Equations (2.17) and (2.25) yield a total force balance equation in the time domain

$$(1-\phi)\nabla\cdot\bar{\mathbf{T}}_s + \phi\nabla\cdot\bar{\mathbf{T}}_f + \mathbf{F} = (1-\phi)\rho_s\ddot{\mathbf{u}}_s + \phi\rho_f\ddot{\mathbf{u}}_f, \qquad (2.123)$$

$$\nabla\cdot\bar{\mathbf{T}} + \mathbf{F} = \rho\ddot{\mathbf{u}}_s + \rho_f\ddot{\mathbf{w}}, \qquad (2.124)$$

where the bulk density and the bulk force are defined by

$$\rho = \phi\rho_f + (1-\phi)\rho_s, \qquad (2.125)$$

$$\mathbf{F} = (1-\phi)\bar{\mathbf{F}}_s + \phi\bar{\mathbf{F}}_f, \qquad (2.126)$$

respectively. An expression in the time domain for the force in terms of source mechanism is given in the following text.

The total stress tensor and the stress tensor on the pore fluid are given by

$$\bar{\mathbf{T}} = K_U(\nabla\cdot\bar{\mathbf{u}}_s)\mathbf{I} + C(\nabla\cdot\bar{\mathbf{w}})\mathbf{I} + 2G_U\otimes\bar{\mathbf{d}}_s + 2C_G\otimes\bar{\mathbf{d}}_w, \qquad (2.127)$$

$$\bar{\mathbf{T}}_f = C(\nabla\cdot\bar{\mathbf{u}}_s)\mathbf{I} + M(\nabla\cdot\bar{\mathbf{w}})\mathbf{I} + 2C_G\otimes\bar{\mathbf{d}}_s + 2M_G\otimes\bar{\mathbf{d}}_w, \qquad (2.128)$$

where the circled cross stands for the Stieltjes convolution product, K_U and G_U are the undrained bulk and shear moduli of the porous medium, C and M are two Biot moduli, and $\bar{\mathbf{d}}_s$ and $\bar{\mathbf{d}}_w$ denote the deviatoric components of the deformation tensors for the solid and fluid, respectively.

2.1.4 The Maxwell equations

As shown by Pride (1994), the local Maxwell equations can be volume averaged to obtain the macroscopic Maxwell equations. With the Donnan model developed by Revil and Linde (2006), the Maxwell equations are

$$\nabla\times\mathbf{E} = -\dot{\mathbf{B}}, \qquad (2.129)$$

$$\nabla\times\mathbf{H} = \mathbf{J} + \dot{\mathbf{D}}, \qquad (2.130)$$

$$\nabla\cdot\mathbf{B} = 0, \qquad (2.131)$$

$$\nabla\cdot\mathbf{D} = \phi\bar{Q}_V, \qquad (2.132)$$

where $\mathbf{H}$ is the magnetic field, $\mathbf{B}$ is the magnetic induction, and $\mathbf{D}$ is the displacement vector. These equations are completed by two electromagnetic (EM) constitutive equations: $\mathbf{D} = \varepsilon\mathbf{E}$ and $\mathbf{B} = \mu\mathbf{H}$ where ε is the permittivity of the medium and μ is the magnetic permeability. If the porous material does not contain magnetized grains, these two material properties are given by (Pride, 1994)

$$\varepsilon = \frac{1}{F}(\varepsilon_f + (F-1)\varepsilon_s), \qquad (2.133)$$

$$\mu = \mu_0, \tag{2.134}$$

where F is the electrical formation factor defined in Chapter 1 for the electrical conductivity and by Equation (2.41); ε_f and ε_s are the dielectric constants of the pore fluid and the solid, respectively; and μ_0 is the magnetic permeability of free space. Note that only two textural properties, Λ and F, are required to describe the influence of the topology of the pore network upon the material properties entering into the transport and EM constitutive equations.

The coupling between the mechanical and the Maxwell equations occurs in the current density, which can be written in the time domain as

$$\mathbf{J} = \sigma\mathbf{E} + \mathbf{J}_S, \tag{2.135}$$

$$\mathbf{J}_S = \hat{Q}_V^0 \dot{\mathbf{w}} = -\frac{k(\omega)\hat{Q}_V^0}{\eta_f}\left(\nabla\bar{p}_f + \rho_f\ddot{\mathbf{u}}_s - \bar{\mathbf{F}}_f\right), \tag{2.136}$$

where $\mathbf{J}_S$ (in A m^{-2}) is the source current density of electrokinetic nature. The body force $\bar{\mathbf{F}}_f$ can be responsible for EM signals that are directly associated with the source (see discussions in Chapter 5). An important point is that we expect a resonance effect in the source current density, and due to the resonance effects in the Darcy velocity, as shown in Figure 2.1 for a capillary. Such resonances associated with harmonic resonance due to bulk flow have never been observed to date.

2.1.5 Analysis of the wave modes

In this section, we solve the seismoelectric model for a poroelastic media saturated by Newtonian or generalized viscoelastic fluids. Specifically, the case of a poroelastic body saturated by a Newtonian fluid is a special case of the theory developed previously. In the following text, we develop the general field equations, and in the process, we show that these equations yield estimates that are consistent with the classical Biot–Frenkel theory in the limit of a Newtonian fluid at low frequencies.

The three equations to solve are given, in the time domain, by Darcy's law plus the combined equations of motion for the solid and the fluid phases:

$$b(t)\otimes\dot{\mathbf{w}} + \tilde{\rho}_f\ddot{\mathbf{w}} = -\left(\nabla\bar{p}_f + \rho_f\ddot{\mathbf{u}}_s - \bar{\mathbf{F}}_f\right), \tag{2.137}$$

$$\rho\ddot{\mathbf{u}}_s + \rho_f\ddot{\mathbf{w}} - \nabla\cdot\bar{\mathbf{T}} = \bar{\mathbf{F}}, \tag{2.138}$$

$$\frac{\bar{p}_f}{M} + \nabla\cdot\bar{\mathbf{w}} + \alpha\nabla\cdot\bar{\mathbf{u}}_s = 0, \tag{2.139}$$

where M is one of the Biot moduli and $b(t) = \eta_f^*(t)/k_0$ where $\eta_f^*(t)$ is the inverse Fourier transform of $\eta_f^*(\omega)$ as defined by Equation (2.16). Equation (2.66) is derived from Equation (2.39) neglecting the electroosmotic effect, and $\tilde{\rho}_f$ denotes the effective fluid density given by $\tilde{\rho}_f = \rho_f F$. Equation (2.138) follows from Equation (2.124), while Equation (2.139) has been derived in Section 2.1.3.1 (see Eq. 2.92) using the relationship $\alpha = C/M$.

Next, we take the time derivative of Equation (2.139) and apply Darcy's law in which we have neglected the electroosmotic contribution. This method is used because we are only interested in describing the seismoelectric coupling and not electroseismic effects in this book. The assumption of whether the electroosmotic component can be neglected in this type of EM–hydromechanical coupling formulation process has been investigated by Revil et al. (1999b). From this formulation process, we obtain, in the frequency domain, the following hydraulic diffusion equation for the pore water:

$$\frac{-i\omega\bar{p}_f}{M} + \nabla\cdot\left[\frac{k(\omega)}{\eta_f}\left(-\nabla\bar{p}_f + \omega^2\rho_f\bar{\mathbf{u}}_s\right)\right] = \alpha\nabla\cdot(i\omega\bar{\mathbf{u}}_s). \tag{2.140}$$

The first term in Equation (2.140) corresponds to the storage term, while the second term corresponds to the divergence of the Darcy velocity. The term on the right-hand side of Equation (2.140) corresponds to a source term for this partial differential equation.

Combining Equations (2.137) and (2.138) yields

$$\left(\rho - \frac{\rho_f}{F}\right)\ddot{\mathbf{u}}_s = \nabla\cdot\bar{\mathbf{T}} - \frac{1}{F}\nabla\cdot\bar{\mathbf{T}}_f + \frac{b(t)}{F}\otimes\dot{\mathbf{w}}, \tag{2.141}$$

$$\left(F - \frac{\rho_f}{\rho}\right)\rho_f\ddot{\mathbf{w}} = \nabla\cdot\bar{\mathbf{T}}_f - \frac{\rho_f}{\rho}\nabla\cdot\bar{\mathbf{T}} - \frac{b(t)}{F}\otimes\dot{\mathbf{w}}. \tag{2.142}$$

The dissipation mechanism occurring in our model stems from the viscoelastic properties of the pore fluid.

In the time domain, the constitutive equations for the total stress tensor and the fluid phase average stress tensor have been derived in Section 2.1.3.4 and are shown in Equations (2.143) and (2.144):

$$\begin{aligned}\bar{\mathbf{T}} = K_U(\nabla\cdot\bar{\mathbf{u}}_s)\mathbf{I} + C(\nabla\cdot\bar{\mathbf{w}})\mathbf{I} + 2G_U(t)\otimes\bar{\mathbf{d}}_s \\ + 2C_G(t)\otimes\bar{\mathbf{d}}_w,\end{aligned} \tag{2.143}$$

$$\bar{\mathbf{T}}_f = C(\nabla\cdot\bar{\mathbf{u}}_s)\mathbf{I} + M(\nabla\cdot\bar{\mathbf{w}})\mathbf{I} + 2C_G(t)\otimes\bar{\mathbf{d}}_s \\ + 2M_G(t)\otimes\bar{\mathbf{d}}_w, \tag{2.144}$$

$$G_U(t) = \text{FT}^{-1}G_U(\omega), \tag{2.145}$$

$$C_G(t) = \text{FT}^{-1}C_G(\omega), \tag{2.146}$$

$$M_G(t) = \text{FT}^{-1}M_G(\omega), \tag{2.147}$$

where the deviators $\bar{\mathbf{d}}_s$ and $\bar{\mathbf{d}}_w$ have been defined previously (see Eqs. 2.53 and 2.55), where $\bar{\mathbf{d}}_s$ is the mean strain deviator of the solid phase, and the coefficients $G_U(t)$, $G_U(\omega)$, $C_G(t)$, $C_G(\omega)$, $M_G(t)$, and $M_G(\omega)$ are defined in Section 2.1.3.1.

We use the classical decomposition of the displacements into dilatational components

$$\nabla\cdot\bar{\mathbf{u}}_s = e, \tag{2.148}$$

$$\nabla\cdot\bar{\mathbf{w}} = \varsigma. \tag{2.149}$$

The dilatational wave propagation is obtained by applying the divergence operator to Equations (2.141) and (2.142) and then inserting Equations (2.148) and (2.149) into the resulting equation. This yields

$$\left(\rho - \frac{\rho_f}{F}\right)\ddot{e} = \nabla^2[H\otimes e + C\varsigma] - \frac{1}{F}\nabla^2(Ce + M\varsigma) \\ + \frac{b(t)}{F}\otimes\dot{\varsigma}, \tag{2.150}$$

$$\left(F - \frac{\rho_f}{\rho}\right)\rho_f\ddot{\varsigma} = \nabla^2[Ce + M\varsigma] - \frac{\rho_f}{\rho}\nabla^2(H\otimes e + C\varsigma) \\ - b(t)\otimes\dot{\varsigma}, \tag{2.151}$$

where H is the stiffness coefficient of the Biot–Frenkel theory $(H = K_U + (4/3)G_U)$. We assume plane wave propagation in the x-direction:

$$e = A_1\exp[i(lx - \omega t)], \tag{2.152}$$

$$\varsigma = A_2\exp[i(lx - \omega t)], \tag{2.153}$$

where ω is the angular frequency and l is the complex wave number. Using Equations (2.152) and (2.153) into (2.150) and (2.151), we obtain

$$\left(\rho - \frac{\rho_f}{F}\right)(-\omega^2 A_1) = H(\omega)(-A_1 l^2) - CA_2 l^2 \\ + \frac{1}{F}(CA_1 l^2 + MA_2 l^2) - i\omega\frac{b(\omega)}{F}A_2, \tag{2.154}$$

$$\left(F - \frac{\rho_f}{\rho}\right)\rho_f(-\omega^2 A_2) = -Cl^2 A_1 - Ml^2 A_2 \\ + \frac{\rho_f}{\rho}(H(\omega)l^2 A_1 + Cl^2 A_2) + i\omega A_2 b(\omega). \tag{2.155}$$

Eliminating A_1 and A_2 between these two equations yields the following equation for the speed: $az_p^4 + bz_p^2 + c = 0$, where $z_p = \omega/l$ is the complex speed and

$$a = \rho - \frac{\rho_f}{F} + i\frac{b(\omega)}{\omega}\left(\frac{\rho}{\tilde{\rho}_f}\right), \tag{2.156}$$

$$b = (-1)\left[\frac{M\rho}{\tilde{\rho}_f} + H(\omega) - \frac{2C}{F} + i\frac{b(\omega)}{\omega\tilde{\rho}_f}H(\omega)\right], \tag{2.157}$$

$$c = \frac{1}{\tilde{\rho}_f}[H(\omega)M - C^2]. \tag{2.158}$$

The two roots of this equation are

$$\left(z_p^I\right)^2 = \frac{-b + \sqrt{b^2 - 4ac}}{2a}, \tag{2.159}$$

$$\left(z_p^{II}\right)^2 = \frac{-b - \sqrt{b^2 - 4ac}}{2a}, \tag{2.160}$$

with $z_p^I > z_p^{II}$. These two waves correspond to the complex wave speeds associated with the fast P-wave when the solid and the fluid move in phase and the slow dilatational wave (slow wave) when the solid and the fluid move out of phase. The phase speeds are given by $1/c_p^I = \text{Re}\left[\left(z_p^I\right)^{-1}\right]$ and $1/c_p^{II} = \text{Re}\left[\left(z_p^{II}\right)^{-1}\right]$, and the inverse quality factors are $1/Q_p^I = \text{Im}\left(z_p^I\right)^2/\text{Re}\left(z_p^I\right)^2$ and $1/Q_p^{II} = \text{Im}\left(z_p^{II}\right)^2/\text{Re}\left(z_p^{II}\right)^2$.

Using the decomposition of the displacements into rotational components

$$\nabla\times\bar{\mathbf{u}}_s = \mathbf{\Omega}, \tag{2.161}$$

$$\nabla\times\bar{\mathbf{w}} = \mathbf{\Psi}, \tag{2.162}$$

applying the curl operator to Equations (2.141) and (2.142), and using Equations (2.161) and (2.162), we obtain

$$\left(\rho - \frac{\rho_f}{F}\right)\ddot{\mathbf{\Omega}} = G_U\otimes\nabla^2\mathbf{\Omega} + C_G\otimes\nabla^2\mathbf{\Psi} - \frac{1}{F}C_G\otimes\nabla^2\mathbf{\Omega} \\ - \frac{1}{F}M_G\otimes\nabla^2\mathbf{\Psi} + \frac{b(t)}{F}\otimes\dot{\mathbf{\Psi}}, \tag{2.163}$$

$$\left(F-\frac{\rho_f}{\rho}\right)\rho_f\ddot{\boldsymbol{\Psi}} = C_G\otimes\nabla^2\boldsymbol{\Omega} + M_G\otimes\nabla^2\boldsymbol{\Psi} - \frac{\rho_f}{\rho}G_U\otimes\nabla^2\boldsymbol{\Omega}$$
$$-\frac{\rho_f}{\rho}C_G\otimes\nabla^2\boldsymbol{\Psi} - b(t)\otimes\dot{\boldsymbol{\Psi}}. \tag{2.164}$$

We assume plane wave propagation in the x-direction:

$$\boldsymbol{\Omega} = B_1\exp[i(lx-\omega t)], \tag{2.165}$$

$$\boldsymbol{\Psi} = B_2\exp[i(lx-\omega t)], \tag{2.166}$$

where ω is the angular frequency and l is the complex wave number. Using these equations, we obtain

$$\left(\rho-\frac{\rho_f}{F}\right)(-\omega^2 B_1) = G_U(\omega)(-B_1 l^2) - C_G B_2 l^2$$
$$+\frac{1}{F}(C_G B_1 l^2 + M_G B_2 l^2) - i\omega\frac{b(\omega)}{F}B_2, \tag{2.167}$$

$$\left(F-\frac{\rho_f}{\rho}\right)\rho_f(-\omega^2 B_2) = -C_G l^2 B_1 - M_G l^2 B_2$$
$$+\frac{\rho_f}{\rho}(G_U l^2 B_1 + C_G l^2 B_2) + i\omega B_2 b(\omega). \tag{2.168}$$

Eliminating the constants B_1 and B_2 between these two equations yields the following quadratic equation for the speed: $a'z_S^4 + b'z_S^2 + c' = 0$, where $z_p = \omega/l$ is the complex speed and

$$a' = \rho - \frac{\rho_f}{F} + i\frac{b(\omega)}{\omega}\left(\frac{\rho}{\tilde{\rho}_f}\right), \tag{2.169}$$

$$b' = (-1)\left[\frac{M_G(\omega)\rho}{\tilde{\rho}_f} + G_U(\omega) - \frac{2C_G(\omega)}{F} + i\frac{b(\omega)}{\omega\tilde{\rho}_f}G_U(\omega)\right], \tag{2.170}$$

$$c' = \frac{1}{\tilde{\rho}_f}[G_U(\omega)M_G(\omega) - C_G(\omega)^2]. \tag{2.171}$$

The two roots of this equation are

$$(z_S^I)^2 = \frac{-b' + \sqrt{b'^2 - 4a'c'}}{2a'}, \tag{2.172}$$

$$(z_S^{II})^2 = \frac{-b' - \sqrt{b'^2 - 4a'c'}}{2a'}, \tag{2.173}$$

with $z_S^I > z_S^{II}$. These two waves correspond to the complex wave speeds associated with a fast rotational S-wave and

a slow rotational S-wave. The fast rotational wave represents the classical shear wave, while the slow rotational wave is related to the fact that the fluid can sustain shear stresses above its resonance frequency. This wave does not exist at low frequencies ($\omega \ll \omega_m$), because the pore fluid behaves as a Newtonian fluid and cannot sustain shear stresses. The phase speeds of these two waves are given by $1/c_S^I = \text{Re}\left[(z_S^I)^{-1}\right]$ and $1/c_S^{II} = \text{Re}\left[(z_S^{II})^{-1}\right]$, and the inverse quality factors are $1/Q_S^I = \text{Im}(z_S^I)^2/\text{Re}(z_S^I)^2$ and $1/Q_S^{II} = \text{Im}(z_S^{II})^2/\text{Re}(z_S^{II})^2$, respectively. To the best of our knowledge, this second type of fast shear wave has been introduced first by Revil and Jardani (2010).

To check the consistency of the model, we can look for the case where $G_f^*(\omega) = 0$ and $G_U = G_{fr}$, $C_G(\omega) = 0$, and $M_G(\omega) = 0$. In this case, we obtain

$$a' = \rho - \frac{\rho_f}{F} + i\frac{b(\omega)}{\omega}\left(\frac{\rho}{\tilde{\rho}_f}\right), \tag{2.174}$$

$$b' = -G_{fr}\left[1 + i\frac{b(\omega)}{\omega\tilde{\rho}_f}\right], \tag{2.175}$$

$$c' = 0, \tag{2.176}$$

and therefore

$$(z_S^I)^2 = 0, \tag{2.177}$$

$$(z_S^{II})^2 = G_{fr}\left(\frac{\tilde{\rho}_f}{\rho}\right)\frac{\tilde{\rho}_f - (\rho_f^2/\rho) + (b^2(\omega)/\omega^2\tilde{\rho}_f) - i(b(\omega)/\omega)(\rho_f/F\rho)}{(\tilde{\rho}_f - (\rho_f^2/\rho))^2 + (b(\omega)/\omega)^2}. \tag{2.178}$$

In the absence of dissipation, this second wave corresponds to the classical shear wave with phase speed given by

$$c_S^2 = \frac{G_{fr}}{\rho - \rho_f/F} \approx \frac{G_{fr}}{\rho}, \tag{2.179}$$

when $F \gg 1$.

The next section will be concerned with a simplification of the present equations to the case of a poroelastic material filled with a Newtonian fluid like liquid water.

2.1.6 Synthetic case studies

To support our analysis, we introduce two 2D numerical experiments (Experiment #1 and Experiment #2) to demonstrate the occurrence of the seismoelectric signals

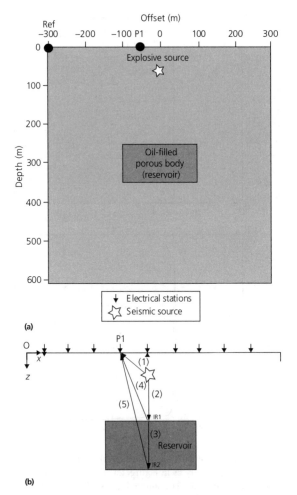

(a)

(b)

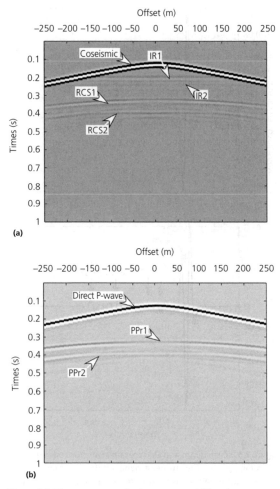

(a)

(b)

Figure 2.3 Sketch of the geometry used for the numerical simulation. **a)** The depth of the oil reservoir is 250 m. Its length and thickness are 200 and 100 m, respectively. The depth of the source is 50 m. The observation station is P_1. **b)** Wavepaths. There is a signal associated with the source itself that diffuses nearly instantaneously from the seismic source to the observation station. We decided not to model this signal. In the early times of the electrogram, various contributions include the seismoelectric conversions of the P-wave reaching the top of the reservoir (2) (labeled IR1) and the bottom of the reservoir (3) (labeled IR2). In addition, coseismic signals are associated with the direct wave (1) (labeled "coseismic") and the reflected waves (4) and (5) (labeled RCS1 and RCS 2).

Figure 2.4 Electrograms and seismograms. **a)** The electrograms show the coseismic electrical potential field associated with the direct wave and the reflections of the P-wave (labeled RCS1 and RCS 2) and the seismoelectric conversions with a smaller amplitude and a flat shape (labeled IR1 and IR2). **b)** The seismograms reconstructed by the geophones show the P-wave direct field and the reflections of the P-waves. The reflections PPr1 and PPr2 correspond to the reflections on the top and bottom of the reservoir, respectively.

that are generated in response to seismic stimulation of a simple oil reservoir geometry.

Experiment #1 corresponds to the simulation of the seismoelectric response of an oil-filled reservoir (see Figure 2.3) in a half-space with the nonwetting oil

considered as a Newtonian fluid. The ground surface corresponds to the top surface of the system where the normal component of the electrical field vanishes. We consider a volumetric source located 50 m below the ground surface generating P-waves. The reservoir is 100 m thick and its top surface is located 250 m below the ground surface (Figure 2.3a). The time dependence of the source $S(t)$ is a Ricker with a dominant frequency of 30 Hz. We consider the arrival of the seismic and

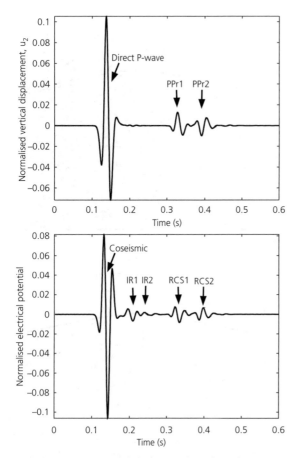

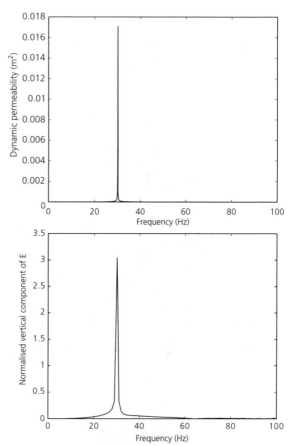

Figure 2.5 Seismogram and electrogram at an electrode (receiver P1 in Figure 2.3). The strongest signal on the electrogram corresponds to the coseismic disturbance associated with the direct wave (see Figure 2.3). RCS1 and RCS2 stand for the coseismic disturbances associated with the reflected P-waves (see Figure 2.3). IR1 and IR2 stand for the two seismoelectric disturbances associated with the seismoelectric conversions at the top and bottom of the reservoir.

Figure 2.6 Results of the second numerical experiment use the same geometry as in Figure 2.3. The fluid of the reservoir is viscoelastic. The source function is a Dirac and we investigate the electrical response in the frequency band (1–100 Hz). The dynamic permeability versus frequency shows the relaxation peak associated with the resonance of the viscoelastic fluid. The peak of the vertical component of the electrical field **E** at a remote self-potential station located at $x = 150$ m and $y = 0$ m.

seismoelectric signals at a station $P_1(-50 \text{ m, } 0 \text{ m})$ corresponding to a single electrode. The reference for the electrical potential recorded at this electrode is located at position Ref(-300 m, 0 m) (see Figure 2.3). The four edges are absorbing boundaries for which we use PML boundary conditions (see Jardani et al., 2010, for more details). To solve numerically the problem, we use the finite element modeling COMSOL Multiphysics 3.5 to simulate both seismograms and electrograms at the ground surface. Much more information regarding the implementation of the partial differential equations in a commercial finite element package will be discussed in

Chapter 4. We only provide numerical results in this section.

In Experiment #1, both seismoelectric conversion and coseismic electrical signals are generated at the reservoir boundaries (Figures 2.3b, 2.4, and 2.5). Figure 2.5 shows two electrograms for station P1. The first type of signals corresponds to the coseismic electrical signal associated with the propagation of the P-wave. It is labeled "Coseismic" in Figures 2.4a and 2.5. Other coseismic signals are associated with the reflected P-waves at the top and bottom interfaces of the reservoir. These coseismic signals are labeled RCS1 and RCS2, respectively, and occur when a seismic wave travels through a porous material, creating a relative displacement between the pore water and the solid phase. The associated current density is balanced by a conduction current density that results an electrical field traveling at the same speed as the seismic wave. The second type of seismoelectric signals correspond to the converted seismoelectric signals associated with the arrival of the P-waves at the top and bottom interface of the oil reservoir. These converted seismoelectric signals are labeled IR1 and IR2, respectively. When crossing an interface between two domains characterized by different properties, a seismic wave generates a time-varying charge separation, which acts as a dipole radiating EM energy. These dipoles oscillate with the waveform of the seismic waves. Because the EM diffusion of the electrical disturbance is very fast (instantaneous in our simulations), the seismoelectric conversions are observed nearly at the same time by all the electrodes but with different amplitudes. The seismoelectric conversions appear therefore as flat lines in the electrograms shown in Figure 2.4a.

Experiment #2 uses the same geometry as Experiment #1, but the fluid in the reservoir is now a viscoelastic wetting oil. The dynamic permeability response of the reservoir filled with the viscoelastic fluid is shown in Figure 2.6a. The time function of the source is a Dirac and we investigate the electrical field response in the frequency band (1–1000 Hz). A plot of the vertical component of the electrical field at an observation point at depth is shown in Figure 2.6b. The maximum at the electrical field occurs at the same frequency as that of the resonance of the viscoelastic fluid. The amplification of the electrical field reaches several orders of magnitude with respect to the DC value. This very strong amplification of the signal could be the basis of a new detection and mapping method of heavy oils and DNAPL in the ground. However, we feel that such theory should be checked first through experimental investigations in the laboratory.

2.1.7 Conclusions

We have developed previously a general extension of the Biot–Frenkel dynamic theory in the case of a porous material saturated by a generalized Maxwell-type fluid that can sustain shear stresses. In this case, the whole Biot–Frenkel theory becomes symmetric in its constitutive equations, and two types of P-waves and two types of S-waves are

expected. In the next section, we are dealing with the classical Biot–Frenkel theory for which the pore fluid is a Newtonian fluid that cannot sustain shear stresses. In this case, we expect two P-waves and one type of S-wave.

2.2 Poroelastic medium filled with a Newtonian fluid

We consider now that the pore fluid is a Newtonian fluid (in the previous section, we were dealing with a more general fluid that was Maxwellian in nature). Water is the best example of a linear Newtonian viscous fluid. We can eventually use the previous theory in the low-frequency behavior of the Maxwell fluid neglecting shear stresses in the fluid. The corresponding theory is the one developed by Biot and is generally called the classical dynamic Biot (or Biot–Frenkel) theory of linear poroelasticity. We present in the following text the fundamental equations of this theory and their application to the seismoelectric problem.

2.2.1 Classical Biot theory

The Biot theory (Biot, 1956a, b, 1962a, b) provides a starting framework to model the propagation of seismic waves in linear poroelastic media saturated by a Newtonian viscous pore fluid like water. The theory predicts the existence of an additional compressional wave by comparison with the P- and S-waves found for purely (nonporous) elastic materials. The existence of this slow P-wave was first confirmed by Plona (1980). The physical interpretations of the elastic constants in the Biot theory were given by Biot and Willis (1957) and Geertsma and Smit (1961). According to the Biot theory, the equations of motion in a 2D statistically isotropic, fully saturated, heterogeneous, porous elastic medium are given, in the frequency domain, by (e.g., Haartsen et al., 1998)

$$-\omega^2(\rho\mathbf{u} + \rho_f\mathbf{w}) = \nabla\cdot\mathbf{T} + \mathbf{F}, \qquad (2.180)$$

$$\mathbf{T} = [\lambda_u\nabla\cdot\mathbf{u} + C\nabla\cdot\mathbf{w}]\mathbf{I} + G[\nabla\mathbf{u} + \nabla\mathbf{u}^T] \qquad (2.181)$$

$$-\omega^2(\rho_f\mathbf{u} + \tilde{\rho}_f\mathbf{w}) - ib\omega\mathbf{w} = -\nabla p, \qquad (2.182)$$

$$-p = C\nabla\cdot\mathbf{u} + M\nabla\cdot\mathbf{w} + S, \qquad (2.183)$$

where $i^2 = -1$; $\mathbf{u}$ is the displacement vector of the solid; $\mathbf{w}$ is the displacement vector of the fluid relative to the solid (called the filtration displacement); $\mathbf{T}$ is the stress tensor; $\mathbf{I}$ is the identity matrix; $\mathbf{F}$ is the body force on the elastic solid phase; S is a pressure source acting on the pore fluid; ρ represents the mass density of the saturated medium; ρ_f and ρ_s are the mass density of the fluid and the solid, respectively; $\tilde{\rho}_f$ is an apparent density of the pore fluid; p is the fluid pressure; $\lambda_U = K_U - (2/3)G$ is the undrained Lamé modulus of the porous material; b is the mobility of the fluid; G is the shear modulus of the porous frame; and C and M are elastic moduli. Equation (2.180) corresponds to Newton's law, while Equation (2.181) represents a constitutive expression for the total stress tensor as a function of the displacement (Hooke's law). This constitutive equation comprises the classical term of linear elasticity plus an additional term related to the expansion/contraction of the porous body to accommodate the flow of the pore fluid relative to a Lagrangian framework attached to the solid phase. Equation (2.110) is the Darcy constitutive equation in which the bulk force acting on the fluid phase has been neglected and Equation (2.183) is one of the classical Biot–Frenkel constitutive equations of poroelasticity. The mass density of the porous material is given by $\rho = \phi \rho_f + (1 - \phi)\rho_s$. These Equations (2.180) through (2.183) can also be derived from equations presented in Section 2.1, assuming that the pore fluid cannot bear shear stresses.

The material properties entering Equations (2.180)–(2.183) are given by Pride (1994) and Rañada Shaw et al. (2000):

$$b = \frac{\eta_f}{k_0}, \tag{2.184}$$

$$\tilde{\rho}_f = \frac{\rho_f \phi}{a}, \tag{2.185}$$

$$\alpha = 1 - \frac{K_{fr}}{K_s} \tag{2.186}$$

$$K_U = \frac{K_f(K_s - K_{fr}) + \phi K_{fr}(K_s - K_f)}{K_f(1 - \phi - K_{fr}/K_s) + \phi K_s}, \tag{2.187}$$

$$C = \frac{K_f(K_s - K_{fr})}{K_f(1 - \phi - K_{fr}/K_s) + \phi K_s}, \tag{2.188}$$

$$M = \frac{C}{\alpha} = \frac{K_f K_s}{K_f(1 - \phi - K_{fr}/K_s) + \phi K_s}, \tag{2.189}$$

where K_U (in Pa) is the bulk modulus of the porous medium, K_{fr} is the bulk modulus of the dry porous frame (skeleton), K_f is the bulk modulus of the fluid, K_s is the

bulk modulus of the solid phase, and α is the Biot–Willis coefficient. Equation (2.187) is the Gassmann equation, η_f is the dynamic viscosity of the pore fluid, k_0 is the permeability of the medium, ϕ is the porosity, and a is the tortuosity. The ratio a/ϕ corresponds to the electrical formation factor F also defined by Archie's law to be $F = \phi^{-m}$ where m is called the cementation exponent. In the following, we consider the tortuosity equal to $\phi^{-1/2}$, which is equivalent to a cementation exponent of 1.5, typical of a pack of spherical grains.

2.2.2 The u–p formulation

The classical formulation described in Equations (2.180)–(2.183) is based on solving partial differential equations for two unknown fields, $\mathbf{u}$ and $\mathbf{w}$. For a 2D discretized problem, four degrees of freedom per node are therefore present. All the papers dealing with the modeling of the seismoelectric problem use this type of formulation (e.g., Haarsten & Pride, 1997; Haartsen et al., 1998; Garambois & Dietrich, 2002). Atalla et al. (1998) introduced an alternative approach using $\mathbf{u}$ and p as unknowns (see also Karpfinger et al., 2009). In 2D, this implies three unknown parameters (u_1, u_2, and p) to be solved at each node.

We start with the Darcy equation where the electroosmotic coupling term neglected:

$$\frac{\eta_f}{k(\omega)}\dot{\mathbf{w}} = -\nabla p - \rho_f \ddot{\mathbf{u}} + \mathbf{F}_f, \tag{2.190}$$

where $\mathbf{F}_f$ is the body force acting on the pore fluid phase. The fact that the electroosmotic term can be safely neglected in this type of formulation has been discussed by a number of authors including Revil et al. (1999b). The dynamic permeability is written as (e.g., Morency & Tromp, 2008)

$$\frac{1}{k(\omega)} \equiv \frac{1 - i(\omega/\omega_c)}{k_0}, \tag{2.191}$$

which is based on the low-frequency approximation to the dynamic permeability given by Pride (1994). The critical frequency is $\omega_c = \eta_f/(k_0 \rho_f F) = b/\tilde{\rho}_f$ where $b = \eta_f/k_0$ and $\tilde{\rho}_f = \rho_f F$. Neglecting the body force acting on the fluid phase, Equation (2.190) can be used to express the filtration displacement, $\mathbf{w}$, as a function of the pore fluid pressure, p, and the displacement of the solid phase, $\mathbf{u}$:

$$\mathbf{w} = k_\omega \left(\nabla p - \omega^2 \rho_f \mathbf{u} \right), \qquad (2.192)$$

where k_ω is defined by

$$k_\omega = \frac{1}{\omega^2 \tilde{\rho}_f + i\omega b}. \qquad (2.193)$$

Equation (2.192) can be used in Newton's law to give

$$-\omega^2 \rho_\omega^s \mathbf{u} - \omega^2 \rho_f k_\omega \nabla p = \nabla \cdot \mathbf{T}, \qquad (2.194)$$

$$\rho_\omega^s = \rho + \omega^2 \rho_f^2 k_\omega, \qquad (2.195)$$

where ρ_ω^s is an apparent mass density for the solid phase. Equation (2.194) is a partial differential equation between $\mathbf{u}$ and p but the stress tensor $\mathbf{T}$ also depends on $\mathbf{w}$. Using the relationships between stress and strain, we obtain the following relationship between the divergence of the filtration displacement, $\nabla \cdot \mathbf{w}$, and the divergence of the displacement of the solid phase, $\nabla \cdot \mathbf{u}$:

$$\nabla \cdot \mathbf{w} = -\frac{1}{M}(p + S) - \alpha \nabla \cdot \mathbf{u}, \qquad (2.196)$$

where α is the classical Biot coefficient of poroelasticity. We use Equation (2.196) in the stress/strain relationships to remove the dependence of the stress tensor on $\mathbf{w}$. This yields

$$\mathbf{T} = \lambda (\nabla \cdot \mathbf{u})\mathbf{I} + G\left[\nabla \mathbf{u} + \nabla \mathbf{u}^T \right] - \alpha p \mathbf{I}, \qquad (2.197)$$

$$\lambda = K - \frac{2}{3}G, \qquad (2.198)$$

where λ is the Lamé modulus of the skeleton. The effective stress tensor is written as

$$\hat{\mathbf{T}} = \lambda (\nabla \cdot \mathbf{u})\mathbf{I} + G\left[\nabla \mathbf{u} + \nabla \mathbf{u}^T \right], \qquad (2.199)$$

$$\mathbf{T} = \hat{\mathbf{T}} - \alpha p \mathbf{I}. \qquad (2.200)$$

The effective stress tensor is the equivalent stress tensor of the skeleton without fluid (in vacuo). Using Equations (2.195)–(2.198) and (2.192), we obtain an equation connecting the solid displacement and the fluid pressure assuming that the Biot coefficient is constant:

$$-\omega^2 \rho_\omega^s \mathbf{u} + \theta_\omega \nabla p = \nabla \cdot \hat{\mathbf{T}}, \qquad (2.201)$$

$$\theta_\omega = \alpha + \rho_f \omega^2 k_\omega, \qquad (2.202)$$

where θ_ω is a volumetric hydromechanical coupling coefficient. Equation (2.201) corresponds to Newton's law for a poroelastic body. This equation is similar to the classical Newton's law for an elastic solid except the coupling term $\theta_\omega \nabla p$, which accounts for the dynamic coupling between the pore fluid and the solid phases.

Regarding the filtration displacement, we obtain the following relationship between $\mathbf{w}$ and $\mathbf{u}$ and p:

$$\frac{1}{M}(p + S) + \nabla \cdot \left\{ k_\omega \left[\nabla p - \omega^2 (\rho_f + \alpha) \mathbf{u} \right] \right\} = 0. \qquad (2.203)$$

Equation (2.203) is the classical diffusion equation for the pore fluid pressure with a source term related to the harmonic change of displacement of the solid phase. This yields

$$-\omega^2 \rho_\omega^s \mathbf{u} + \theta_\omega \nabla p = \nabla \cdot \hat{\mathbf{T}}, \qquad (2.204)$$

$$\frac{1}{M}(p + S) + \nabla \cdot \left[k_\omega \left(\nabla p - \omega^2 (\rho_f + \alpha) \mathbf{u} \right) \right] = 0. \qquad (2.205)$$

Therefore, in summary, the equations of motion can be written in terms of the two new unknown fields $(\mathbf{u}, p)$ as

$$-\omega^2 \rho_\omega^s \mathbf{u} + \theta_\omega \nabla p = \nabla \cdot \hat{\mathbf{T}} + \mathbf{F}, \qquad (2.206)$$

$$\hat{\mathbf{T}} = \lambda (\nabla \cdot \mathbf{u})\mathbf{I} + G\left[\nabla \mathbf{u} + \nabla \mathbf{u}^T \right], \qquad (2.207)$$

$$\mathbf{T} = \hat{\mathbf{T}} - \alpha p \mathbf{I}, \qquad (2.208)$$

$$\frac{1}{M}(p + S) + \nabla \cdot \left\{ k_\omega \left[\nabla p - \omega^2 (\rho_f + \alpha) \mathbf{u} \right] \right\} = 0. \qquad (2.209)$$

Equation (2.206) corresponds to Newton's law applied to the solid skeleton of the porous material. This equation is similar to Newton's equation of elastic bodies except for the coupling term, $\theta_\omega \nabla p$, which represents the coupling between the solid and fluid phases. The stress tensor defined by Equation (2.207) corresponds to the stress tensor with the porous material in vacuum (i.e., it corresponds to the stress acting on the solid phase if the pore fluid is replaced by vacuum). Equation (2.208) describes the relationship between the total stress tensor and the effective stress tensor. The material properties entering into Equations (2.206)–(2.209) are given by

$$k_\omega = \frac{1}{\omega^2 \tilde{\rho}_f + i\omega b}, \qquad (2.210)$$

$$\lambda = K - \frac{2}{3}G, \qquad (2.211)$$

Table 2.3 Material properties used in the seismoelectric forward model.

Parameter	Value	Units	Reference
ρ_s	2650	kg m^{-3}	Mavko et al. (1998)
ρ_w	1000	kg m^{-3}	Mavko et al. (1998)
ρ_o	900	kg m^{-3}	Karaoulis et al. (2012)
K_s	36.5	GPa	Mavko et al. (1998)
K_{fr}	18.2	GPa	Mavko et al. (1998)
G	13.8	GPa	Mavko et al. (1998)
K_w	2.25	GPa	Jardani et al. (2010)
K_o	1.50	GPa	Charoenwongsa et al. (2010)
η_w	1×10^{-3}	Pa s	Jardani et al. (2010)
η_o	50×10^{-3}	Pa s	Light motor oil

$$\rho_\omega^s = \rho + \omega^2 \rho_f^2 k_\omega, \tag{2.212}$$

$$\theta_\omega = \alpha + \omega^2 \rho_f k_\omega, \tag{2.213}$$

where k_ω is not the dynamic permeability of the porous material, $\tilde{\rho}_f$ is an effective fluid density, λ is the Lamé coefficient, and ρ_ω^s corresponds to the apparent mass density of the solid phase at a given frequency ω. Typical material properties can be found in Table 2.3.

2.2.3 Description of the electrokinetic coupling

As explained in Chapter 1, the electrokinetic coupling at work in the seismoelectric response is the result of the relative displacement of the pore water with respect to the solid phase. The drag of a fraction of the charge density contained in the pore water is responsible for a net source current density in a framework attached to the solid phase (Pride, 1994; Revil et al., 1999a; Leroy & Revil, 2004). The model developed by Pride (1994) is an extension of the classical streaming potential theory and takes the form of the Biot–Frenkel equations coupled to the Maxwell equations via the source current density. However, this formulation has a drawback: it required knowledge of the zeta potential, a microscopic potential of the electrical double layer at the pore water/solid interface.

In the following text, we also consider the resulting EM disturbances in the quasistatic limit of the Maxwell equations, like the works on self-potential signals by Suski et al. (2006) and Jardani et al. (2007). This assumption is valid because the target is assumed to be close enough (less than a kilometer) from the receivers (such as antennas, nonpolarizing electrodes, and magnetometers).

In such a case, we can neglect the time required by the EM disturbances to diffuse between the reservoir and the receivers (see Revil et al., 2003, for a discussion of the diffusion time associated with the diffusion of low-frequency EM disturbances).

Using the constitutive equation derived in Chapter 1, we can model the problem by solving only the quasistatic Poisson-type problem:

$$\nabla \cdot (\sigma \nabla \psi) = \nabla \cdot \mathbf{J}_S, \tag{2.214}$$

$$\mathbf{J}_S = \hat{Q}_V^0 \dot{\mathbf{w}} = -i\omega \hat{Q}_V^0 k_\omega \left(\nabla p - \omega^2 \rho_f \mathbf{u} \right), \tag{2.215}$$

where ψ is the electrostatic potential ($\mathbf{E} = -\nabla \psi$ in the quasistatic limit of the Maxwell equations), σ is the electrical conductivity of the porous medium, $\mathbf{J}_S$ is the source current density of the electrokinetic kind, ρ_f is the fluid density, and $\hat{Q}_V^0$ is the effective excess charge (of the diffuse layer) per unit pore volume (in C m^{-3}). For saturated rocks, $\hat{Q}_V^0$ can be directly computed from the low-frequency permeability k_0 through the semiempirical formula shown in Figure 1.9.

2.3 Experimental approach and data

2.3.1 Measuring key properties
2.3.1.1 Measuring the cation exchange capacity and the specific surface area
Seismoelectric phenomena are related to the properties of the interface between the mineral phase and the pore water phase. Two of the most important parameters characterizing this interface are the cation exchange capacity (CEC) and the specific surface area. The CEC can be measured by the titration of the mineral surface with a cation that has a strong affinity for the active sites populating this surface (e.g., copper ammonium, cobalt). Knowing the amount of sorbed charge and the mass of the grains, we directly measure the charge per unit mass of solid grains, which is defined as the CEC (expressed therefore in the International System of Unit in Coulomb kg^{-1} or C kg^{-1}). Key references include Chapman (1965), Worthington (1973), Gillman (1979), Thomas (1982), and Gillman and Sumpter (1986).

The specific surface area of a material can be obtained using the so-called Brunauer–Emmett–Teller (BET) method (Brunauer et al., 1938). A dry sample is used and its surface is covered by a monolayer of nitrogen following the so-called BET isotherm. Such types of experiment have usually two parts. In the first, known volumes of nitrogen are allowed to expand into the apparatus and the resulting pressure is recorded. From this information, the volume of the chamber can be computed. Then, any gas adsorbed on the surface of the solid grains is removed by heating a sample of the porous material in a vacuum. Once the surface has been stripped of adsorbed gas molecules, known volumes of nitrogen are fed in and allowed to reach equilibrium with the surface of the mineral grains. By measuring the resulting pressure in the apparatus, the amount of nitrogen that remains unadsorbed can be found, and therefore, the amount adsorbed on the solid can also be computed. Knowing the size of a nitrogen molecule, it is possible to determine the surface area per unit mass of grains (expressed therefore in $m^2 \, kg^{-1}$ in the International System of Units).

2.3.1.2 Measuring the complex conductivity

The experimental setup used to measure the complex conductivity of a rock sample is shown in Figure 2.7. It consists of an impedance meter able to measure the complex conductance or impedance in a broad range of frequencies (1 mHz to 45 kHz for the equipment described in Figure 2.7b). A harmonic current is imposed on the sample (through the current electrodes A and B; see Figure 2.7a), and the harmonic electrical field is measured with the pairs of electrodes M and N. The phase angle between the current and the voltage is generally very small (less than 30 mrad), and measuring it adequately used to be quite challenging. Nowadays, however, we can easily reach a phase accuracy of 0.1 mrad from 1 mHz to 10 kHz by using a high sampling rate. The amplitude of the measured conductance can be transformed into the amplitude of the electrical conductivity using a geometrical factor, which depends on the geometry of the sample measurement system. This includes the position of the electrodes and the boundary conditions for the electrical current and the electric potential. The geometrical factor can be determined using two approaches: (1) using a benchmark with a material of same shape as the test sample but of known conductivity and (2) by solving numerically the Laplace equation (see Jougnot et al., 2010, for additional details).

One example of an experimental data set performed at different salinities but at a single frequency is shown in Figure 2.8. We see that both the in-phase conductivity

and the quadrature conductivity depend on the conductivity of the pore water. The formation factor and the surface conductivity can be found by using the in-phase conductivity data performed at the different salinities by fitting the core sample conductivity versus pore water conductivity using an electrical conductivity equation such as Equation (1.91) (the surface conductivity corresponding to the last term of this equation). In this case, we have two unknowns to fit the formation factor and the surface conductivity, and such fitting (using a least-square approach) is shown in Figure 2.8.

2.3.1.3 Measuring the streaming potential coupling coefficient

The quasistatic streaming potential coupling coefficient is defined by

$$C_0 = \lim_{\omega \to 0} \left(\frac{\partial \psi}{\partial p} \right)_{\mathbf{J} = 0}. \qquad (2.216)$$

Measuring this fundamental coefficient, used in the whole electrokinetic theory, is actually pretty simple. Figure 2.9a shows a very simple experimental setup in which a gradient in hydraulic head $h = p/(g\rho_f)$ (g denotes the acceleration of the gravity, $9.81 \, m \, s^{-1}$) is applied to the core sample. Figure 2.9b shows that the measured potential drops $\partial \psi$, at the end faces of the cylindrical core sample, are proportional to the imposed head differences. This result is confirmed in Figure 2.10 by directly plotting the potential differences as a function of the hydraulic head differences. The quasistatic coupling coefficient is determined from the slope of the linear trends shown in Figure 2.10. The permeability can be measured from the same setup by measuring the flow rate of water through the core sample and applying Darcy's law to determine either the hydraulic conductivity or the permeability. If the electrical conductivity of the core sample has also been determined, we can use the expression of the streaming potential coupling coefficient to determine the effective charge density $\hat{Q}_V^0$. An example is provided in the next section.

2.3.2 Streaming potential dependence on salinity

First, we investigate the effect of salinity upon the streaming potential coupling coefficient C. At low frequencies ($\omega \tau_k \ll 1$), the streaming potential coupling coefficient is given by

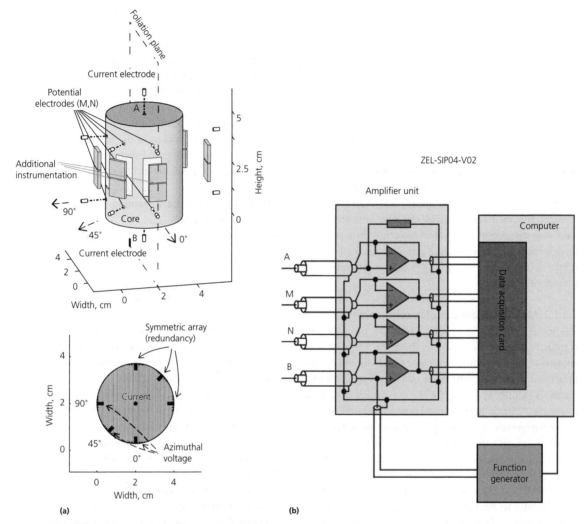

Figure 2.7 Experimental setup to measure complex conductivity. **a)** Position of the electrodes on the core sample. A and B denote the current electrodes used to inject a harmonic current, while the electrodes M and N denote the voltage electrodes used to record the harmonic electrical field. **b)** ZEL-SIP04-V02 impedance meter built by Egon Zimmermann in Germany (Zimmermann et al., 2008). The data acquisition system operates in the frequency range from 1 mHz to 45 kHz with a phase accuracy close to 0.1 mrad below 1 kHz. This instrument can be used to measure the complex conductivity (amplitude and phase) in the frequency range 1 mHz to 45 kHz. (*See insert for color representation of the figure.*)

$$C_0 \approx \frac{k_0 \hat{Q}_V^0}{\eta_w \left(\frac{\sigma_w}{F} + \sigma_S \right)}. \qquad (2.217)$$

We can use Equation (2.217) to fit the values of the static coupling coefficient displayed in Figure 2.11, for the Berea sandstone. Knowing the permeability and the formation factor of the core sample, the pore water conductivity, and the viscosity of the pore water, the two unknowns that remain are the charge per unit pore volume, $\hat{Q}_V^0$, and the surface conductivity, σ_S. After fitting the model to the data, we obtain $\sigma_S = (1.2 \pm 0.3) \times 10^{-3}$ S m^{-1} and $\hat{Q}_V^0 = 1.4 \pm 0.2$ C m^{-3}. These two values can be independently confirmed using our model: (1) the value of $\hat{Q}_V^0$ can be independently obtained by the empirical model of Figure 1.9, which yields $\hat{Q}_V^0 = 2.0$ Cm^{-3}. (2) The surface conductivity can be compared to the estimate made by Moore et al. (2004) using electrical

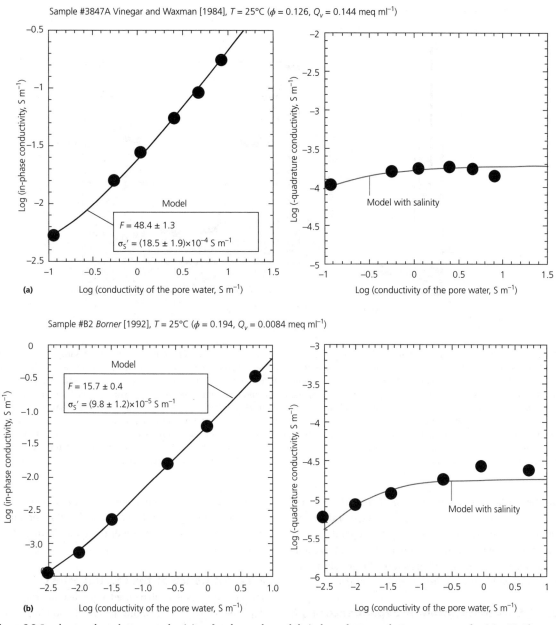

Figure 2.8 In-phase and quadrature conductivity of rock samples and their dependence on the pore water conductivity (NaCl solutions). **a)** Experimental data for a sample from the Vinegar and Waxman (1984) database (clayey sand). The error bars are smaller than the size of the symbols. **b)** The same for one core sample (sandstone) from the database of Börner (1992). In both cases, the quadrature conductivity model (materialized by the plain line) is the one described in Revil (2012). The in-phase conductivity data are fitted using a conductivity model providing the values of the formation factor and the surface conductivity.

conductivity measurements. They found $\sigma_S = 2.7 \times 10^{-3}$ S m^{-1}. Therefore, there is a fair agreement between the present theory and the published experiment data. From the surface conductivity ($\sigma_S = 1.2 \times 10^{-3}$ S m^{-1}), the

formation factor ($F = 18$), and the value of the mobility of the counterions in the diffuse layer [β(Na$^+$, 25°C) = 5.2×10^{-8} m^2 s^{-1} V^{-1}], we can estimate the value of the high-frequency charge density $\hat{Q}_V^\infty$ using the expression

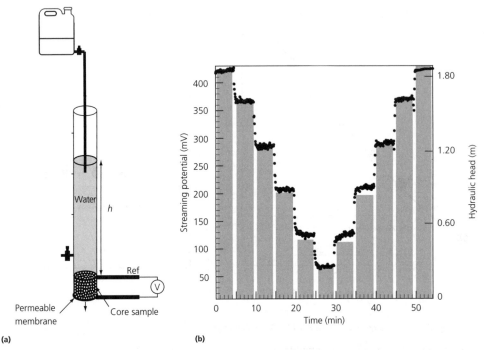

(a)   **(b)**

Figure 2.9 Simple streaming potential coupling coefficient measurement setup and experimental results. **a)** The sample is packed at the bottom of a Plexiglas tube and is maintained in the tube by a permeable membrane with a coarse mesh (the mesh is, however, finer than the diameter of the grains). The record of the self-potential during the flow of the electrolyte through the sample is done with Ag/AgCl$_2$ electrodes ("Ref" is the reference electrode) attached to the end faces of the sample. The hydraulic heads are maintained constant at different levels as the streaming potentials are recorded at the various levels. **b)** Example of a typical run for sample S3 (grain size of 150–212 µm) and a water conductivity of 10^{-3} S m^{-1}. The filled circles correspond to the measurements of the streaming potential at the two end faces of the sample, while the gray columns correspond to the measurement of the hydraulic heads. From this, it can be seen that the measured streaming potentials are proportional to the imposed hydraulic heads. The results are reproducible. This means that there is no drift of the electrical potential of the electrodes for the duration of the experiment.

of the surface conductivity. We obtain $\hat{Q}_V^\infty = 4 \times 10^5$ Cm^{-3}. We check that $\hat{Q}_V^\infty \gg \hat{Q}_V^0$, in the case of the Berea sandstone (porosity 0.23, permeability 450 mD, NaCl). This is expected because the Berea sandstone has pretty large pores (6 to 9 µm), and therefore, the double layer is very thin with respect to the size of the pores.

2.3.3 Streaming potential dependence on pH

Because the ζ-potential of the surface of silicates and aluminosilicates is pH dependent, we expect that the seismoelectric current generated through the seismoelectric effect should be also pH dependent. Dukhin et al. (2010) performed seismoelectric current measurements at different pH for three sandstones including the Berea sandstone. Their data, shown in Figure 2.12, is showing a clear dependence with the pH. The Berea sandstone has some kaolinite, and the observed pH dependence seems

controlled by the pH dependence of the zeta potential of kaolinite. This pH dependence can be therefore predicted from the double layer theory discussed in Chapter 1.

2.3.4 Influence of the inertial effect

The previous results will be used to predict the frequency dependence of the streaming potential coupling coefficient in the Berea sandstone. In Figure 2.13, we compare the prediction of our model with the recent measurements of the dynamic streaming potential coupling coefficient from Zhu and Toksöz (2013) (the values reported at 1 kHz are actually the static values). Our model is able to reproduce these data very well up to 100 kHz for five different salinities. The decrease of the coupling coefficient at high frequency is due to the effect of the inertial term in the Navier–Stokes equation.

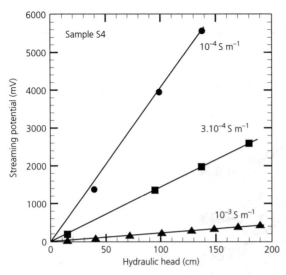

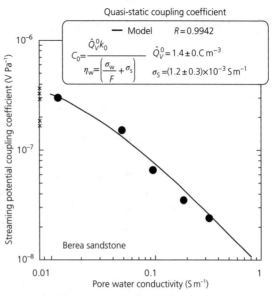

Figure 2.10 Example of three typical runs for a core sample (in the present case, glass beads with grain size of 212–300 µm) at three pore water conductivities. We observe linear relationships between the variation of the streaming potentials and the variation of the hydraulic heads at these different salinities. At each salinity, the quasistatic streaming potential coupling coefficient C_0 is obtained from the slope of these linear trends.

Figure 2.11 Low-frequency streaming potential coupling coefficient C_0. The black circles correspond to the measurements by Zhu and Toksöz (2013) (Berea sandstone, porosity 0.23, permeability 450 mD, NaCl). The crosses correspond to the laboratory measurements by Moore et al. (2004) (Berea sandstone, porosity 0.19, water). The plain line corresponds to the fit of the proposed model to the data of Zhu and Toksöz (2013) only.

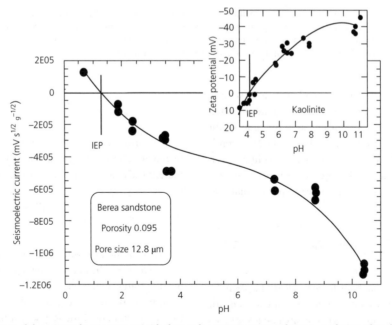

Figure 2.12 Dependence of the seismoelectric current with the pH for a water-saturated Berea sandstone (data from Dukhin et al., 2010). The plain line is a guide for the eyes. IEP corresponds to the isoelectric point for which the zeta potential of the mineral is equal to zero. The Berea sandstone is mainly made of silica grains with some kaolinite. The insert corresponds to zeta potential data reported by Leroy and Revil (2004) for kaolinite (NaCl, 2×10^{-3} Mol l^{-1}). The plain line corresponds to the prediction of a triple layer model.

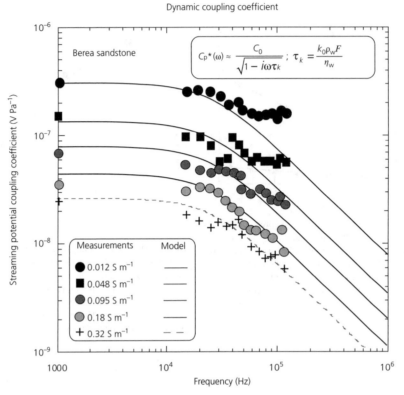

Dynamic coupling coefficient

$$C_P{}^*(\omega) \approx \frac{C_0}{\sqrt{1 - i\omega\tau_k}} \;;\; \tau_k = \frac{k_0\rho_w F}{\eta_w}$$

Figure 2.13 Dynamic streaming potential coupling coefficient (the values reported at 1 kHz are the static values). The data are from Zhu and Toksöz (2013) for the same Berea sandstone (porosity 0.23, permeability 450 mD, NaCl). The relaxation is due to the transition between the viscous laminar flow regime at low frequency and the inertial laminar flow regime at high frequencies. Below 10 kHz, the streaming potential coupling coefficient can be considered independent of frequency.

Another way to look at the dependency of the streaming potential coupling coefficient with the inertial effect is to plot this coupling coefficient as a function of the Reynolds number. Teng and Zhao (2000) recently derived a generalized Darcy equation by volume averaging the local Navier–Stokes momentum equation over a representative elementary volume of a porous material, given by

$$\rho_f \frac{d\dot{w}}{dt} + \frac{1 + \mathrm{Re}}{k_0}\eta_f \dot{w} = -\nabla p + \mathbf{F}, \qquad (2.218)$$

where **F** is a macroscopic body force and Re is the Reynolds number, a key dimensionless number that expresses the ratio of inertial to viscous forces in the Navier–Stokes equation (Batchelor, 1972). For a capillary of radius R, U being the strength of the seepage velocity, the Reynolds number is then defined by (e.g., Batchelor, 1972)

$$\mathrm{Re} = \frac{\rho_f \dot{w} R}{\eta_f}. \qquad (2.219)$$

In a porous material, the radius of the capillary should be replaced by a corresponding length scale of the porous material. The Reynolds number is defined by

$$\mathrm{Re} = \frac{\rho_f \dot{w} \Lambda}{\eta_f}, \qquad (2.220)$$

where Λ is a characteristic length of the flow (for capillaries $\Lambda = R$ where R is the radius of the capillary). If we replace $\dot{w}$ by the Darcy equation (neglecting the electroosmotic contribution), we can use the following approximation to get an estimate of the Reynolds number:

$$\mathrm{Re} = \frac{\rho_f k_0 \Lambda}{\eta_f^2 (1 + \mathrm{Re})}\frac{p}{L}, \qquad (2.221)$$

where p is the pore fluid pressure and L is the length of the cylindrical core. For a granular medium with a unimodal particle size distribution, an expression to determine the length scale Λ is given by (Revil, 2002)

$$\Lambda = \frac{d_0}{2m(F-1)}. \qquad (2.222)$$

From Equations (2.221) and (2.222), the Reynolds number is the solution of the following equation:

$$\mathrm{Re}^2 + \mathrm{Re} - \frac{\rho_f^2 g}{2m\alpha\eta_f^2}\frac{d_0^3}{F(F-1)^3}\left(\frac{h}{L}\right) = 0. \qquad (2.223)$$

The positive root of Equation (2.223) is

$$\mathrm{Re} = \frac{1}{2}\left(\sqrt{1+c}-1\right), \qquad (2.224)$$

$$c = \frac{\beta\rho_f}{\eta_f^2}\frac{d_0^3}{F(F-1)^3}\left(\frac{p}{L}\right), \qquad (2.225)$$

where $\beta \approx 2.25 \times 10^{-3}$ is a numerical constant (determined from the constants given earlier). Equation (2.224) is a new equation that has a strong practical value since it can be easily used to determine the Reynolds number in a porous material from the knowledge of the pressure gradient.

In the present case, the macroscopic body force corresponds to the electrostatic force associated with the excess of electrical charge per unit pore volume. Therefore, the generalized Darcy equation, Equation (2.218), can be written as

$$\dot{\mathbf{w}} = -\frac{k}{\eta_f}\nabla p - \hat{Q}_V^0 \nabla\psi, \qquad (2.226)$$

where k is an apparent permeability that is related to the Reynolds number by

$$\frac{k}{k_0} = \frac{1}{1+\mathrm{Re}}, \qquad (2.227)$$

$$\lim_{\mathrm{Re}\to 0}\frac{k}{k_0} = 1, \qquad (2.228)$$

where k_0 is the permeability in viscous laminar flow conditions.

The influence of inertial flow upon electrokinetic coupling has been the subject of very few publications (see

Watanabe and Katagishi, 2006, and references therein). Gorelik (2004) used dimensional analysis to demonstrate that the effect of the Reynolds number corresponds to a multiplication of the Helmholtz–Smoluchowski equation by an unspecified function of the Reynolds number. In this chapter, we look for an explicit (quantitative) relationship between the streaming potential coupling coefficient and the Reynolds number. At the scale of a representative elementary volume, the current density is given by

$$\mathbf{J} = -\sigma\nabla\psi + \hat{Q}_V^0 \dot{\mathbf{w}}, \qquad (2.229)$$

$$\mathbf{J} = -\sigma\nabla\psi - \frac{k\hat{Q}_V^0}{\eta_f}\nabla p, \qquad (2.230)$$

where k is the apparent permeability defined earlier. The streaming potential coupling coefficient can be related to the excess charge of the diffuse layer per unit pore volume, $\hat{Q}_V^0$, by $C_0 = k_0\hat{Q}_V^0/\eta_f\sigma$. Equation (2.229) expresses the fact that the source current density is equal to the excess of charge of the pore fluid $\hat{Q}_V^0$ times the Darcy velocity $\dot{\mathbf{w}}$. As the seepage velocity is influenced by the increase of the Reynolds number (for Re >0.1), the Reynolds number also influences the value of the streaming potential coupling coefficient. Following Equations (2.227) and (2.230), the streaming potential coupling coefficient is related to the Reynolds number by

$$\frac{C}{C_0} = \frac{1}{1+\mathrm{Re}}, \qquad (2.231)$$

$$\lim_{\mathrm{Re}\to 0}\frac{C}{C_0} = 1, \qquad (2.232)$$

where C_0 is the streaming potential coupling coefficient in viscous laminar flow conditions and C is the measured coupling coefficient. A comparison between Equations (2.227) and (2.232) with experimental data obtained by Bolève et al. (2007) is shown in Figure 2.14. We see a clear decrease of the streaming potential coupling coefficient with the Reynolds number that is well reproduced by the model.

2.4 Conclusions

In this chapter, we have developed a complete theory of wave propagation in poroelastic media. In the first case, the porous material is considered to be saturated with a viscoelastic fluid able to sustain shear stresses. In this case,

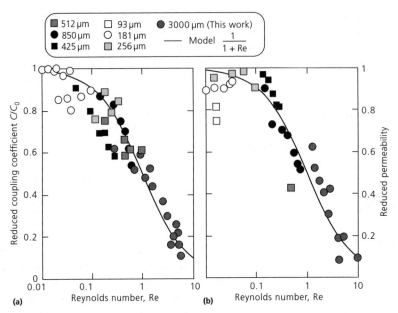

Figure 2.14 Influence of the Reynolds number. **a)** Influence of the Reynolds number upon the relative coupling coefficient C/C_0 (C is the measured apparent streaming potential coupling coefficient and C_0 is determined in the viscous laminar flow regime for Re $\ll$ 1 where Re denotes the Reynolds number). **b)** Influence of the Reynolds number upon the relative permeability k/k_0 (k denotes the measured apparent permeability using Darcy's law, while k_0 is determined in the viscous laminar flow regime for Re $\ll$ 1) (Data from Bolève et al., 2007).

we found four types of waves (2 compressional P-waves and 2 shear (S-)waves). Alternatively, this theory can be simplified (at low frequencies) to recover the classical Biot theory for a poroelastic material saturated with a Newtonian fluid. In this case, there are two compressional P-waves and one shear (S-)wave mode (the fluid is not able to sustain shear stresses). In all these cases, we determined the source current density associated with the effective charge per unit volume and the Darcy velocity.

A comparison between the model and the experimental data showed that our model was able to explain both the streaming potential coupling coefficient at low frequencies and its dependency with salinity and frequency up to 100 kHz. We also discussed the effect of the Reynolds number upon the streaming potential coupling coefficient, which corresponds to an alternative way to see the transition between the laminar viscous and laminar inertial flow regimes.

In Chapter 4, we will show how to solve the macroscopic equations (valid at the scale of the representative elementary volume) developed in this chapter, and we will develop inverse algorithms to determine key properties of the subsurface such as facies boundaries and material properties.

References

Alkafeef, S.F. & Smith, A.L. (2005) Asphaltene adsorption isotherm in the pores of reservoir rock cores, Paper 93188 presented at the SPE International Symposium on Oil field Chemistry, SPE, Houston.

Alkafeef, S., Gochin, R.J. & Smith, A.L. (2001) The effect of double layer overlap on measured streaming currents for toluene flowing through sandstone cores. *Colloids and Surfaces A: Physiochemistry and Engineering Aspects*, **195**, 77–80.

Alkafeef, S.F., Algharaib, M.K. & Alajmi, A.F. (2006), Hydrodynamic thickness of petroleum oil adsorbed layers in the pores of reservoir rocks. *Journal of Colloid and Interface Science*, **298**, 13–19.

Atalla, N., Panneton, R. & Debergue, P. (1998) A mixed displacement-pressure formulation for poroelastic materials. *Journal of the Acoustical Society of America*, **104**, 1444–1452.

Batchelor, G.K. (1972) *An Introduction to Fluid Dynamics*, Cambridge University Press, Cambridge, UK

Behura, J., Batzle, M., Hofmann, R., & Dorgan, J. (2007) Heavy oils: their shear story. *Geophysics*, **72(5)**, E175–E183.

Biot, M. (1956a) Theory of propagation of elastic waves in a fluid-saturated porous solid. I. Low-frequency range. *Journal of the Acoustical Society of America*, **28**, 168–178.

Biot, M. (1956b) Theory of propagation of elastic waves in a fluid-saturated porous solid, II. Higher-frequency range. *Journal of the Acoustical Society of America*, **28**, 178–191.

Biot, M.A. (1962a) Generalized theory of acoustic propagation in porous dissipative media. *Journal of the Acoustical Society of America*, **34**(9), 1254–1264.

Biot, M.A. (1962b) Mechanics of deformation and acoustic propagation in porous media. *Journal of Applied Physics*, **33**(4), 1482–1498.

Biot, M.A. & Willis, D.G. (1957) The elastic coefficients of the theory of consolidation. *Journal of Applied Mechanics*, **24**, 594–601.

Bolève, A., Crespy, A., Revil, A., Janod, F., & Mattiuzzo J.L. (2007) Streaming potentials of granular media: Influence of the Dukhin and Reynolds numbers. *Journal of Geophysical Research*, **112**, B08204, doi:10.1029/2006JB004673.

Börner, F.D. (1992) Complex conductivity measurements of reservoir properties. *Proceedings of the Third European Core Analysis Symposium*, Paris, pp. 359–386.

Brunauer, P., Emmett, H., & Teller, E. (1938) Adsorption of Gases in Multimolecular Layers. *Journal of the American Chemical Society*, **60**(2), 309–319, Doi: 10.1021/ja01269a023.

Buckley, J.S. & Liu, Y. (1998) Some mechanisms of crude oil/brine/solid interactions. *Journal of Petroleum Science and Engineering*, **20**, 155–160.

Castrejon-Pita, J.R., Del Río, J.A., & Huelsz, G. (2003) Experimental observations of dramatic differences in the dynamic response of Newtonian and Maxwellian fluids. *Physical Review E*, **68**, 046301.

Chapman, H.D. (1965) Cation-exchange capacity. *In:* C. A. Black (ed.) *Methods of soil analysis—Chemical and microbiological properties. Agronomy*, **9**, pp. 891–901. American Society of Agronomy, Inc., Madison, WI.

Charoenwongsa, S., Kazemi, H., Miskimins, J., & Fakcharoenpol, P. (2010) A fully-coupled geomechanics and flow model for hydraulic fracturing and reservoir engineering applications. CSUG/SPE SPE-137497, 1–31, Canadian Unconventional Resources and International Petroleum Conference, Calgary, Alberta, Canada.

Cole, K.S. & Cole, R.H. (1941) Dispersion and absorption in dielectrics. *Journal of Chemical Physics*, **9**, 341–351.

Dante, R.C, Geffroy, E., & Chavez, A.E. (2007) Viscoelastic models for Mexican heavy crude oil and comparison with a mixture of heptadecane and eicosane. Part II. *Fuel*, **86**, 2403–2409.

De Groot, S.R., & Mazur, P. (1984) *Non-equilibrium Thermodynamics*, 510 pp. Dover, New York.

Delgado, A.V., González-Caballero, F., Hunter, R.J., Koopal, L. K., & Lyklema J. (2007) Measurement and interpretation of electrokinetic phenomena. *Journal of Colloid and Interface Science*, **309**, 194–224.

Dukhin, A.S., Goetz, P.J., & Thommes, M. (2010) Seismoelectric effect: a non-isochoric streaming current. 1. Experiment. *Journal of Colloid and Interface Science*, **345**, 547–553.

Garambois, S., & Dietrich, M. (2002) Full waveform numerical simulations of seismoelectromagnetic wave conversions in fluid-saturated stratified porous media. *Journal of Geophysical Research*, **107**(B7), 10.1029/2001JB000316.

Gassmann, F. (1951) Über die Elastizität poröser Medien. *Vierteljahrschrift der Naturforschenden Gesellschaft in Zurich*, **96**, 1–23.

Geertsma, J. & Smit, D.C. (1961) Some aspects of elastic wave propagation in fluid-saturated porous solids. *Geophysics*, **26**, 169–181.

Gillman, G.P. (1979) A proposed method for the measurement of exchange properties of highly weathered soils. *Australian Journal of Soil Research*, **17**, 129–139.

Gillman, G.P. & Sumpter, E.A. (1986), Modification to the compulsive exchange method for measuring exchange characteristics of soils. *Australian Journal of Soil Research*, **24**, 61–66.

Gorelik, L.V. (2004) Investigation of dynamic streaming potential by dimensional analysis. *J. Colloid and Interface Science*, **274**, 695–700.

Haartsen, M.W. & Pride, S.R. (1997) Electroseismic waves from point sources in layered media. *Journal of Geophysical Research*, **102**(B11), 24745–24769.

Haartsen, M., Dong, W. & Toksoz, M.N. (1998), Dynamic streaming currents from seismic point sources in homogeneous poroelastic media. *Geophysical Journal International*, **132**, 256–263.

Jardani, A., Revil, A., Santos, F., Fauchard, C. & Dupont, J.P. (2007) Detection of preferential infiltration pathways in sinkholes using joint inversion of self-potential and EM-34 conductivity data. *Geophysical Prospecting*, **55**, 1–11, doi:10.1111/j.1365-2478.2007.00638.x

Jardani, A., Revil, A., Slob, E. & Sollner, W. (2010) Stochastic joint inversion of 2D seismic and seismoelectric signals in linear poroelastic materials. *Geophysics*, **75**(1), N19–N31, doi: 10.1190/1.3279833.

Johnson D.L. (1986) Recent developments in the acoustic properties of porous media. In: Sette D, ed. *Proceedings of the international school of physics « Enrico Fermi » course XCIII Frontiers in physical acoustics*, pp. 255–290. North-Holland, Amsterdam.

Jougnot, D., Ghorbani, A., Revil, A., Leroy, P., & Cosenza, P. (2010) Spectral induced polarization of partially saturated clay-rocks: a mechanistic approach. *Geophysical Journal International*, **180**(1), 210–224, doi: 10.1111/j.1365-246X.2009.04426.x.

Karaoulis, M., Revil, A., Zhang, J., & Werkema, D.D. (2012) Time-lapse cross-gradient joint inversion of cross-well DC resistivity and seismic data: a numerical investigation. *Geophysics*, **77**, D141–D157, doi: 10.1190/GBO2012-0011.1.

Karpfinger, F., Muller, T.M., & Gurevich, B. (2009) Green's function and radiation pattern in poroelastic solid revisited. *Geophysical Journal International*, **178**, 327–337.

Leroy, P. & Revil, A. (2004) A triple layer model of the surface electrochemical properties of clay minerals. *Journal of Colloid and Interface Science*, **270**(2), 371–380.

Mavko, G., Mukerji, T., & Dvorkin, J. (1998) *The Rock Physics Handbook*, pp. 289–309, Cambridge University Press, Cambridge, UK.

Moore, J.R., Glaser, S.D., Morrison, H.F., & Hoversten G.M. (2004) The streaming potential of liquid carbon dioxide in Berea sandstone. *Geophysical Research Letters*, **31**, L17610, doi:10.1029/2004GL020774.

Morency, C. & Tromp, J. (2008) Spectral-element simulations of wave propagation in porous media. *Geophysical Journal International*, **175**, 301–345.

Plona, T.J. (1980) Observation of a second bulk compressional wave in a porous medium at ultrasonic frequencies. *Applied Physics Letters*, **36**, 259–261.

Pride, S. (1994) Governing equations for the coupled electromagnetics and acoustics of porous media. *Physical Review B*, **50**, 15678–15696.

Rañada Shaw, A., Denneman, A.I.M., & Wapenaar, C.P.A. (2000) Porosity and permeability effects on seismo-electric reflection. In: *Proceedings of the EAGE conference*. Paris, France.

Revil, A. (2002) The hydroelectric problem of porous rocks: thermodynamic approach and introduction of a percolation threshold. *Geophysical Journal International*, **151**(3), 944–949.

Revil, A. (2012) Spectral induced polarization of shaly sands: Influence of the electrical double layer. *Water Resources Research*, **48**, **W02517**, doi:10.1029/2011WR011260.

Revil, A. & Jardani, A. (2010) Seismoelectric response of heavy oil reservoirs. Theory and numerical modelling. *Geophysical Journal International*, **180**, 781–797, doi: 10.1111/j.1365-246X.2009.04439.x.

Revil, A. & Linde, N. (2006) Chemico-electromechanical coupling in microporous media. *Journal of Colloid and Interface Science*, **302**, 682–694.

Revil, A., Schwaeger, H., Cathles, L.M. & Manhardt, P. (1999a) Streaming potential in porous media. 2. Theory and application to geothermal systems. *Journal of Geophysical Research*, **104**(**B9**), 20,033–20,048.

Revil, A., Pezard, P.A., & Glover, P.W.J. (1999b) Streaming potential in porous media. 1. Theory of the zeta-potential. *Journal of Geophysical Research*, **104**(**B9**), 20021–20031.

Revil, A., Naudet, V., Nouzaret, J., & Pessel, M. (2003) Principles of electrography applied to self-potential electrokinetic sources and hydrogeological applications. *Water Resources Research*, **39**(**5**), 1114, doi: 10.1029/2001WR000916.

Revil, A., Leroy, P., Ghorbani, A., Florsch, N., & Niemeijer, A.R. (2006) Compaction of quartz sands by pressure solution using a Cole-Cole distribution of relaxation times. *Journal of Geophysical Research*, **111**, B09205, doi:10.1029/2005JB004151.

von Smoluchowski, M. (1906) Zur kinetischen Theorie der Brownschen Molekularbewegung und der Suspensionen (in German). *Annalen der Physik*, **326**(**14**), 756–780.

Suski, B., Revil, A., Titov, K., et al. (2006) Monitoring of an infiltration experiment using the self-potential method. *Water Resources Research*, **42**, W08418, doi:10.1029/2005WR004840

Teng, H. & Zhao, T.S. (2000) An extension of Darcy's law to non-Stokes flow in porous media. *Chemical Engineering Science*, **55**, 2727–2735.

Thomas, G.W. (1982) Exchangeable cations. *In:* A.L. Page, Ed. *Methods of soil analysis, Part 2 Chemical and microbiological properties*, 2nd edition, *Agronomy*, **9**, 159–165. American Society of Agronomy, Madison, WI.

Vinegar, H.J. & Waxman, M.H. (1984) Induced polarization of Shaly sands. *Geophysics*, **49**, 1267–1287.

Watanabe, T. & Katagishi, Y. (2006) Deviation of linear relation between streaming potential and pore fluid pressure difference in granular material at relatively high Reynolds numbers. *Earth, Planets and Space*, **58**(**8**), 1045–1051.

Worthington, A.E. (1973) An automated method for the measurement of cation exchange capacity of rocks. *Geophysics*, **38**(**1**), 140–153.

Yasufuku, S., Ise, T., Inoue, Y., & Ishioka, Y. (1977) Electrokinetic phenomena in electrical insulating oil/impregnated cellulosic pressboard systems, in *«Electrical Insulation »*, *IEEE Transactions*, **EI-12**(**5**), 370–375, Oxford, UK.

Zhu, Z. & Toksöz, M.N. (2013) Experimental measurements of the streaming potential and seismoelectric conversion in Berea sandstone. *Geophysical Prospecting*, **61**(**3**), 688–700, doi: 10.1111/j.1365-2478.2012.01110.

Zimmermann, E., Kemnan, A., Berwix, J., Glaas, W., Munch, H.M., & Huisman, J.A. (2008) A high-accuracy impedance spectrometer for measuring sediments with low polarizability. *Measurement Science and Technology*, 19, doi:10.1088/0957-0233/19/10/105603.

CHAPTER 3

Seismoelectric theory in partially saturated conditions

Now that we have developed a complete theory in the saturated case (both with the poroelastic and acoustic approximations), we consider in this chapter some extensions to two cases. The first extension concerns the unsaturated case for which the material is partially saturated in water (considered to be the wetting phase for the solid grains) but the nonwetting phase (air) is at the atmospheric pressure. The second extension corresponds to the case where there are two immiscible Newtonian fluids present in the pore space. In this case, we need to account explicitly for the capillary pressure. Finally, we reinvestigate the acoustic approximation extending its validity in partially saturated conditions. We end the chapter by comparing some of the predictions of our model with available experimental data.

3.1 Extension to the unsaturated case

3.1.1 Generalized constitutive equations

In this section, we develop a new theory of the seismo-electric effect in unsaturated porous material. This theory can be used to understand the electromagnetic signals associated with seismic wave propagation in saturated and unsaturated porous media. We consider in the following text an isotropic representative elementary volume of a porous material with connected pores. The surfaces of the minerals in contact with the pore water are negatively charged. We consider therefore an excess of (positive) charges in the pore space of the porous material in the electrical double layer. As discussed in Chapter 1, the electrical double layer coating the surface of the grains is made of two layers: (1) a layer of (counter) ions sorbed onto the mineral surface and (2) a diffuse layer where the electrostatic (Coulomb) force prevails. In the following, we will use the subscript a to describe the properties of air (the nonwetting phase), and subscripts w and s will be used to describe the properties of the water and solid phases, respectively. Water is assumed to be the wetting phase. The term skeleton will be used to describe the assemblage of grains alone without the two fluid phases in the pore network.

Another set of assumptions used later pertains to the capillary pressure curve. Hysteretic behavior will be neglected, and therefore, the porous material will be characterized by a unique set of hydraulic functions (for instance, the Brooks & Corey, 1964). We will work also with the Richards model, which makes the assumption that the air pressure is constant and equal to the atmospheric pressure. This implies in turn that the air phase is infinitely mobile and connected to the atmosphere. This represents a simplification of the problem that makes the analysis more tractable. The capillary pressure p_c (in Pa) is defined as the pressure of the nonwetting phase minus the pressure of the wetting phase (Bear & Verruijt, 1987):

$$p_c = p_a - p_w \qquad (3.1)$$

The Seismoelectric Method: Theory and Applications, First Edition. André Revil, Abderrahim Jardani, Paul Sava and Allan Haas.
© 2015 John Wiley & Sons, Ltd. Published 2015 by John Wiley & Sons, Ltd.

where p_a and p_w denote the average air and water pressures (in Pa), respectively. The capillary head (suction), Ψ, is defined as

$$\Psi = -\frac{p_c}{\rho_w g} = \frac{p_w - p_a}{\rho_w g} \qquad (3.2)$$

In unsaturated flow conditions, the gradient of the capillary head is given by

$$\nabla\Psi = \frac{1}{\rho_w g}\nabla p_w \qquad (3.3)$$

This assumption is used to avoid dealing with the flow of the air phase, simplifying the problem solution. In unsaturated conditions, the capillary pressure is positive, the capillary head is negative, and the pressure of the water phase is smaller than the atmospheric pressure. The total head h includes the gravity force and is defined by $\Psi + z$ where z denotes the elevation head (gravity effect). Our model will be restricted to the capillary regime, which has a saturation level that is above the irreducible water saturation level. There are also many mechanisms of electrical polarization in porous media including the diffusion polarization, the polarization of the Stern layer coating the surface of the mineral grains, and the polarization of the metallic particles (e.g., pyrite and magnetite) acting as semiconductors. At low frequencies (<1 MHz) and in the absence of metallic particles, the so-called α-polarization prevails (Revil, 2013a, b), which is due to the polarization of the Stern layer. Finally, attenuation of the seismic waves associated with squirt-flow dissipation mechanisms will be neglected despite the fact that this mechanism is known to control the attenuation of seismic waves in the frequency band usually used in the field for seismic investigations (see Rubino & Holliger, 2012). Neglecting squirt flow makes the problem simpler and more tractable.

In saturated conditions, the (averaged) filtration displacement is defined as (Morency & Tromp, 2008)

$$\mathbf{w} = \phi(\mathbf{u}_w - \mathbf{u}) \qquad (3.4)$$

where ϕ denotes the connected porosity and $\mathbf{u}_w$ and $\mathbf{u}$ correspond to the averaged displacement of the water and solid phases, respectively. All the disturbances considered in the following text will be harmonic, $\exp(-i\omega t)$, where $\omega = 2\pi f$ denotes the angular frequency (in rad s^{-1}) and $f = \omega/(2\pi)$ is the frequency (in Hertz).

In the fully saturated condition, the Darcy velocity (also called the filtration velocity) is defined as the time derivative of the filtration displacement. In these conditions, the generalized Darcy's law is given by (Jardani et al., 2010)

$$\dot{\mathbf{w}} = -\frac{k_0}{\eta_f}\left(\nabla p + \rho_f \ddot{\mathbf{u}} + \rho_f \frac{\alpha}{\phi}\ddot{\mathbf{w}} - \mathbf{F}_f\right) \qquad (3.5)$$

where k_0 denotes the quasistatic permeability of the porous material (in m^2), $F = \alpha/\phi$ denotes the electrical formation factor (dimensionless), which is the ratio of the bulk tortuosity α of the pore space to the connected porosity, and $\mathbf{F}_f$ denotes the body force applied to the pore water phase (in N m^{-3}, e.g., the gravitational body force or the electrical force acting on the excess of electrical charges of the pore water). For the following equations, to keep the notation as light as possible, we will not distinguish the variables expressed in the time domain from those in the frequency domain. However, it will be easy enough to recognize the domain of the formulations when the equations are written in the frequency domain versus the time domain. The switch from one domain to the other is done by a simple Fourier transform or its associated inverse Fourier transform.

In unsaturated conditions, the filtration displacement and the mass density of the fluid (subscript f) phase are given by

$$\mathbf{w}_w = s_w\phi(\mathbf{u}_w - \mathbf{u}) \qquad (3.6)$$

$$\rho_f = (1 - s_w)\rho_g + s_w\rho_w \qquad (3.7)$$

respectively. In these equations, s_w denotes the degree of water saturation ($s_w = 1$ at full saturation). The mass density of the gas phase can be neglected, and therefore, $\rho_f \approx s_w\rho_w$. From Equation (3.4), the porosity can be replaced by $s_w\phi$ (of course, terms in $(1 - \phi)$, dealing with the solid phase, remain unchanged). The Darcy velocity associated with the water phase is given by

$$\dot{\mathbf{w}}_w = -\frac{k_r k_0}{\eta_w}\left(\nabla p_w + \rho_w s_w \ddot{\mathbf{u}} + \rho_w s_w \frac{\alpha_w}{\phi}F\ddot{\mathbf{w}}_w - \mathbf{F}_w\right) \qquad (3.8)$$

where k_r denotes the relative permeability (dimensionless), η_w denotes the dynamic viscosity of the pore water (in Pa s), and p_w is the pressure of the water phase (it will be replaced later by the suction head defined previously by Eq. 3.2). The ratio α_w/ϕ corresponds to the bulk tortuosity of the water phase, α_w, divided by the connected porosity. This ratio can be replaced by Archie's second law Fs_w^n where n is called the second Archie's exponent ($n > 1$, dimensionless) (see, for instance, Archie, 1942, and Revil et al., 2007). From now on, the constitutive equations are described in the frequency domain. Therefore, $\ddot{\mathbf{w}}_w$ is replaced by $-i\omega\dot{\mathbf{w}}_w$ and so on. The Darcy equation, Equation (3.8), can be rewritten as

$$-i\omega\mathbf{w}_w = -\frac{k^*(\omega)}{\eta_w}\left(\nabla p_w - \omega^2\rho_w s_w\mathbf{u} - \mathbf{F}_w\right) \qquad (3.9)$$

where $k^*(\omega)$ is a complex-valued permeability given by

$$k^*(\omega) = \frac{k_r k_0}{1 - i\omega\tau_k} \qquad (3.10)$$

and where the relaxation time is given by

$$\tau_k = \frac{k_r k_0 \rho_w F}{\eta_w} s_w^{1-n} \qquad (3.11)$$

The relaxation time τ_k represents the transition between the viscous laminar flow regime and the inertial laminar flow regime. The critical frequency associated with this relaxation time is given by

$$f_k = \frac{1}{2\pi\tau_k} = \frac{\eta_w}{2\pi k_r k_0 \rho_w F} s_w^{n-1} \qquad (3.12)$$

Note that because $n \geq 1$ (Archie, 1942), $n - 1 \geq 0$. Note that the measurement of this relaxation frequency can be used to estimate the permeability of the material.

In this analysis, low frequencies ($\omega\tau_k \ll 1$, $f \ll f_k$) correspond to the viscous laminar flow regime where the flow in a cylindrical pore obeys Poiseuille law. High frequencies ($\omega\tau_k \gg 1$, $f \gg f_k$) correspond to the inertial laminar flow regime for which the pore water flow represents a potential-flow problem. Note that some authors (like Biot in his earlier works) prefer to include the inertial effect in an apparent (or effective) dynamic viscosity $\eta^*(\omega)$ rather than defining an apparent (or effective) permeability $k^*(\omega)$. This choice (arbitrary) was abandoned later on.

In poroelasticity, it is customary to define the following two variables (e.g., Morency & Tromp, 2008):

$$b = \frac{\eta_w}{k_0} \qquad (3.13)$$

$$\tilde{\rho}_w = \rho_w F \qquad (3.14)$$

where $\tilde{\rho}_w$ denotes an apparent pore water mass density. The relationship between the permeability and the water saturation can be expressed with the Brooks and Corey (1964) relationship:

$$k_r = s_w^{\frac{2+3\lambda}{\lambda}} \qquad (3.15)$$

where λ is termed the Brook and Corey exponent. Using Equation (3.11) and Equations (3.13)–(3.15), the following relationship between the relaxation time and the saturation is obtained:

$$\tau_k = s_w^{\frac{2+3\lambda}{\lambda}+1-n} \frac{\tilde{\rho}_w}{b} \qquad (3.16)$$

In order to write a hydrodynamic equation coupled with the electrical field, the body force $\mathbf{F}_w$ entering Equation (3.9) should be expressed by Coulomb's law:

$$\mathbf{F}_w = \hat{Q}_V^*(\omega)\mathbf{E} \qquad (3.17)$$

where $\hat{Q}_V^*(\omega)$ denotes the frequency-dependent (effective) excess charge that can be dragged by the flow of pore water through the pore space of the material (dynamic excess charge density of the pore space) and $\mathbf{E}$ denotes the electrical field, in V m^{-1}. The charge density $\hat{Q}_V^*(\omega)$ is frequency dependent because there are more charges dragged in the inertial laminar flow regime than in the viscous laminar flow regime, in agreement with the model of Pride (1994). In the following, the parameters $\hat{Q}_V^0$ and $\hat{Q}_V^\infty$ are the volumetric charge density dragged in the low ($\omega\tau_k \ll 1$) and high ($\omega\tau_k \gg 1$) frequency regimes, respectively. Because the transition between low- and high-frequency regimes is governed by the relaxation time, τ_k, the following functional can be used to compute the effective charge density as a function of the frequency:

$$\frac{1}{\hat{Q}_V^*(\omega)} = \frac{1}{\hat{Q}_V^\infty} + \left(\frac{1}{\hat{Q}_V^0} - \frac{1}{\hat{Q}_V^\infty}\right)\frac{1}{\sqrt{1-i\omega\tau_k}} \qquad (3.18)$$

The form of this function is derived and explained further in Revil and Mahardika (2013). We need to find expressions for the low- and high-frequency charge densities, $\hat{Q}_V^0$ and $\hat{Q}_V^\infty$, respectively. We note that:

1 At low frequencies, only a small fraction of the counterions of the diffuse layer are dragged by the flow of the pore water, and therefore, $\hat{Q}_V^0 \ll \hat{Q}_V^\infty$. An expression to compute $\hat{Q}_V^0$ from the low-frequency permeability, k_0, is discussed further later.

2 At high frequencies, all the charge density existing in the pores is uniformly dragged along the pore water flow, and therefore, the charge density $\hat{Q}_V^\infty$ is also equal to the volumetric charge density of the diffuse layer. An expression to compute $\hat{Q}_V^\infty$ from the cation exchange capacity is discussed further later.

Depending on the size of the electrical double layer with respect to the size of the pores, two cases can be considered:

1 In the thick double layer approximation, $\hat{Q}_V^\infty \approx \hat{Q}_V^0$ (all the counterions of the diffuse layer are dragged by the flow whatever the frequency), and therefore,

$$\hat{Q}_V^*(\omega, s_w) \approx \frac{\hat{Q}_V^0}{s_w} \qquad (3.19)$$

2 In the thin double layer approximation (see Chapter 1), one can expect $\hat{Q}_V^\infty \gg \hat{Q}_V^0$. Therefore, we have

$$\hat{Q}_V^*(\omega, s_w) \approx \hat{Q}_V^0 s_w^{-1} \sqrt{1 - i\omega\tau_k(s_w)} \qquad (3.20)$$

Introducing Coulomb's law, Equation (3.17), into the Darcy equation, Equation (3.9), yields the following form of Darcy's law:

$$-i\omega\mathbf{w}_w = -\frac{k^*(\omega)}{\eta_w}\left(\nabla p_w - \omega^2\rho_w s_w \mathbf{u}\right) + \frac{k^*(\omega)\hat{Q}_V^*(\omega)}{\eta_w}\mathbf{E} \qquad (3.21)$$

This equation shows the influence of three forcing terms on the Darcy velocity: (i) the pore fluid pressure gradient, (ii) the displacement of the solid framework, and (iii) the electrical field through electroosmosis.

Now, we turn to investigate the macroscopic electrical current density, $\mathbf{J}$. The first contribution to $\mathbf{J}$ is the conduction current density given by Ohm's law:

$$\mathbf{J}_c = \sigma^*(\omega)\mathbf{E} \qquad (3.22)$$

where the conductivity $\sigma^*(\omega)$ denotes the complex conductivity discussed in Chapter 1 and section 2.3.1.2 of Chapter 2.

The second contribution to the total current density corresponds to the advective drag of the excess of charge

of the pore space by the flow of the pore water (contribution of advective nature). If the Darcy velocity associated with the poromechanical contribution is written as $\dot{\mathbf{w}}_w^m$, the second contribution to the current density is given by

$$\mathbf{J}_m = -i\omega\hat{Q}_V^*(\omega)\mathbf{w}_w^m \qquad (3.23)$$

The mechanical contribution to the filtration displacement is given by the generalized Darcy's law derived previously:

$$-i\omega\mathbf{w}_w^m = -\frac{k^*(\omega)}{\eta_w}\left(\nabla p_w - \omega^2\rho_w s_w \mathbf{u}\right) \qquad (3.24)$$

The total current density is therefore given by the sum of the conductive and advective contributions, which yields the following generalized Ohm's law:

$$\mathbf{J} = \sigma^*(\omega)\mathbf{E} + \frac{k^*(\omega)\hat{Q}_V^*(\omega)}{\eta_w}\left(\nabla p_w - \omega^2\rho_w s_w \mathbf{u}\right) \qquad (3.25)$$

The two constitutive equations for the generalized Ohm's and Darcy's laws are written as two coupled equations:

$$\begin{bmatrix} \mathbf{J} \\ -i\omega\mathbf{w}_w \end{bmatrix} = \begin{bmatrix} \sigma^* & L^*(\omega, s_w) \\ L^*(\omega, s_w) & \dfrac{k^*(\omega, s_w)}{\eta_w} \end{bmatrix} \begin{bmatrix} \mathbf{E} \\ -(\nabla p_w - \omega^2\rho_w s_w \mathbf{u}) \end{bmatrix} \qquad (3.26)$$

where the coefficient $L^*(\omega)$ is defined as

$$L^*(\omega, s_w) = \frac{k^*(\omega, s_w)\hat{Q}_V^*(\omega, s_w)}{\eta_w} \qquad (3.27)$$

The generalized streaming potential coupling coefficient is defined by the following equations in the quasistatic limit of the Maxwell equations:

$$\nabla \times \mathbf{E} = 0 = \rhd \mathbf{E} = -\nabla\psi \qquad (3.28)$$

$$C_p^*(\omega, s_w) = \left(\frac{\partial\psi}{\partial p_w}\right)_{\mathbf{j}=0; \ddot{\mathbf{u}}=0} = -\frac{L^*(\omega, s_w)}{\sigma^*(\omega, s_w)} \qquad (3.29)$$

$$C_p^*(\omega, s_w) = -\frac{k^*(\omega, s_w)\hat{Q}_V^*(\omega, s_w)}{\eta_w\sigma^*(\omega, s_w)} \qquad (3.30)$$

More explicitly, in the thin double layer approximation, the streaming potential coupling coefficient is given by

$$C_p^*(\omega, s_w) \approx \frac{C_0(s_w)}{\sqrt{1 - i\omega\tau_k}} \qquad (3.31)$$

$$C_0 = \lim_{\omega \to 0} C_p(\omega, s_w) \approx \frac{k_r k_0 \hat{Q}_V^0}{\eta_w s_w \left(\frac{1}{F} s_w{}^n \sigma_f + \sigma_S\right)} \qquad (3.32)$$

Similarly, the generalized electroosmotic coupling coefficient is defined in the quasistatic limit of the Maxwell equations. This definition is facilitated by the following constraints: we allow for the change of the pore fluid pressure while the skeleton is at rest with an absence of pore fluid flow influenced by other mechanisms. With these constraints, the electroosmotic coupling coefficient can then be simply defined by the ratio between the gradient of pore fluid pressure divided by the gradient in electrical potential. For the thin double layer case, this yields

$$C_{os}^*(\omega, s_w) = \left(\frac{\partial p_w}{\partial \varphi}\right)_{\dot{w}=0;\,\ddot{u}=0} = -\frac{L^*(\omega, s_w)\eta_w}{k^*(\omega, s_w)} \qquad (3.33)$$

$$C_{os}^*(\omega, s_w) = -\hat{Q}_V^*(\omega) \qquad (3.34)$$

$$C_{os}^*(\omega, s_w) \approx -\hat{Q}_V^0 s_w^{-1} \sqrt{1 - i\omega\tau_k} \qquad (3.35)$$

For the thick double layer case, the electroosmotic coupling coefficient is given by $C_{os}^*(\omega) \approx -\hat{Q}_V^0/s_w$. Therefore, the electroosmotic coupling coefficient is simply a measure of the excess of charges that can be moved by the flow of the pore water caused by an electric field. In unsaturated flow conditions, it is customary to use the capillary head gradient $\nabla\Psi = \nabla p_w/\rho_w g$ instead of the pore water pressure gradient. Below the relaxation frequency separating the low-frequency viscous laminar flow regime from the high-frequency inertial laminar flow regime, Equation (3.26) can be rewritten in the time domain using the pressure head, including the gravitational field in the hydraulic driving force. This yields two coupled equations:

$$\begin{bmatrix} \mathbf{J} \\ \dot{\mathbf{w}}_w \end{bmatrix} = -\begin{bmatrix} \sigma^* & L^*(s_w) \\ L^*(s_w) & \dfrac{k_r(s_w)k_0}{\eta_w} \end{bmatrix} \begin{bmatrix} \nabla\psi \\ \rho_w g\nabla(\Psi + z) + \rho_w s_w \ddot{\mathbf{u}} \end{bmatrix} \qquad (3.36)$$

where the quasistatic approximation coupling coefficient L^* is given by

$$L^*(s_w) = \frac{k_r(s_w)k_0\hat{Q}_V^0}{\eta_w s_w} \qquad (3.37)$$

Therefore, Equations (3.36) and (3.37) provide a simple streaming potential and electroosmosis model for porous media. This simple model can be used to characterize the occurrence of the streaming potential and electroosmosis in porous media systems.

3.1.2 Description of the hydromechanical model

The development of the unsaturated hydromechanical model starts with the following Biot constitutive equation in saturated conditions:

$$-p = C_s\nabla \cdot \mathbf{u} + M_s\nabla \cdot \mathbf{w} \qquad (3.38)$$

Equation (3.38) is also often written as

$$-p = C_s\varepsilon_{kk} - M_s\zeta \qquad (3.39)$$

where $\varepsilon_{kk} = \nabla \cdot \mathbf{u} = \delta V/V$ (where V denotes the representative elementary volume) represents the volumetric strain of the porous body and $\zeta = -\nabla \cdot \mathbf{w} = \phi(\nabla \cdot \mathbf{u} - \nabla \cdot \mathbf{u}_w)$ denotes the linearized increment of fluid content (e.g., Lo et al., 2002). The parameter ζ represents the fractional volume of water flowing in or out of the representative volume of the skeleton in response to an applied stress. The bulk moduli, M_s and C_s, in saturated conditions, are defined as

$$\frac{1}{C_s} \equiv -\left(\frac{d\varepsilon_{kk}}{dp_w}\right)_\zeta = \frac{1 + \Delta}{K_f + K_S\Delta} \qquad (3.40)$$

$$\frac{1}{M_s} \equiv \left(\frac{d\zeta}{dp_w}\right)_{\varepsilon_{kk}} = \phi\left(\frac{1 + \Delta}{K_f}\right) \qquad (3.41)$$

where $\alpha \equiv 1 - K_{fr}/K_S$ denotes the Biot coefficient in the saturated state and Δ is defined by $\Delta = (K_f/\phi K_S^2)[(1 - \phi)K_S - K_{fr}]$. The Biot modulus, M, corresponds to the inverse of the poroelastic component of the specific storage and is defined as the increase of the amount of fluid (per unit volume of rock) as a result of a unit increase of pore pressure under constant volumetric strain. From that, the following relationship is obtained: $C_s = \alpha M_s$.

To extend these equations to the unsaturated case, we apply the classical change of variables discussed previously, that is, $\rho_f \Rightarrow s_w\rho_w$, $p \Rightarrow p_w$, $\phi \Rightarrow s_w\phi$, and $\mathbf{w} \Rightarrow \mathbf{w}_w = s_w\phi(\mathbf{u}_w - \mathbf{u})$). This yields

$$\frac{1}{C} = \frac{1+\Delta}{K_f + K_S \Delta} \tag{3.42}$$

$$\frac{1}{M} = \theta\left(\frac{1+\Delta}{K_f}\right) \tag{3.43}$$

where $\theta = s_w\phi$ denotes the water content (dimensionless) and where, in unsaturated conditions, the fluid increment is defined by

$$\zeta = -\nabla \cdot \mathbf{w}_w = \phi s_w (\nabla \cdot \mathbf{u} - \nabla \cdot \mathbf{u}_w) \tag{3.44}$$

Note that the term $\Delta = \left(K_f/\phi K_S^2\right)\left[(1-\phi)K_S - K_{fr}\right]$ depends on the saturation because the compressibility of the fluid is given by the Wood formula $(1/K_f) = (1-s_w)/K_a + s_w/K_w$ (Wood, 1955; Teja & Rice, 1981) where K_a and K_w represent the bulk moduli of air and water, respectively ($K_a = 0.145$ MPa and $K_w = 2.25$ GPa; see Lo et al., 2005). In unsaturated conditions, from Equations (3.42) and (3.43), $C = a_w M$, and the Biot coefficient, a_w, is given by $a_w = s_w\alpha$ where α denotes the Biot coefficient in saturated conditions. That said, there should be no exchange of water below the irreducible water saturation, and therefore, the scaling should be adjusted such that $a_w = s_e\alpha$ rather than $a_w = s_w\alpha$. In this case, s_e denotes the reduced water saturation or the effective saturation that is related to the saturation of the water phase by $s_e = (s_w - s_r)/(1 - s_r)$. In this formulation adjustment, s_r denotes the irreducible water saturation. In other words, $(1/M) \rightarrow 0$ as $s_w \rightarrow s_r$.

Generalizing Equation (3.38) to unsaturated conditions yields

$$\left(\frac{1}{M} + \frac{\partial\theta}{\partial p_w}\right)(p_a - p_w) = \frac{C}{M}\nabla \cdot \mathbf{u} + \nabla \cdot \mathbf{w}_w \tag{3.45}$$

where $\partial\theta/\partial p_w$ denotes the specific moisture capacity, which is determined from the derivative of the capillary pressure with respect to the water content (in unsaturated flow, the air pressure is kept constant). The filtration displacement of the water phase $-i\omega\mathbf{w}_w$ is given by

$$-i\omega\mathbf{w}_w = -\frac{k^*(\omega)}{\eta_w}\left(\nabla p_w - \omega^2\rho_w s_w \mathbf{u} - \mathbf{F}_w\right) \tag{3.46}$$

Therefore, the bulk filtration displacement of the water phase is given by

$$\mathbf{w}_w = \frac{k^*(\omega)}{i\omega\eta_w}\left(\nabla p_w - \omega^2\rho_w s_w \mathbf{u} - \mathbf{F}_w\right) \tag{3.47}$$

Combining Equations (3.45) and (3.47) yields

$$\left(\frac{1}{M} + \frac{\partial\theta}{\partial p_w}\right)(p_a - p_w)$$
$$= a_w\nabla \cdot \mathbf{u} + \nabla \cdot \left[\frac{k^*(\omega)}{i\omega\eta_w}\left(\nabla p_w - \omega^2\rho_w s_w \mathbf{u} - \mathbf{F}_w\right)\right] \tag{3.48}$$

where the relationship $C/M = a_w$ has been used. Equation (3.48) is a nonlinear diffusion equation for the fluid pressure. For this to be obvious, the terms of this equation need to be reworked. Multiplying all the terms by $(i\omega\eta_w/k^*(\omega))$, separating the pressure terms in the left-hand side from the source term in the right-hand side, and taking into consideration that the air pressure is constant (unsaturated flow assumption), the following nonlinear hydraulic diffusion equation is obtained:

$$\nabla \cdot \left[\frac{k^*(\omega)}{\eta_w}\nabla p_w\right] + i\omega\left(\frac{1}{M} + \frac{\partial\theta}{\partial p_w}\right)(p_w - p_a)$$
$$= \nabla \cdot \left[\frac{k^*(\omega)}{\eta_w}\left(\omega^2\rho_w s_w \mathbf{u}\right)\right] - i\omega a_w\nabla \cdot \mathbf{u} + \nabla \cdot \left(\frac{k^*(\omega)}{\eta_w}\mathbf{F}_w\right) \tag{3.49}$$

This equation may be written in the time domain by applying an inverse Fourier transform. Assuming that the permeability is given by its low-frequency asymptotic limit (which is correct below 10 kHz), and using Coulomb's law, plus the gravity force acting as the body force (the frequency-dependent volumetric charge density is also taken in its low-frequency limit) yields

$$\nabla \cdot \left[\frac{k_r k_0}{\eta_w}\nabla p_w\right] - \left(\frac{1}{M} + \frac{\partial\theta}{\partial p_w}\right)\dot{p}_w$$
$$= -\nabla \cdot \left[\frac{k_r k_0 \rho_w s_w}{\eta_w}\ddot{\mathbf{u}}\right] + a_w\nabla \cdot \dot{\mathbf{u}} + \nabla \cdot \left[\frac{k_r k_0}{\eta_w}\left(\frac{\hat{Q}_V^0}{s_w}\mathbf{E} + \rho_w\mathbf{g}\right)\right] \tag{3.50}$$

where $\dot{\Theta} = \partial\Theta/\partial t$ denotes the time derivative of a parameter Θ (whatever it is) and t represents time. The origins of the three forcing terms on the right-hand side of this equation are now clearly established: the first term is related to the acceleration of the seismic wave acting on the skeleton of the material, the second term is due to the velocity of the seismic wave, and the third term (at constant gravity acceleration) corresponds to the pore water flow associated with the electroosmotic forcing associated with the drag of the pore water by the electromigration of the excess of charge contained in the pore space of the material.

Another possibility is to write a generalized Richards equation (Richards, 1931) that shows the influence of the forcing term (we assume again that the air pressure is constant). Starting with Equation (3.50) and replacing the water pressure by the capillary head defined by $\Psi = (p_w - p_a)/\rho_w g$ (in m), the following generalized Richards equation is obtained:

$$
\left(\frac{\rho_w g}{M} + \frac{\partial \theta}{\partial \Psi} \right) \dot{\Psi} + \nabla \cdot [-K_h \nabla (\Psi + z)]
$$
$$
= \nabla \cdot \left[K_h \frac{s_w}{g} \ddot{\mathbf{u}} \right] - \alpha_w \nabla \cdot \dot{\mathbf{u}} - \nabla \cdot \left(\frac{K_h}{\rho_w g} \frac{\hat{Q}_V^0}{s_w} \mathbf{E} \right) \tag{3.51}
$$

$$
C_e \frac{\partial \Psi}{\partial t} + \nabla \cdot [-K_h (\nabla \Psi + 1)]
$$
$$
= -\alpha_w \nabla \cdot \dot{\mathbf{u}} + \nabla \cdot \left[K_h \left(\frac{s_w}{g} \ddot{\mathbf{u}} - \frac{\hat{Q}_V^0}{s_w \rho_w g} \mathbf{E} \right) \right] \tag{3.52}
$$

$$
K_h = \frac{k_r k_0 \rho_w g}{\eta_w} = k_r K_s \tag{3.53}
$$

where $\alpha_w = s_e \alpha = s_e (1 - K_{fr}/K_S)$ and $K = K_{fr}$, denoting the bulk modulus of the skeleton (drained bulk modulus). Here, K_h denotes the hydraulic conductivity (in m s^{-1}), K_s denotes the hydraulic conductivity at saturation, z denotes the elevation above a datum, and C_e denotes the specific storage term. This storage term is the sum of the specific moisture capacity (in m^{-1}) (also called the water capacity function) and the specific storage corresponding to the poroelastic deformation of the material. This yields

$$
C_e = \frac{\partial \theta}{\partial \Psi} + \frac{\rho_w g}{M} \tag{3.54}
$$

Usually, in unsaturated conditions, the poroelastic term is much smaller than the specific moisture capacity, but the poroelastic term should not be neglected so that we have a formulation that remains consistent with the saturated state. The unsaturated hydraulic conductivity is related to the relative permeability, k_r, and to K_0, the hydraulic conductivity at saturation. With the Brooks and Corey (1964) model, the porous material is saturated when the fluid pressure reaches the atmospheric pressure ($\psi = 0$ at the water table). The effective saturation, the specific moisture capacity, the relative permeability, and the water content are defined by

$$
s_e = \begin{cases} (\alpha_b \Psi)^{-\lambda}, & \Psi \le 1/\alpha_b \\ 1, & \Psi > 1/\alpha_b \end{cases} \tag{3.55}
$$

$$
\frac{\partial \theta}{\partial \Psi} = \begin{cases} -\lambda \alpha_b (\phi - \theta_r)(\alpha_b \Psi)^{-(\lambda+1)}, & \Psi \le 1/\alpha_b \\ 0, & \Psi > 1/\alpha_b \end{cases} \tag{3.56}
$$

$$
k_r = \begin{cases} s_e^{\frac{2+3\lambda}{\lambda}} = (\alpha_b \Psi)^{-(2+3\lambda)}, & \Psi \le 1/\alpha_b \\ 1, & \Psi > 1/\alpha_b \end{cases} \tag{3.57}
$$

$$
\theta = \begin{cases} \theta_r + s_e(\phi - \theta_r), & \Psi \le 1/\alpha_b \\ \phi, & \Psi > 1/\alpha_b \end{cases} \tag{3.58}
$$

respectively, where α_b denotes the inverse of the capillary entry pressure, which is related to the matric suction at which pore fluid begins to leave a drying soil water system. The pore size distribution index is represented by λ (a textural parameter), and θ_r represents the residual water content ($\theta_r = s_r \phi$). Sometimes, the residual water saturation is not accounted for, and the capillary pressure curve and the relative permeability are written as

$$
s_w = \begin{cases} \left(\frac{p_c}{p_e} \right)^{-\lambda}, & p_c > p_e \\ 1, & p_c \le p_e \end{cases} \tag{3.59}
$$

$$
k_r = \begin{cases} s_w^{\frac{2+3\lambda}{\lambda}}, & p_c > p_e \\ 1, & p_c \le p_e \end{cases} \tag{3.60}
$$

Because $\alpha_w = s_e \alpha = s_e (1 - K_{fr}/K_S)$, when the water saturation reaches the irreducible water saturation, the two source terms on the right-hand side of Equation (3.52) are null. Therefore, there is no possible excitation below the irreducible water saturation. In reality, this is not necessarily true, and the model should be completed by including film flow below the irreducible water saturation.

The hydromechanical equations are defined in terms of an effective stress tensor. As explained in details in Jardani et al. (2010) and Revil and Jardani (2010), there is a computational advantage in expressing the coupled hydromechanical problem in terms of the fluid pressure and displacement of the solid phase (4 unknowns in total) rather than using the displacement of the solid and filtration displacement (6 unknowns in total).

In saturated conditions, Newton's law (which is a momentum conservation equation for the skeleton, partially filled with pore water) is written as

$$\nabla \cdot \bar{\bar{T}} + \mathbf{F} = -\omega^2 (\rho \mathbf{u} + \rho_f \mathbf{w}) \tag{3.61}$$

where $\bar{\bar{T}}$ is the total stress tensor (positive normal stress implies tension; see Detournay & Cheng, 1993) and $\mathbf{F}$ denotes the total body force applied to the porous material. In unsaturated conditions, Newton's law is written as

$$\nabla \cdot \bar{\bar{T}} + \mathbf{F} = -\omega^2 (\rho \mathbf{u} + s_w \rho_w \mathbf{w}_w) \tag{3.62}$$

where the filtration displacement of the pore water phase is given by

$$\mathbf{w}_w = \frac{k^*(\omega)}{i\omega\eta_w}\left(\nabla p_w - \omega^2 \rho_w s_w \mathbf{u} - \mathbf{F}_w\right) \tag{3.63}$$

Combining Equations (3.62) and (3.63) yields

$$\nabla \cdot \bar{\bar{T}} + \mathbf{F} = -\omega^2 \left[\rho \mathbf{u} + s_w \rho_w \frac{k^*(\omega)}{i\omega\eta_w}\left(\nabla p_w - \omega^2 \rho_w s_w \mathbf{u} - \mathbf{F}_w\right)\right] \tag{3.64}$$

$$\nabla \cdot \bar{\bar{T}} + \mathbf{F} = -\omega^2 \rho_\omega^S \mathbf{u} - \omega^2 \left[s_w \rho_w \frac{k^*(\omega)}{i\omega\eta_w}\left(\nabla p_w - \mathbf{F}_w\right)\right] \tag{3.65}$$

where

$$\rho_\omega^S = \rho - (s_w \rho_w)^2 \frac{k^*(\omega)}{i\omega\eta_w}\omega^2 \tag{3.66}$$

The effective stress in unsaturated conditions is taken as

$$\bar{\bar{T}}_{eff} = \bar{\bar{T}} + p_a \bar{\bar{I}} + s_e \alpha (p_w - p_a) \bar{\bar{I}} \tag{3.67}$$

which is consistent with the Bishop effective stress principle in unsaturated conditions and the Biot stress principle in saturated conditions. The confining pressure and the effective confining pressure are defined as

$$P = -\frac{1}{3}\mathrm{Trace}\,\bar{\bar{T}} \tag{3.68}$$

$$P_{eff} = -\frac{1}{3}\mathrm{Trace}\,\bar{\bar{T}}_{eff} \tag{3.69}$$

respectively. This yields $P_{eff} = P - p_a - s_e\alpha(p_w - p_a)$. Equations (3.65) and (3.67) yield

$$\nabla \cdot \bar{\bar{T}}_{eff} + \mathbf{F} - \omega^2 s_w \rho_w \frac{k^*(\omega)}{i\omega\eta_w}\mathbf{F}_w = -\omega^2 \rho_\omega^S \mathbf{u} + \theta_\omega \nabla p_w \tag{3.70}$$

where the hydromechanical coupling term, θ_ω, is defined by

$$\theta_\omega = s_w \left(\alpha - \omega^2 \rho_w \frac{k^*(\omega)}{i\omega\eta_w}\right) \tag{3.71}$$

The last fundamental constitutive equation needed to complete the hydromechanical model in unsaturated conditions is a relationship between the total stress tensor (or the effective stress tensor) and the displacement of the solid phase and filtration displacement of the pore water phase. This equation is Hooke's law, which, in linear poroelasticity and for saturated conditions, is given by (Biot, 1962a, b)

$$\bar{\bar{T}} = (\lambda_u \nabla \cdot \mathbf{u} + C\nabla \cdot \mathbf{w})\bar{\bar{I}} + G(\nabla \mathbf{u} + \nabla \mathbf{u}^T), \tag{3.72}$$

where $\bar{\bar{\varepsilon}} = (1/2)(\nabla \mathbf{u} + \nabla \mathbf{u}^T)$ denotes the deformation tensor, $G = G_{fr}$ denotes the shear modulus that is equal to the shear modulus of the skeleton (frame), and $\lambda_u = K_u - (2/3)G$ denotes the Lamé modulus in undrained conditions (K_u denotes the undrained bulk modulus). In unsaturated conditions and accounting for the air pressure, Equation (3.70) can be written as

$$\bar{\bar{T}} + p_a \bar{\bar{I}} = (\lambda_u \nabla \cdot \mathbf{u} + C\nabla \cdot \mathbf{w}_w)\bar{\bar{I}} + G(\nabla \mathbf{u} + \nabla \mathbf{u}^T) \tag{3.73}$$

From Equation (3.45), the linearized increment of fluid content is given by

$$-\nabla \cdot \mathbf{w}_w = \frac{1}{M}(p_w - p_a) + \frac{C}{M}\nabla \cdot \mathbf{u} \tag{3.74}$$

Combining Equations (3.73) and (3.74) yields

$$\bar{\bar{T}} + p_a\bar{\bar{I}} = \left[\left(K_u - \frac{2}{3}G\right)\nabla \cdot \mathbf{u} + C\left(-\frac{1}{M}(p_w - p_a) - \alpha_w \nabla \cdot \mathbf{u}\right)\right]$$
$$\bar{\bar{I}} + G(\nabla \mathbf{u} + \nabla \mathbf{u}^T) \tag{3.75}$$

$$\bar{\bar{T}} + p_a\bar{\bar{I}} = \left[\left(K_u - \alpha_w C - \frac{2}{3}G\right)\nabla \cdot \mathbf{u} - \alpha_w(p_w - p_a)\right]$$
$$\bar{\bar{I}} + G(\nabla \mathbf{u} + \nabla \mathbf{u}^T) \tag{3.76}$$

where the following expression, derived previously, has been used for the Biot coefficient in unsaturated conditions:

$$\frac{C}{M} = \alpha_w = s_e\alpha \tag{3.77}$$

In addition, the bulk modulus is given by $K = K_u - \alpha_w C$ and the Lamé modulus is given by $\lambda = K - (2/3)G$.

Equation (3.76) yields the following Hooke's law for the effective stress given by

$$\overline{\overline{T}} + p_a\overline{\overline{I}} + \alpha_w(p_w - p_a)\overline{\overline{I}} = \lambda(\nabla \cdot \mathbf{u})\overline{\overline{I}} + G(\nabla\mathbf{u} + \nabla\mathbf{u}^T) \quad (3.78)$$

$$\overline{\overline{T}}_{\text{eff}} = \lambda(\nabla \cdot \mathbf{u})\overline{\overline{I}} + G(\nabla\mathbf{u} + \nabla\mathbf{u}^T) \quad (3.79)$$

where the effective stress is given by Equation (3.67). Note that the effective stress is only related to the deformation of the skeleton of the porous material by definition.

3.1.3 Maxwell equations in unsaturated conditions

Pride (1994) volume averaged the local Maxwell equations to obtain a set of macroscopic Maxwell equations in the thin double layer limit (i.e., assuming that the thickness of the diffuse layer is much smaller than the size of the throat controlling the flow of the pore water). The same equations were obtained by Revil and Linde (2006) for the thick double layer case. The general form of these macroscopic Maxwell equations is

$$\nabla \times \mathbf{E} = -\dot{\mathbf{B}} \quad (3.80)$$

$$\nabla \times \mathbf{H} = \mathbf{J} + \dot{\mathbf{D}} \quad (3.81)$$

$$\nabla \cdot \mathbf{B} = 0, \; \mathbf{B} = \nabla \times \mathbf{A} \quad (3.82)$$

$$\nabla \cdot \mathbf{D} = \phi \hat{Q}_V^{\infty} \quad (3.83)$$

where $\mathbf{B}$ is the magnetic induction vector, $\mathbf{H}$ is the magnetic field (in $A\,m^{-1}$), $\mathbf{D}$ is the current displacement vector (in $C\,m^{-2}$), $\mathbf{A}$ is the magnetic potential vector, and $\mathbf{E} = -\nabla\psi - \dot{\mathbf{A}}$ where ψ denotes the electrostatic potential (in V). These equations are completed by two electromagnetic constitutive equations: $\mathbf{D} = \varepsilon\mathbf{E}$ and $\mathbf{B} = \mu\mathbf{H}$ where ε is the permittivity of the material and μ denotes its magnetic permeability. In the absence of magnetized grains, $\mu = \mu_0$ where μ_0 denotes the magnetic permeability of free space.

When the harmonic electrical field is written as $\mathbf{E} = \mathbf{E}_0 \exp(-i\omega t)$, and ω is the angular frequency with $\mathbf{E}_0$ being a constant electrical field magnitude and direction, the displacement current density vector is given by $\mathbf{J}_d = \dot{\mathbf{D}} = -i\omega\varepsilon\mathbf{E}$. The total current density, $\mathbf{J}_T$, entering Ampère's law,

$$\nabla \times \mathbf{H} = \mathbf{J}_T \quad (3.84)$$

is given by

$$\mathbf{J}_T = \mathbf{J} + \mathbf{J}_d \quad (3.85)$$

$$\mathbf{J}_T = \sigma^*(\omega)\mathbf{E} + \frac{k^*(\omega)\hat{Q}_V}{\eta_w}(\nabla p_w - \omega^2\rho_w s_w\mathbf{u}) - i\omega\varepsilon\mathbf{E} \quad (3.86)$$

Equation (3.86) can be written as an equation in which we have a generalized Ohm's law appearing on the right-hand side of the equation together with an electrokinetic coupling term:

$$\mathbf{J}_T = (\sigma^*(\omega) - i\omega\varepsilon)\mathbf{E} + \frac{k^*(\omega)\hat{Q}_V}{\eta_w}(\nabla p_w - \omega^2\rho_w s_w\mathbf{u}) \quad (3.87)$$

An effective conductivity can be introduced, such as $\sigma_{\text{eff}}^* = \sigma^*(\omega) - i\omega\varepsilon$, where $\sigma_{\text{eff}}^* = \sigma_{\text{eff}} + i\omega\varepsilon_{\text{eff}}$ is the effective or apparent complex conductivity and σ_{eff} and ε_{eff} are real scalars that are dependent upon frequency. These effective parameters are those that are measured during an experiment in the laboratory or in the field, but these terms contain both electromigration and true dielectric polarization effects. They are given by $\sigma_{\text{eff}} = \sigma'$ and $\varepsilon_{\text{eff}} = \sigma''/\omega - \varepsilon$.

3.2 Extension to two-phase flow

3.2.1 Generalization of the Biot theory in two-phase flow conditions

For the generalization of the Biot–Frenkel theory in two-phase flow conditions, we consider, in the following, an isotropic porous elastic body wherein two Newtonian fluid phases fill the full of the pore space of the solid matrix. In this development, wave propagation through a porous medium containing two immiscible fluids, in the space–time domain, is governed by coupled partial differential equations. Here, Lo et al. (2005) presented a formulation that used the continuum mechanics of mixtures theory (Lo et al., 2002). This theory was derived in an Eulerian framework and describes the propagation of three (compressional) P-waves and one S-wave (shear). These waves have dissipation terms that result from the momentum transfer or interaction terms of the solid and fluids within the porous medium. Using numerical simulations, Santos et al. (1990, 2006) concluded that both of the compressional waves are analogs to the Biot fast and slow P-waves in a single-fluid-saturated porous material and that a third P-wave that is developed is a diffusive wave related to fluctuations in the pressure difference (capillary pressure) between the fluid phases. The problem consisted of solving a set coupled equation with three unknown vectors, $\mathbf{u}$ (the displacement vector field of the solid phase), and $\mathbf{u}_1$ and $\mathbf{u}_2$, the displacement vector fields of the two immiscible fluid phases.

For the two fluid phases, the momentum conservation equation is written as (Lo et al., 2005)

$$
\rho_i \theta_i \frac{\partial^2 \mathbf{u}_i}{\partial t^2} + \theta_i \nabla p_i - R_{ii} \left(\frac{\partial \mathbf{u}_i}{\partial t} - \frac{\partial \mathbf{u}}{\partial t} \right)
$$
$$
- \sum_{j=1}^{2} A_{ij} \left(\frac{\partial^2 \mathbf{u}_j}{\partial t^2} - \frac{\partial^2 \mathbf{u}}{\partial t^2} \right) = 0, \quad (i = 1, 2),
\tag{3.88}
$$

while for the solid phase, we have

$$
\rho_s \theta_s \frac{\partial^2 \mathbf{u}}{\partial t^2} + \sum_{i=1}^{2} p_i \nabla \theta_i + \sum_{i=1}^{2} R_{ii} \left(\frac{\partial \mathbf{u}_i}{\partial t} - \frac{\partial \mathbf{u}}{\partial t} \right)
$$
$$
+ \sum_{i=1}^{2} \sum_{j=1}^{2} A_{ij} \left(\frac{\partial^2 \mathbf{u}_j}{\partial t^2} - \frac{\partial^2 \mathbf{u}}{\partial t^2} \right) - \nabla \cdot \mathbf{T} = 0
\tag{3.89}
$$

The stress tensor is given by

$$
\mathbf{T} = \left[\left(a_{11} - \frac{2}{3}G \right) \nabla \cdot \mathbf{u} + a_{12} \nabla \cdot \mathbf{u}_1 + a_{13} \nabla \cdot \mathbf{u}_2 \right] \mathbf{I} + 2G\mathbf{e}
\tag{3.90}
$$

In these equations, the nonwetting fluid corresponds to $i = 1$ (fluid 1), and the wetting fluid corresponds to $i = 2$ (fluid 2), $\mathbf{I}$ is the identity matrix, and p_i is the fluid pressure of the fluid phases i. The A_{ij} coefficients are constitutive coefficients that account for the effect of inertial coupling, while the cross-coupling effects caused by inertial drag are neglected ($A_{12} = A_{21} = 0$). The R_{ii} coefficients denote the coefficients related to viscous drag, and ρ_i and ρ_s are the mass densities of the fluid i and the solid, respectively. In Equations (3.89) and (3.90), the parameter θ_s is the volume fraction of the solid phase, and the parameter θ_i is the volume fraction of the ith fluid phase. Note that the volume fraction of the fluid phase may be written in terms of the porosity ϕ and the fluid saturation as $\theta_i = \phi S_i$. The tensor $\mathbf{T}$ denotes the stress tensors associated with the solid phase, and a_{nm} ($n, m = 1, 2, 3$) are elastic coefficients where their cross terms are symmetric, that is, $a_{nm} = a_{mn}$. The solid phase strain tensor of the elastic skeleton is defined by $\mathbf{e} = (1/2)[\nabla \mathbf{u} + \nabla \mathbf{u}^T]$, where the

superscript "T" denotes the transpose of the matrix and G denotes the shear modulus of the porous medium frame.

Following the Biot–Frenkel theory for a single phase, the linear stress–strain relations for an elastic porous medium bearing two immiscible Newtonian fluids have been generalized by Tuncay and Corapcioglu (1997) using microscopic volume averaging:

$$
-\theta_1 p_1 = a_{12} \nabla \cdot \mathbf{u} + a_{22} \nabla \cdot \mathbf{u}_1 + a_{23} \nabla \cdot \mathbf{u}_2
\tag{3.91}
$$

$$
-\theta_2 p_2 = a_{13} \nabla \cdot \mathbf{u} + a_{23} \nabla \cdot \mathbf{u}_1 + a_{33} \nabla \cdot \mathbf{u}_2
\tag{3.92}
$$

where $\mathbf{u}_1$ and $\mathbf{u}_2$ denote the displacements of the respective fluids. In this formulation, p_1 and p_2 represent the infinitesimal changes of the nonwetting and wetting pressures, respectively, where their linear relationships with the volumetric strains (dilatations) of the solid and two fluid phases depend on the elastic properties of the solid matrix and on the properties of the two fluids. To simplify the previous set of equations, we neglect the inertial drag ($A_{12} = A_{21} = 0$) of the interactions of a fluid with an adjacent fluid, and we introduce the average displacement vector of a fluid relative to the solid phase instead of the displacement of the fluid phase i given by $\mathbf{w}_i = \theta_i (\mathbf{u}_i - \mathbf{u})$.

For the two fluid phases, we have

$$
\rho_i \frac{\partial^2 \mathbf{u}}{\partial t^2} + \left(\frac{\rho_i}{\theta_i} - \frac{A_{ii}}{\theta_i^2} \right) \frac{\partial^2 \mathbf{w}_i}{\partial t^2} - \frac{R_{ii}}{\theta_i^2} \frac{\partial \mathbf{w}_i}{\partial t} = -\nabla p_i, \quad (i = 1, 2),
\tag{3.93}
$$

while for the solid phase, we have

$$
\rho_s \theta_s \frac{\partial^2 \mathbf{u}}{\partial t^2} + \sum_{i=1}^{2} \frac{A_{ii}}{\theta_i} \frac{\partial^2 \mathbf{w}_i}{\partial t^2} + \sum_{i=1}^{2} \frac{R_{ii}}{\theta_i} \frac{\partial \mathbf{w}_i}{\partial t} = \nabla \cdot \mathbf{T}
\tag{3.94}
$$

We use Equation (3.93) to rewrite Equation (3.94) as

$$
\rho_T \frac{\partial^2 \mathbf{u}}{\partial t^2} + \sum_{i=1}^{2} \rho_i \frac{\partial^2 \mathbf{w}_i}{\partial t^2} + \sum_{i=1}^{2} \theta_i \nabla p_i = \nabla \cdot \mathbf{T}
\tag{3.95}
$$

$$
\mathbf{T} = \left[\left(\bar{a}_{1s} - \frac{2}{3}G \right) \nabla \cdot \mathbf{u} + \frac{a_{12}}{\theta_1} \nabla \cdot \mathbf{w}_1 + \frac{a_{13}}{\theta_2} \nabla \cdot \mathbf{w}_2 \right] \mathbf{I} + 2G\mathbf{e}
\tag{3.96}
$$

with

$$
\bar{a}_{1s} = a_{11} + a_{12} + a_{13}
\tag{3.97}
$$

and where $\rho_T = \rho_s(1-\phi) + \rho_1\theta_1 + \rho_2\theta_2$ denotes the averaged density of the mixture. The material properties entering Equations (3.91)–(3.92) are given by Lo et al. (2005) where

$$a_{11} = K_s(1-\phi-\delta_s), \quad \text{with} \tag{3.98}$$

$$a_{12} = a_{21} = -K_s\delta_1 \tag{3.99}$$

$$a_{13} = a_{31} = -K_s\delta_2 \tag{3.100}$$

$$a_{22} = -\frac{1}{M_1}\left[\left(K_1K_2\frac{dS_1}{dp_c} + \frac{K_1K_2S_1}{1-S_1}\frac{dS_1}{dp_c} + K_1S_1\right)\delta_1 \right.$$
$$\left. + \frac{K_1K_2\phi S_1}{1-S_1}\frac{dS_1}{dp_c} + K_1S_1\phi\right] \tag{3.101}$$

$$a_{23} = a_{32} = -\left[\frac{\delta_1\delta_2}{\delta_s}K_s + \frac{K_1K_2\phi}{M_1}\frac{dS_1}{dp_c}\right] \tag{3.102}$$

$$a_{33} = -\frac{1}{M_1}\left[\begin{array}{l}\left(K_1K_2\dfrac{dS_1}{dp_c} + \dfrac{K_1K_2(1-S_1)}{S_1}\dfrac{dS_1}{dp_c} + K_2(1-S_1)\right)\delta_2 \\ + \dfrac{K_1K_2\phi(1-S_1)}{S_1}\dfrac{dS_1}{dp_c} + K_1(1-S_1)\phi\end{array}\right], \tag{3.103}$$

$$R_{11} = -\frac{\theta_1\eta_1}{k_s k_{r1}} \tag{3.104}$$

$$R_{22} = -\frac{\theta_2\eta_2}{k_s k_{r2}} \tag{3.105}$$

$$A_{11} = (1-\alpha_s)\rho_1\theta_1 \tag{3.106}$$

$$A_{22} = (1-\alpha_s)\rho_2\theta_2 \tag{3.107}$$

$$\delta_s = \frac{(1-\phi-(K_b/K_s))K_s}{K_s + (M_1/M_2)((K_b/K_s)+\phi-1)} \tag{3.108}$$

$$\delta_1 = -\frac{K_1(S_1 + K_2(dS_1/dp_c) + (K_2S_1/(1-S_1))(dS_1/dp_c))(1-\phi-(K_b/K_s))}{K_sM_1 + M_2((K_b/K_s)+\phi-1)} \tag{3.109}$$

$$\delta_2 = \frac{K_2(1-S_1 + (K_1/S_1)(dS_1/dp_c))(1-\phi-(K_b/K_s))}{K_sM_1 + M_2((K_b/K_s)+\phi-1)} \tag{3.110}$$

$$M_1 = -\left(\frac{K_1}{S_1}\frac{dS_1}{dp_c} + \frac{K_2}{(1-S_1)}\frac{dS_1}{dp_c} + 1\right) \tag{3.111}$$

$$M_2 = \left(\frac{K_1K_2}{\phi S_1(1-S_1)}\frac{dS_1}{dp_c} + \frac{K_1S_1}{\phi} + \frac{K_2(1-S_1)}{\phi}\right) \tag{3.112}$$

In these formulae, K_s denotes the bulk modulus of solid phase, K_1 stands for the bulk modulus of the nonwetting fluid phase, K_2 is the bulk modulus of the wetting fluid phase, ϕ represents the porosity, $S_1 = \theta_1/\phi$ is the relative saturation of the nonwetting fluid, and $p_c = p_1 - p_2$ refers to the capillary pressure. Equations (3.104) and (3.105) are the classical formulations used to infer the viscous coefficients of the two fluids that are governed by Darcy's law. The variable η_i is the dynamic viscosity of the pore fluid i, k_s is the permeability of the porous medium, and k_{ri} is the relative permeability of fluid phase i that can be deduced from the degree of saturation in the media. In this, tortuosity is represented by a_s (a_s/ϕ corresponds to the formation factor). The elastic parameters are mainly dependent on changes in saturation; hence, we use the van Genuchten (1980) approach to determine (dp_c/dS_1) and the relative permeabilities (Lo et al., 2005):

$$\frac{dp_c}{dS_1} = \frac{\rho_2 g}{m_v n_v \chi}\left((1-S_1)^{-\frac{n_v}{n_v-1}}-1\right)^{\frac{1-n_v}{n_v}}(1-S_1)^{-\left(\frac{2n_v-1}{n_v-1}\right)} \tag{3.113}$$

$$k_{r1}(S_2) = (1-S_2)^\lambda\left(1-(S_2)^{\frac{1}{m_v}}\right)^{2m_v} \tag{3.114}$$

$$k_{r2}(S_2) = (S_2)^\lambda\left(1-\left[1-(S_2)^{\frac{1}{m_v}}\right]^{m_v}\right)^{2m_v} \tag{3.115}$$

where $m_v = 1 - 1/n_v$, n_v, χ, and λ are the van Genuchten parameters (m_v and n_v should not be confused with the m and n of Archie's laws).

3.2.2 The u–p formulation for two-phase flow problems

The classical formulation described previously, using Equations (3.95) to (3.96), is based on solving partial differential equations for three unknown fields, $\mathbf{u}$ (solid displacement field) and $\mathbf{w}_1$ and $\mathbf{w}_2$, corresponding to the two filtration displacements for the two fluids. A two-dimensional discretized problem results in six degrees of freedom per node that must be solved to compute the wave propagation problem. We seek to simplify the problem by reducing the number of degrees of freedom that need to be solved; therefore, we introduce an alternate formulation that is simpler to solve numerically. In this alternate formulation, we eliminate the unknowns $\mathbf{w}_1$ and $\mathbf{w}_2$ and only use $\mathbf{u}$ and p_i as the unknown parameters we solve for. This will reduce the problem to three unknown parameters (u_1, u_2, and p_i) that we have to solve at each node. To get to this alternate formulation, we use a two-step technique. In the first step, we apply the Fourier transform to the set of equations to convert them into the frequency domain, and then, in the second step, we conduct algebraic manipulations to derive the formulations that are usable for numerical processing. We start with the following formulae, in the frequency domain:

$$-\omega^2(\rho_i\mathbf{u}_s + \tilde{\rho}_i\mathbf{w}_i) + i\omega\frac{R_{ii}}{\theta_i^2}\mathbf{w}_i = -\nabla p_i \quad (3.116)$$

$$\tilde{\rho}_i = \left(\frac{\rho_i}{\theta_i} - \frac{A_{ii}}{\theta_i^2}\right) \quad (3.117)$$

Equation (3.116) can be used to express the filtration displacement of each fluid, $\mathbf{w}_i$, as a function of the pore fluid pressure p_i and the displacement of the solid phase $\mathbf{u}$:

$$\dot{\mathbf{w}}_i = -k_i(\omega)\left(\nabla p_i - \omega^2\rho_i\mathbf{u}\right) \quad (3.118)$$

$$k_i(\omega) = -\frac{1}{i\omega\tilde{\rho}_i + \frac{R_{ii}}{\theta_i^2}} \quad (3.119)$$

which can be written as $\mathbf{w}_i = \tilde{k}_i(\nabla p_i - \omega^2\rho_i\mathbf{u}_s)$ where $k_i(\omega)$ is the frequency-dependent apparent permeability that is similar to the one defined in Jardani et al. (2010) for the seismoelectric problem in saturated conditions represented by k_ω (see Chapter 2). We can write

$$\tilde{k}_i = \frac{k_i(\omega)}{i\omega} = \frac{1}{\omega^2\tilde{\rho}_i - i\omega\frac{R_{ii}}{\theta_i^2}} \quad (3.120)$$

For the solid phase, the momentum conservation equation is written as

$$-\omega^2\left(\rho_T\mathbf{u} + \sum_{i=1}^{2}\rho_i\mathbf{w}_i\right) + \sum_{i=1}^{2}\theta_i\nabla p_i = \nabla\cdot\mathbf{T} \quad (3.121)$$

We now discuss the elimination of the unknown variables, $\mathbf{w}_i$, from these equations. We use the momentum conservation equations for the two fluid phases written in the frequency domain as follows:

$$-\omega^2(\rho_1\mathbf{u} + \tilde{\rho}_1\mathbf{w}_1) + i\omega\frac{R_{11}}{\theta_1^2}\mathbf{w}_1 = -\nabla p_1 \quad (3.122)$$

$$-\omega^2(\rho_2\mathbf{u} + \tilde{\rho}_2\mathbf{w}_2) + i\omega\frac{R_{22}}{\theta_2^2}\mathbf{w}_2 = -\nabla p_2 \quad (3.123)$$

For the solid phase, the momentum conservation equation (Newton's law) and the total stress are expressed by

$$-\omega^2(\rho_T\mathbf{u} + \rho_1\mathbf{w}_1 + \rho_2\mathbf{w}_2) + \theta_1\nabla p_1 + \theta_2\nabla p_2 = \nabla\cdot\mathbf{T} \quad (3.124)$$

$$\mathbf{T} = \left[\left(\bar{a}_{1s} - \frac{2}{3}G\right)\nabla\cdot\mathbf{u} + \frac{a_{12}}{\theta_1}\nabla\cdot\mathbf{w}_1 + \frac{a_{13}}{\theta_2}\nabla\cdot\mathbf{w}_2\right]\mathbf{I} + 2G\mathbf{e} \quad (3.125)$$

where

$$\bar{a}_{1s} = a_{11} + a_{12} + a_{13} \quad (3.126)$$

Following Biot (1962a, b), we can express the stress–strain relations in an isotropic porous medium as

$$-\theta_1 p_1 = a_{2s}\nabla\cdot\mathbf{u} + \frac{a_{22}}{\theta_1}\nabla\cdot\mathbf{w}_1 + \frac{a_{23}}{\theta_2}\nabla\cdot\mathbf{w}_2 \quad (3.127)$$

$$-\theta_2 p_2 = a_{3s}\nabla\cdot\mathbf{u} + \frac{a_{23}}{\theta_1}\nabla\cdot\mathbf{w}_1 + \frac{a_{33}}{\theta_2}\nabla\cdot\mathbf{w}_2 \quad (3.128)$$

Solving for $\nabla\cdot\mathbf{w}_1$ and $\nabla\cdot\mathbf{w}_2$,

$$\nabla\cdot\mathbf{w}_1 = \alpha_{s1}\nabla\cdot\mathbf{u} + M_{11}p_1 - M_{12}p_2 \quad (3.129)$$

$$\nabla\cdot\mathbf{w}_2 = \alpha_{s2}\nabla\cdot\mathbf{u} - M_{12}p_1 + M_{22}p_2 \quad (3.130)$$

Subsequently, we can rewrite the equations of motion in terms of the three new unknown fields ($\mathbf{u}$, p_1, and p_2) as

$$-\omega^2\tilde{\rho}_T\mathbf{u} + \tilde{\theta}_1\nabla p_1 + \tilde{\theta}_2\nabla p_2 = \nabla\cdot\hat{\mathbf{T}} \quad (3.131)$$

$$\hat{\mathbf{T}} = \bar{\lambda}_s(\nabla\cdot\mathbf{u})\mathbf{I} + G\left[\nabla\mathbf{u} + \nabla\mathbf{u}^T\right] \quad (3.132)$$

$$M_{11}p_1 - \nabla\cdot\left\{\tilde{k}_1\left[\nabla p_1 - \omega^2\rho_1\mathbf{u}\right]\right\} = M_{12}p_2 - \bar{\alpha}_{s1}\nabla\cdot\mathbf{u} \quad (3.133)$$

$$M_{22}p_2 - \nabla\cdot\left\{\tilde{k}_2\left[\nabla p_2 - \omega^2\rho_2\mathbf{u}\right]\right\} = M_{12}p_1 - \bar{\alpha}_{s2}\nabla\cdot\mathbf{u} \quad (3.134)$$

Equation (3.131) corresponds to Newton's law applied to the solid skeleton of the porous material. This equation is similar to Newton's equation of elastic bodies except for the coupling terms $\tilde{\theta}_i\nabla p_i$, which represents the coupling between the solid and fluid phases. The stress tensor defined by Equation (3.132) corresponds to the stress tensor with the porous material in vacuum (i.e., it corresponds to the stress acting on the solid phase if the pore fluid is replaced by vacuum). Equations (3.133) and (3.134) are the nonlinear diffusion equations for the pore fluid pressure of each fluid in which the effect of the compression of the solid phase on the pore fluid pressure and the hydraulic pressure exercised mutually between the two fluids are taken in the account.

The parameters used into Equations (3.131)–(3.134) are given by

$$\bar{\lambda}_s = \left(\bar{a}_{1s} + \frac{a_{12}}{\theta_1}\bar{\alpha}_{s1} + \frac{a_{13}}{\theta_2}\bar{\alpha}_{s2} - \frac{2}{3}G \right) \quad (3.135)$$

$$\tilde{\rho}_i = \left(\frac{\rho_i}{\theta_i} - \frac{A_{ii}}{\theta_i^2} \right) \quad (3.136)$$

$$k_i(\omega) = -\frac{1}{i\omega\tilde{\rho}_i + \dfrac{R_{ii}}{\theta_i^2}} \quad (3.137)$$

$$\tilde{k}_i = \frac{k_i(\omega)}{i\omega} = \frac{1}{\omega^2\tilde{\rho}_i - i\omega\dfrac{R_{ii}}{\theta_i^2}}, \quad (3.138)$$

$$\tilde{\rho}_T = \rho_T - \omega^2\rho_1^2\tilde{k}_1 - \omega^2\rho_2^2\tilde{k}_2 \quad (3.139)$$

$$\tilde{\theta}_1 = -\alpha_{s1} - \omega^2\rho_1\tilde{k}_1 \quad (3.140)$$

$$\tilde{\theta}_2 = -\alpha_{s2} - \omega^2\rho_2\tilde{k}_2 \quad (3.141)$$

$$\alpha_{s1} = \frac{(\bar{a}_{2s}a_{33} - \bar{a}_{3s}a_{23})\theta_1}{(a_{23}^2 - a_{33}a_{22})} \quad (3.142)$$

$$\alpha_{s2} = \frac{(\bar{a}_{3s}a_{22} - \bar{a}_{2s}a_{23})\theta_2}{(a_{23}^2 - a_{33}a_{22})} \quad (3.143)$$

$$M_{11} = \frac{a_{33}\theta_1^2}{(a_{23}^2 - a_{33}a_{22})} \quad (3.144)$$

$$M_{12} = \frac{a_{23}\theta_1\theta_2}{(a_{23}^2 - a_{33}a_{22})} \quad (3.145)$$

$$M_{22} = \frac{a_{22}\theta_2^2}{(a_{23}^2 - a_{33}a_{22})} \quad (3.146)$$

$$\bar{a}_{1s} = a_{11} + a_{12} + a_{13} \quad (3.147)$$

$$\bar{a}_{2s} = a_{12} + a_{22} + a_{23} \quad (3.148)$$

$$\bar{a}_{3s} = a_{13} + a_{23} + a_{33} \quad (3.149)$$

where $\tilde{\theta}_1$ and $\tilde{\theta}_2$ are the volumetric hydromechanical coupling coefficients of the each fluid, $\bar{\lambda}_s$ is the Lamé coefficient of the solid phase, $\tilde{k}_1$ and $\tilde{k}_2$ are the dynamic permeability of the each fluid, $\tilde{\rho}_T$ is an apparent mass density for the solid phase at a given frequency ω, and M_{11} and M_{22} are the storativity coefficients for the nonwetting and wetting fluids, respectively. The parameter $\tilde{\rho}_i$ denotes the inertial drag interactions between the solid and fluid phase i.

3.2.3 Seismoelectric conversion in two-phase flow

To describe the electromagnetic response in the quasistatic limit, we consider that the target is relatively close (less than 1000 m) to the sensors (antennas, nonpolarizing electrodes, and magnetometers). In this case, we can neglect the time required for the electromagnetic disturbances to diffuse between the reservoir and the electromagnetic sensors. In this case, we can model the problem by solving only the quasistatic electromagnetic problem using the Poisson equation. We note that this seismoelectric signal is a result of the relative displacement of the fluids generated by seismic source. Hence, the conversion of the mechanical energy into an electrical signal is due to the electrokinetic coupling where the drag of the charge density contained in the pore fluids is responsible for the polarization of the medium (Revil et al., 1999a; Leroy & Revil, 2004). The hydroelectric problem is coupled to hydromechanical via the current density term that is colinear with velocity of the relative displacement of the two fluid phases. The electric problem can be written in frequency domain as

$$\nabla \cdot (\sigma\nabla\psi) = \nabla \cdot \mathbf{J}_S \quad (3.150)$$

$$\mathbf{J}_S = -i\omega\left(\frac{\hat{Q}_1^0}{S_1}\mathbf{w}_1 + \frac{\hat{Q}_2^0}{S_2}\mathbf{w}_2 \right) \quad (3.151)$$

$$\mathbf{J}_S = -i\omega\frac{\hat{Q}_1^0}{S_1}\tilde{k}_1\left(\nabla p_1 - \omega^2\rho_1\mathbf{u}\right) - i\omega\frac{\hat{Q}_2^0}{S_2}\tilde{k}_2\left(\nabla p_2 - \omega^2\rho_2\mathbf{u}\right)$$

$$(3.152)$$

where ψ is the electrostatic potential ($\mathbf{E} = -\nabla\psi$ in the quasistatic limit of the Maxwell equations). The electrical conductivity of the medium is σ, and it can be inferred from electrical resistivity tomography or other electromagnetic techniques. The medium conductivity can also be expressed as a function of the pore water conductivity, σ_f, as

$$\sigma = \frac{1}{F}S_2^n\sigma_f + S_2^{n-1}\sigma_S \quad (3.153)$$

where F denotes the formation factor, introduced previously (ratio of the pore space tortuosity by the connected porosity), n (≥ 1) denotes the second Archie's exponent, σ_S denotes the surface conductivity associated with electrical conduction in the electrical double layer, and $\mathbf{J}_S$ is the source current density of an electrokinetic nature.

The parameters $\hat{Q}_1^0$ and $\hat{Q}_2^0$ denote the effective excess charge (of the diffuse layer) per unit pore volume (in $C\,m^{-3}$) of the nonwetting and wetting fluids, respectively. For water-saturated rocks, $\hat{Q}_2^0$ can be directly computed from the low-frequency permeability, k_0, of the porous material. However, we assume that the excess charge $\hat{Q}_1^0 = 0\ C\,m^{-3}$ for the nonwetting phase, and therefore, the seismoelectric conversion originates only from the wetting phase. In this case, we write the current density source as

$$\mathbf{J_S} = -i\omega\frac{\hat{Q}_2^0}{S_2}\tilde{k}_2\left(\nabla p_2 - \omega^2\rho_2\mathbf{u}\right). \tag{3.154}$$

The effect of water saturation on the excess electrical charges per unit volume is developed by extending the empirical relationship between the two charge density parameters in unsaturated conditions. This is done by substituting $\hat{Q}_2^0$ in water-saturated conditions for $\hat{Q}_2^0/S_2$ in unsaturated conditions (see Linde et al., 2007; Revil et al., 2007).

3.2.4 The effect of water content on the coseismic waves

Pride and Haartsen (1996) proposed a relationship of the proportionality between electromagnetic fields and the horizontal displacements of the grains. This proportionality was observed in a set of seismic and seismoelectric data acquired in the field by Garambois and Dietrich (2001). To establish this relationship, the authors defined the problem in the diffusive regime, a valid assumption given the frequencies encountered in field measurements. In the fully saturated conditions, the transfer function for the P-wave was derived from the eigenvalue response of an isotropic and homogeneous medium and expressed in the low-frequency regime under the following form:

$$\frac{E_x}{-\omega^2 u_x} = -\frac{\rho_f\varepsilon_f\zeta}{\sigma_f\mu_f}\left(1-\frac{\rho}{\rho_f}\frac{C}{H}\right) \tag{3.155}$$

where E_x and u_x are the amplitudes of the horizontal components of the electrical field and solid phase displacement, respectively, ε_f is the fluid's dielectric permittivity, σ_f is the fluid's electrical conductivity, μ_f is the fluid's viscosity, and ζ is the zeta potential. Parameters C and H are the elastic moduli (as defined by Biot, 1962a, b) that are related to the undrained bulk modulus and the Skempton coefficient as shown here:

$$C = BK_u \tag{3.156}$$

$$H = K_u + \frac{4}{3}G \tag{3.157}$$

Equation (3.155) links various physical properties to the amplitudes of coseismic electric fields that travel along with seismic P-wave.

In this section, we extend this proportionality for a medium saturated with two fluid phases. For this reason, we solve the set of equations that govern seismoelectric coupling by considering a compressional, pure harmonic plane wave in 1D, in a homogeneous and isotropic material:

$$\mathbf{u} = \mathbf{U}\ \exp[i(kx-\omega t)] \tag{3.158}$$

$$\mathbf{w}_1 = \mathbf{W}_1\ \exp[i(kx-\omega t)] \tag{3.159}$$

$$\mathbf{w}_2 = \mathbf{W}_2\ \exp[i(kx-\omega t)] \tag{3.160}$$

$$\mathbf{E} = \mathbf{E}_0\ \exp[i(kx-\omega t)] \tag{3.161}$$

where $\mathbf{U}$ and $\mathbf{W}_{1,2}$ are wave amplitudes, ω is the angular wave excitation frequency, and $k = k_r + ik_i$ is a complex wave number that includes an attenuation coefficient as the phase slowness that can be defined as $s = k_r/\omega$. The complete set of governing equations that described the poroelastic problem (Eqs. 3.93, 3.99) may be reduced to three vector equations as follows:

$$\left\{-\omega^2\begin{bmatrix}\rho_s\theta_s & \dfrac{A_{11}}{\theta_1} & \dfrac{A_{22}}{\theta_2} \\[2mm] \rho_1\theta_1 & \tilde{\rho}_1\theta_1 & 0 \\[2mm] \rho_2\theta_2 & 0 & \tilde{\rho}_2\theta_2\end{bmatrix} + i\omega\begin{bmatrix}0 & -\dfrac{R_{11}}{\theta_1} & -\dfrac{R_{22}}{\theta_2} \\[2mm] 0 & \dfrac{R_{11}}{\theta_1} & 0 \\[2mm] 0 & 0 & \dfrac{R_{22}}{\theta_1}\end{bmatrix}\right.$$
$$\left.+ k^2\begin{bmatrix}\tilde{a}_{11} & \dfrac{a_{12}}{\theta_1} & \dfrac{a_{13}}{\theta_2} \\[2mm] a_{2s} & \dfrac{a_{22}}{\theta_1} & \dfrac{a_{23}}{\theta_2} \\[2mm] a_{3s} & \dfrac{a_{23}}{\theta_1} & \dfrac{a_{33}}{\theta_2}\end{bmatrix}\right\}\begin{bmatrix}U \\ W_1 \\ W_2\end{bmatrix} = \begin{bmatrix}0 \\ 0 \\ 0\end{bmatrix} \tag{3.162}$$

The existence of a solution to this equation requires that the determinant of its coefficients must vanish. The determinant can be formulated as

$$Q_3\left(\frac{\omega^2}{k^2}\right)^3 + Q_2\left(\frac{\omega^2}{k^2}\right)^2 + Q_1\left(\frac{\omega^2}{k^2}\right) + Q_0 = 0 \tag{3.163}$$

For a given frequency, the polynomial Equation (3.163) of the dispersion has three complex roots, and the wave number has six roots. However, only three of these roots are physically possible. This implies the

Table 3.1 Petrophysical properties of the soil used to solve the seismoelectric coupling.

Property	Symbol	Value
Bulk modulus of air	K_1 (MPa)	0.145
Bulk modulus of framework	K_b (GPa)	1.02
Bulk modulus of solid	K_s (GPa)	35
Bulk modulus of water	K_2 (GPa)	2.25
Fitting parameter	n (—)	2.03
Fitting parameter	η	0.5
Fitting parameter	χ (—)	2.39
Material density of air	ρ_1 (kg m^{-3})	1.1
Material density of solid	ρ_s (kg m^{-3})	2650
Material density of water	ρ_2 (kg m^{-3})	997
Permeability	k (m^2)	10^{-12}
Porosity	ϕ (—)	0.23
Shear modulus of framework	G (GPa)	1.44
Viscosity of air	η_1 (Ns m^{-2})	18×10^{-6}
Viscosity of water	η_2 (Ns m^{-2})	0.001

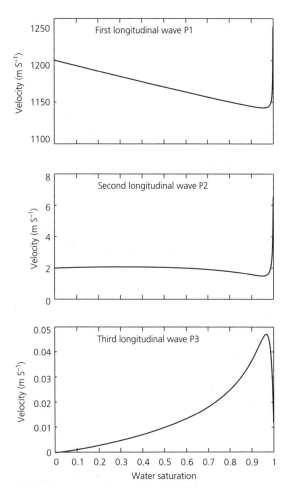

Figure 3.1 Effect of water saturation on the three-phase velocities of the longitudinal mode wave propagation as derived from the real component of the roots of cubic Equation (3.163) with 10 Hz as the source frequency used in the numerical model.

existence of three compressional waves in a poroelastic medium saturated by two immiscible fluids. According to the numerical analysis conducted by Vasco and Minkoff (2012), Lo et al. (2005), Santos et al. (1990, 2006), and Tuncay and Corapcioglu (1997), the P1 and P2 waves correspond to the fast and slow compressional waves in the Biot (1962a, b) model. The third wave (P3) is a diffusive wave that is due to the pressure difference between the fluid phases, and its phase speed depends on the slope of the capillary pressure effect. We solved the polynomial (Eq. 3.163) in the low-frequency region for an air–water mixture, to study the effect of water saturation on the velocity of the three wave modes using the poroelastic parameters noted in Table 3.1. The results obtained are reported in Figure 3.1 and show that for the first, fast P-wave, there is a decrease in compressional wave velocity with water saturation. The velocity decreases with increasing water saturation because total density increases (Lo et al., 2005). At very high water saturation, the compressional velocity is more controlled by the fluid bulk modulus H than bulk density. The velocity of the first wave, P1, in the low-frequency condition can be determined from the following formulation (Lo et al., 2005):

$$c_{pI} = \sqrt{\frac{H}{\rho_T}} \qquad (3.164)$$

where $H = \tilde{a}_{11} + a_{22} + a_{33} + 2(a_{12} + a_{13} + a_{23})$. The second dilatational wave (P2) propagates as a diffusion wave (Biot, 1956a; b; Santos et al., 1990; Tuncay & Corapcioglu, 1997) in which its velocity increases for a fully saturated soil. The third mode of propagation P3 is shown in Figure 3.1 to have the lowest velocity and is caused by the pressure difference between the two fluid phases. The third wave does not occur in systems with a single fluid. The attenuation coefficients derived from the imaginary part of the velocity for the three wave modes are plotted in Figure 3.2 where it is shown that the attenuation increases with water saturation for the P1 wave; however, its attenuation is much lower relative to the

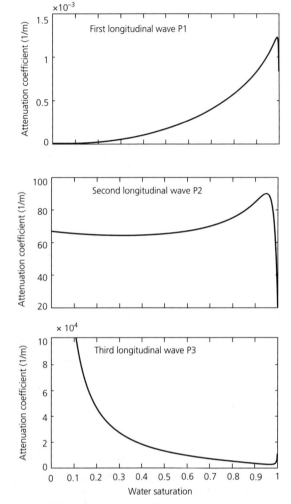

Figure 3.2 Effect of water saturation on the attenuation of the three P-wave modes calculated from the imaginary components of the roots of the cubic Equation (3.163).

strong attenuation of the second and third modes, making the experimental detection and verification of the two modes extremely difficult.

Next, we use the set of the relations introduced in the matrix previously to reformulate the amplitude of the relative displacement of the each of the fluid phases, $\mathbf{W}_{1,2}$, as a function of the amplitude of the solid phase $\mathbf{U}$:

$$\mathbf{w}_1 = \beta^I_{p_1,p_2,p_3} \mathbf{U} \, \exp[i(kx - \omega t)] \tag{3.165}$$

$$\mathbf{w}_2 = \beta^{II}_{p_1,p_2,p_3} \mathbf{U} \, \exp[i(kx - \omega t)] \tag{3.166}$$

using

$$\beta^I_{p_1,p_2,p_3} = \frac{\begin{aligned}&\left(s^2_{p_1,p_2,p_3} a_{2s} - \rho_1\theta_1\right)\left(s^2_{p_1,p_2,p_3}\frac{a_{13}}{\theta_2} - \Gamma_2\right)\\&- \left(s^2_{p_1,p_2,p_3}\tilde{a}_{11} - \rho_s\theta_s\right)\frac{a_{23}}{\theta_2}s^2_{p_1,p_2,p_3}\end{aligned}}{\begin{aligned}&\frac{a_{23}}{\theta_2}s^2_{p_1,p_2,p_3}\left(s^2_{p_1,p_2,p_3}\frac{a_{12}}{\theta_1} - \Gamma_1\right)\\&- \left(s^2_{p_1,p_2,p_3}\frac{a_{22}}{\theta_1} - \Lambda_1\right)\left(s^2_{p_1,p_2,p_3}\frac{a_{13}}{\theta_2} - \Gamma_2\right)\end{aligned}} \tag{3.167}$$

$$\beta^{II}_{p_1,p_2,p_3} = \frac{\begin{aligned}&\left(s^2_{p_1,p_2,p_3} a_{2s} - \rho_1\theta_1\right)\left(s^2_{p_1,p_2,p_3}\frac{a_{12}}{\theta_1} - \Gamma_1\right)\\&- \left(s^2_{p_1,p_2,p_3}\tilde{a}_{11} - \rho_s\theta_s\right)\left(s^2_{p_1,p_2,p_3}\frac{a_{22}}{\theta_1} - \Lambda_1\right)\end{aligned}}{\begin{aligned}&\left(s^2_{p_1,p_2,p_3}\frac{a_{22}}{\theta_1} - \Lambda_1\right)\left(s^2_{p_1,p_2,p_3}\frac{a_{13}}{\theta_2} - \Gamma_2\right)\\&- \left(s^2_{p_1,p_2,p_3}\frac{a_{12}}{\theta_1} - \Gamma_1\right)\frac{a_{23}}{\theta_2}s^2_{p_1,p_2,p_3}\end{aligned}} \tag{3.168}$$

with

$$\Gamma_1 = \frac{A_{11}}{\theta_1} - \frac{R_{11}}{i\omega\theta_1} \tag{3.169}$$

$$\Gamma_2 = \frac{A_{22}}{\theta_2} - \frac{R_{22}}{i\omega\theta_2} \tag{3.170}$$

$$\Lambda_1 = \tilde{\rho}_1\theta_1 + \frac{R_{11}}{i\omega\theta_1} \tag{3.171}$$

where $\beta^I_{p_1,p_2,p_3}$ and $\beta^{II}_{p_1,p_2,p_3}$ are the factors of the three compressional waves that link their amplitude in the two fluid phases with the amplitude in the solid frame. This shows that the amplitudes are dependent on the velocity of each type of wave. The electric field corresponding to the propagation of the three compressional wave modes in a homogeneous medium drives a conduction current that exactly balances the streaming current. Therefore,

$$\mathbf{J} = \sigma^*(\omega)\mathbf{E} + \frac{\hat{Q}^0_2}{S_2}\dot{\mathbf{w}}_2 = 0 \tag{3.172}$$

and

$$\mathbf{E} = i\omega\frac{\hat{Q}_{V2}(\omega)}{\sigma^*(\omega)S_2}\beta^{II}_{p_1,p_2,p_3}(\omega)\mathbf{U} \, \exp[i(kx - \omega t)] \tag{3.173}$$

Here, we reformulated the different parameters at low frequency where the angular frequency of the wave motions is much smaller than the critical frequency. The critical frequency is equal to the inverse of the intrinsic time scale in the two fluid system where $\omega < \omega_c = \eta_1\eta_2/[k_s(\rho_1 k_{r1}\eta_1 + \rho_2 k_{r2}\eta_2)]$:

$$\hat{Q}_{V2}(\omega) \approx \hat{Q}^0_{V2} \tag{3.174}$$

$$\sigma^*(\omega) = \sigma_0 \qquad (3.175)$$

$$\beta_{p_1}^{II}(\omega) \approx \frac{i\omega k_0 k_r \rho_2}{\eta_2} \qquad (3.176)$$

Therefore, the electrical field can be written as

$$\mathbf{E} = \frac{\hat{Q}_{V2}}{\sigma_0 S_2} K_r \ddot{\boldsymbol{u}} \qquad (3.177)$$

$$K_r = \frac{k_0 k_r \rho_2}{\eta_2} \qquad (3.178)$$

Equation (3.177) connects the electrical field recorded between two electrodes and the seismic signal in terms of the acceleration of the seismic wave. We conclude that this proportionality depends both on the saturation state and the hydraulic and electrical conductivities of the medium, which are in turn related to the water content. The ratio of the acceleration, $\ddot{\boldsymbol{u}}_x$, and the electrical field is plotted in Figure 3.3 using Equation (3.177). This result reveals that an increase in the amplitude of the coseismic electrical disturbance is due to an increase of the water content. Strasher (2006) expressed this transfer equation for the normalized seismoelectric field as a power law of the effective saturation $\mathbf{E} \propto C_0 S_e^{(0.42\,\pm\,0.25)\alpha} \rho_2 \ddot{\boldsymbol{u}}$ where C_0 denotes the steady-state streaming potential coupling in the saturated conditions. To check this proportionality between the seismic and electric fields, a numerical simulation, based on the poroelastic and electric coupling Equation (3.177) for a homogeneous and isotropic medium that is saturated by two immiscible fluid phases, was coded in MATLAB and COMSOL. The elasticity, hydraulic, and electrical parameters required to solve the problems are reported in Table 3.1. The size of the model is $400 \times 400\ \text{m}^2$ with a PML of $50\ \text{m}$ surrounding the studied domain. The fluid mixtures of air and water are investigated, with two values of the water saturation $S_2 = 0.2$ and $S_2 = 0.9$, in which the horizontal component of the electrical field associated with the seismic wave propagation is perfectly proportional to the horizontal acceleration $\ddot{\boldsymbol{u}}_x$, in accordance with the transfer function in Figure 3.3 (see Figure 3.4). The advantage of this ratio between seismic and electric fields is its usefulness to help solve for important hydraulic and electrical parameters. Because the origin of the seismoelectric response is due to the existence of water in the porous medium, we also studied the behavior of the coseismic amplitudes with water saturation for the fluid mixtures of air and water. We numerically simulated different values of water saturation on the coseismic signal. The amplitude variations of the electrical potential as recorded at the surface of the earth and its maximums versus water saturation were reported in Figure 3.5. The results obtained show an increase in the amplitude of the coseismic electrical

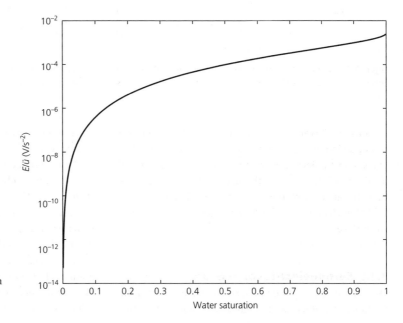

Figure 3.3 The transfer function linking the ratio of the seismoelectric field to the acceleration of the grains versus water saturation. This plot shows that as water saturation increases, the coupling between the acceleration of the grains and the seismoelectric field also increases.

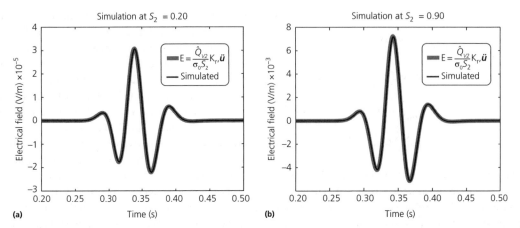

(a)   Time (s)

(b)   Time (s)

Figure 3.4 A numerical test to confirm the validity of the transfer function $E_x/\ddot{u}_x$ at two values of water saturation (20 and 90%). **a)** Simulation at 20% water saturation. **b)** Simulation at 90% water saturation. In both case, there is a perfect agreement between the numerical simulation and the analytical formula.

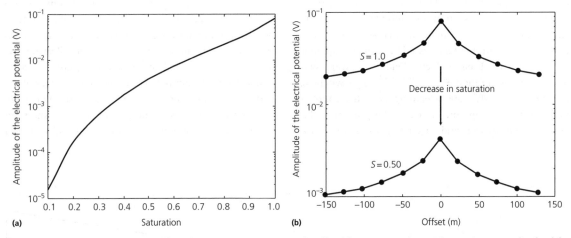

(a)   Saturation

(b)   Offset (m)

Figure 3.5 The effect of water saturation on the coseismic response. **a)** The effect of water saturation on the maximum amplitude of the coseismic response. **b)** The effect of water saturation on the coseismic response as recorded at the simulated electrodes.

response with water saturation. Consequently, a perfect reproduction of the coseismic signal during seismoelectric surveys reflects the saturation of the medium and its absence in the case of the dry conditions. We think that the analysis of the coseismic signal may provide key information on moisture conditions of the subsurface.

3.2.5 Seismoelectric conversion

We dedicate this section to the study of the ability of the seismoelectric method to detect saturation contrasts.

In this sense, we divided the $400\,\text{m} \times 400\,\text{m}$ domain in the previous problem vertically into two tabular layers: the first is a shallow unsaturated layer that represents the vadose zone, and the second is a fully saturated layer (Figure 3.6). The petrophysical parameters used to simulate the seismoelectric data are identical, except for the water saturation difference between the layers. That means that all other petrophysical parameters in the entire domain are homogeneous throughout the entire domain of the model (perfect matching layers excluded). The simulations were realized with three values of saturation for the vadose zone (the upper layer) $S_2 = 0.1$,

$S_2 = 0.5$, and $S_2 = 0.8$. The results obtained point out that the seismoelectric conversion response can be detected when there is a saturation contrast between the vadose and saturated zones. The seismoelectric conversion response can be identified in the first and the second cases (Figure 3.7) by the increase of the amplitude of the coseismic signature with the increase of the water saturation in the vadose zone. However, for the third case, when the saturation of the vadose zone is $S_2 = 0.8$, the amplitude of the seismoelectric conversion response is

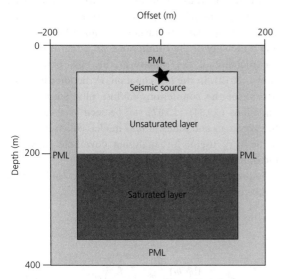

Figure 3.6 Sketch of the model used for the simulations. The electrodes are collocated at $z = 60$ m. All the electrodes are assumed to be connected to a reference electrode.

very weak relative to the coseismic signal (Figure 3.7). We conclude that a dry vadose zone is the best condition to identify the groundwater level, because the coseismic signal is low relative to the seismoelectric conversion response. This phenomenon was proven during several seismoelectric surveys carried out at different fields with different water content conditions by Strahser (2006).

3.3 Extension of the acoustic approximation

In unsaturated conditions, the density of the pore fluid, ρ_f, the bulk modulus of the pore fluid, K_f, and the dynamic viscosity of the pore fluid, η_f, are related to the properties of the nonwetting (subscript "nw") and wetting (subscript "w") phase by

$$\rho_f = (1 - s_w)\rho_{nw} + s_w\rho_w \tag{3.179}$$

$$\frac{1}{K_f} = \frac{(1 - s_w)}{K_{nw}} + \frac{s_w}{K_w} \tag{3.180}$$

$$\eta_f = \eta_{nw}\left(\frac{\eta_w}{\eta_{nw}}\right)^{s_w} \tag{3.181}$$

$$\hat{Q}_V(s_w) = \frac{\hat{Q}_V}{s_w} \tag{3.182}$$

This models the effect of saturation on the permeability (through the effect of saturation on the relative

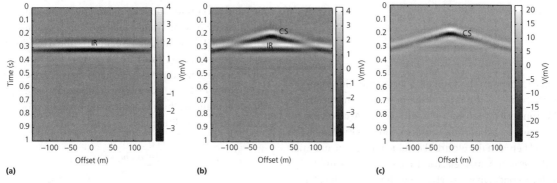

Figure 3.7 The electric potential generated with our modeling code for three values of the water saturation of the vadose zone $S_2 = 0.1$, $S_2 = 0.5$, and $S_2 = 0.8$. **a)** Case corresponding to $S_2 = 0.1$. **b)** Case corresponding to $S_2 = 0.5$. In both cases, the interface response at 0.28 s can be distinguished when the contrast between the saturated and vadose zones is strong enough. **c)** Case corresponding to $S_2 = 0.8$. In this situation, the contrast in saturation between the vadose zone and the aquifer is very weak, and consequently, it is difficult to detect the interface response (seismoelectric conversion). IR, interface response; CS, coseismic field.

permeability) and the effect of saturation upon the electrical conductivity of the material (using Archie's second law plus the effect of saturation on the surface conductivity). We assume that the two fluid phases are immiscible and that the wetting phase is water, while the nonwetting phase has insulating properties (e.g., air or nonpolar oil).

3.4 Complex conductivity in partially saturated conditions

If we want to account for the low-frequency polarization effects, we need to provide a generalization of the equations presented in Chapter 1 for the complex conductivity. As discussed in Chapter 1, the complex conductivity is written as

$$\sigma^* = \sigma' + i\sigma'' = |\sigma|\exp(i\varphi) \tag{3.183}$$

where $|\sigma^*| = \left(\sigma'^2 + \sigma''^2\right)^{1/2}$ represents the magnitude of the conductivity and $\varphi = \mathrm{atan}(\sigma''/\sigma')$ is the phase lag between the electrical current and the resulting electrical field. For clayey materials, the frequency dependence of the complex conductivity is usually very small and can be neglected in the frequency range of 0.1 Hz to 10 kHz (see Vinegar & Waxman, 1984; Revil, 2012, for some extensive discussions). This is important for field applications, because the good signal transport properties of porous media in this frequency range make field operations useful. The following linear model is used to describe the overall inphase electrical conductivity as a function of the pore water electrical conductivity, σ_f:

$$\sigma' = \frac{1}{F}s_w^n \sigma_f + \sigma_S \tag{3.184}$$

where F denotes the formation factor introduced previously (which is the ratio of the pore space tortuosity to the connected porosity), n is the second Archie's exponent, and σ_S denotes the surface conductivity. If a linear model between the electrical conductivity of the material and the pore water electrical conductivity is used, the surface conductivity is given by the model developed recently by Revil (2012, 2013a, b):

$$\sigma_S = \frac{A(\phi,m)}{F}s_w^{n-1}\beta_{(+)}\hat{Q}_V^\infty \tag{3.185}$$

$$\sigma_S = \frac{A(\phi,m)}{F}s_w^{n-1}\beta_{(+)}(1-f)Q_V \tag{3.186}$$

where f represents the fraction of counterions in the Stern layer (attached to the grains) and $A(\phi, m)$ is defined by

$$A(\phi,m) = m(F-1)\frac{2}{3}\left(\frac{\phi}{1-\phi}\right) \tag{3.187}$$

Equation (3.186) means that the surface conductivity is controlled by the charge density in the diffuse layer, $\bar{Q}_V = (1-f)Q_V$, with its fraction of counterions $(1 - f)$, and a mobility of the counterions, $\beta_{(+)}$, that is equal to the mobility of the cations in the bulk pore water (Revil, 2012). In the last equation, Q_V denotes the total density of the counterions (diffuse plus Stern layer). Equations (3.184)–(3.187) can be seen as a variant of the Waxman and Smits (1968) model, which is known to be very useful for analyzing downhole resistivity measurements in shaly sand reservoirs.

Following Revil (2012), the quadrature conductivity can be expressed as

$$\sigma'' = -\frac{A(\phi,m)}{F}s_w^p \beta_{(+)}^S f Q_V \tag{3.188}$$

$$\sigma'' = -\frac{A(\phi,m)}{F}s_w^p \beta_{(+)}^S \left(\frac{f}{1-f}\right)\hat{Q}_V^\infty \tag{3.189}$$

where $p = n-1$ and $\beta_{(+)}^S$ denotes the mobility of the counterions in the Stern layer. These equations provide a simple and accurate model to describe the complex conductivity of shaly sands and soils and generalize the Vinegar and Waxman (1984) model. As noted by Vinegar and Waxman (1984) and Revil (2012), the frequency dependence of the quadrature conductivity is not explicit in Equation (3.189). It should be mentioned that for quasistatic conditions, the quadrature conductivity should go to zero in DC conditions:

$$\lim_{\omega \to 0} \sigma''(\omega) = 0 \tag{3.190}$$

The typical frequency in the following text, which the quadrature conductivity becomes frequency dependent, is typically smaller than 0.1 Hz (see Revil, 2012) and therefore is not relevant to the seismoelectric problem.

3.5 Comparison with experimental data

3.5.1 The effect of saturation

We consider an unsaturated material with two immiscible fluid phases: (1) water that is considered to be the wetting phase for the solid and (2) a nonwetting and insulating fluid phase like air or oil. The saturation can be related to the capillary pressure by the Brook and Corey (1964) relationship:

$$s_w = \begin{cases} s_r + (1-s_r)\left(\dfrac{p_e}{p_c}\right)^{\lambda}, & p_c \geq p_e \\ 1, & p_c < p_e \end{cases} \tag{3.191}$$

or alternatively $p_c = p_e s_e^{-1/\lambda}$ (or $s_e = (p_c/p_e)^{-\lambda}, p_c \geq p_e$; see equation 12 of Brooks & Corey, 1964) where $s_e = (s_w - s_r)/(1-s_r)$ denotes the effective or reduced water saturation, s_r is the irreducible water saturation, and p_c represents the capillary pressure. The capillary entry pressure, p_e, is the critical pressure needed to displace the water phase by the gas phase when the porous material is fully water saturated. The capillary entry pressure is related to the media permeability in the following full saturation condition.

The saturation dependence of the dynamic mass density is given by

$$\tilde{\rho}_f(s_w) = \frac{1}{F} s_w^n \left[(1-s_w)\rho_g + s_w \rho_w\right] \tag{3.192}$$

The saturation dependence of the permeability-related constant, k_{ω}, and the coefficients, ρ_{ω}^S, can be easily calculated. The dependence of the Biot coefficient on saturation can be found in Revil and Mahardika (2013).

Now, we need to determine the effect of the saturation on the electrical conductivity. We investigate here two models:

$$\sigma = \frac{1}{F} s_w^n \left(\sigma_w + \beta_S \frac{\bar{Q}_V}{s_w}\right) \text{(Model A)}, \tag{3.193}$$

$$\sigma = \frac{1}{F} s_w^n \left(\sigma_w + \beta_S \frac{\bar{Q}_V}{s_w^n}\right) \text{(Model B)}, \tag{3.194}$$

where n is called the saturation exponent (Archie, 1942). Model A is discussed in detail in Revil (2013a) and was used for the seismoelectric problem by Revil and Mahardika (2013). It is also consistent with the Waxman

and Smits (1968) model for unsaturated siliciclastic materials. The saturation dependence of Model B was recently proposed for the seismoelectric coupling in unsaturated conditions by Warden et al. (2013; see equation 13). Later, we check the consistency of these two models with respect to the available electrokinetic data.

The effect of saturation on the excess of electrical charges per unit volume is developed by extending the empirical relationship between these two parameters in unsaturated conditions (see Linde et al., 2007; Revil et al., 2007):

$$\hat{Q}_V(s_w) = \frac{\hat{Q}_V}{s_w} \tag{3.195}$$

This equation has been recently challenged by Jougnot et al. (2012) but seems to be consistent with experimental data as checked by Linde et al. (2007), Revil et al. (2007), and Mboh et al. (2012). We will show later that this model is also consistent with the data from Guichet et al. (2003), Revil et al. (2011), and Vinogradov and Jackson (2011).

3.5.2 Additional scaling relationships

In this section, we develop a unified set of scaling relationships between the hydraulic and electrical properties in order to reduce the number of input parameters. Three scaling laws are developed later, one for the relative water permeability, one for the capillary entry pressure, and one for the streaming potential coupling coefficient. For each new scaling law, we will show that it is in agreement with existing experimental data or empirical scaling laws based on fitting experimental data.

Johnson (1986) developed the following equation for the electrical conductivity for a water-saturated rock:

$$\sigma = \frac{1}{F}\left(\sigma_w + \frac{2}{\Lambda}\Sigma_S\right) \tag{3.196}$$

where Λ is a characteristic length scale of the pore space. A comparison of Equations (3.192) and (3.194) with Equation (3.196) implies the following scaling laws for the dependence of the formation factor and length scale, Λ, with the relative water saturation:

$$F \Leftrightarrow F s_w^{-n}, \tag{3.197}$$

$$\Lambda \Leftrightarrow \Lambda s_w \quad \text{(Model A)}, \tag{3.198}$$

$$\Lambda \Leftrightarrow \Lambda s_{\mathrm{w}}^{n} \quad (\text{Model B}), \qquad (3.199)$$

where n denotes the second Archie's exponent (Archie, 1942). The left side of Equations (3.197)–(3.199) indicates the parameters used to compute the electrical conductivity in fully saturated conditions, and on the right side, we have the scaling of the same parameters with saturation for unsaturated materials.

The permeability, k, is related to the formation factor, F, and the dynamic pore radius, Λ, by (see Johnson, 1986, for the saturated case)

$$k_0(s_{\mathrm{w}}) = \frac{\Lambda(s_{\mathrm{w}})^2}{8F(s_{\mathrm{w}})} \qquad (3.200)$$

Therefore, the permeability should scale with the water saturation as

$$k_0(s_{\mathrm{w}}) = \frac{\Lambda^2}{8F} s_w^{2+n} \quad (\text{Model A}), \qquad (3.201)$$

$$k_0(s_{\mathrm{w}}) = \frac{\Lambda^2}{8F} s_w^{3n} \quad (\text{Model B}). \qquad (3.202)$$

Therefore, according to this scaling process, the permeability can be computed as the product of the permeability at saturation, $k_{\mathrm{S}} = \Lambda^2/8F$, and a relative permeability that depends only on the relative water saturation:

$$k_0(s_{\mathrm{w}}) = k_{\mathrm{S}} k_{\mathrm{r}}(s_{\mathrm{w}}) \qquad (3.203)$$

with

$$k_{\mathrm{r}}(s_{\mathrm{w}}) = s_e^{n+2} \quad (\text{Model A}), \qquad (3.204)$$

$$k_{\mathrm{r}}(s_{\mathrm{w}}) = s_e^{3n} \quad (\text{Model B}). \qquad (3.205)$$

Here, we used the effective water saturation, s_e, rather than the water saturation, to enforce the fact that the relative permeability is null for the irreducible water saturation case. These equations can be compared to the one proposed by Li and Horne (2005; equations 5 and 6) $k_{\mathrm{r}}(s_{\mathrm{w}}) = s_e s_w^n$. In the Brooks and Corey (1964) model, the relative permeability is also given by a power law relationship (see also Purcell, 1949):

$$k_{\mathrm{r}}(s_{\mathrm{w}}) = \left(\frac{p_{\mathrm{c}}}{p_{\mathrm{e}}}\right)^{-2-3\lambda} = s_e^{\mathrm{r}} \qquad (3.206)$$

$$r = \frac{2+3\lambda}{\lambda} \qquad (3.207)$$

where λ is called the pore size distribution index. Therefore, we identify the following equality:

$$\lambda = \frac{2}{n-1}, \quad r = n+2 \quad (\text{Model A}), \qquad (3.208)$$

$$\lambda = \frac{2}{3(n-1)}, \quad r = 3n \quad (\text{Model B}). \qquad (3.209)$$

This result is very important because it provides an explicit relationship between a hydraulic parameter and an electrical parameter. To our knowledge, this is the first time that these relationships are proposed, despite some attempts by others to connect the resistivity index and the capillary pressure curves (see, for instance, Li & Horne, 2005).

Figure 3.8 shows that Model A seems to agree better than Model B with experimental data but more data

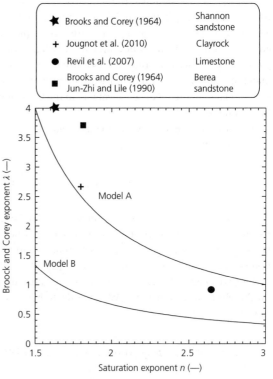

Figure 3.8 Comparison between Model A and Model B used to predict the value of the Books and Corey exponent, λ, from the saturation exponent, n. Data from Jougnot et al. (2010), Revil et al. (2007), Brooks and Corey (1964), and Jun-Zhi and Lile (1990). The Shannon sandstone is also known under the term "Hygiene sandstone" in the literature. The data seem to favor Model A indicating therefore a saturation dependence of surface conductivity.

are needed to check the predictive capabilities of the two models. These new relationships also mean that the measurement of the second Archie's exponent can be used to predict the capillary pressure curve. The hysteretic behavior seen in the capillary pressure curve implies that a hysteretic behavior should be present in the value of n.

We now turn our attention to the capillary entry pressure, which can be related to the length scale, Λ, by

$$p_e = \frac{2\gamma}{r_c} \qquad (3.210)$$

where γ represents the surface tension between water and air $(71.99 \pm 0.05) \times 10^{-3}\,\mathrm{N\,m^{-1}}$ and r_c represents the smallest pore out of the set of pores involved in the percolation of water through the porous material (Katz & Thompson, 1987). Also, Katz and Thompson (1987) developed a relationship between the permeability and the percolation length scale, r_c, using percolation principles: $k_S = r_c^2/(226F)$. A comparison with $k_S = \Lambda^2/8F$ yields $r_c \approx 5.3\,\Lambda$. Using the relationship between the length scale, r_c, and the permeability at saturation, we obtain

$$p_e = \frac{2\gamma}{\sqrt{226F}} k_S^{-0.5} \qquad (3.211)$$

This equation can be compared to the empirical equation derived by Thomas et al. (1968), $p_e = 52 k_S^{-0.43}$, with p_e expressed in kPa and k_S in mD. Archie's law can be used to compute F. Taking $m = 2.0$ (a default value for soils), and a porosity of $\phi = 0.20$ (a reasonable average value), we obtain $p_e = a k_S^{-0.5}$ with $a = 61$ where p_e is expressed in kPa and k_S in mD. We also obtain $p_e = b\phi/\sqrt{k_S}$. The proportionality between the capillary entry pressure and the ratio, $\phi/\sqrt{k_S}$, is checked in Figure 3.9. Therefore, Equation (3.199) is able to represent data but accounts explicitly for the porosity dependence, which is not the case of the empirical equation developed by Thomas et al. (1968). Note that in mercury porosimetry, mercury is the nonwetting phase, while water is the wetting phase.

3.5.3 Relative coupling coefficient with the Brooks and Corey model

The last coefficient we want to test is the streaming potential coupling coefficient, which can be defined for

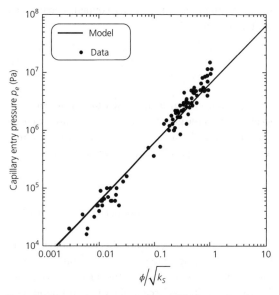

Figure 3.9 Comparison between the model proposed in the main text for the capillary entry pressure, assuming that the formation factor is related to the porosity by $F = \phi^{-2}$ (classical Archie's law, Archie, 1942). The experimental data are from Huet et al. (2005). They correspond to 89 sets of mercury injection (Hg–air) capillary pressure data. Core samples include both carbonate and sandstone lithologies. The permeability is expressed in mD.

a fully water-saturated material (Revil & Mahardika, 2013, and see Chapter 1) as

$$C_0 = \lim_{\omega \to 0} \left(\frac{\partial \psi}{\partial p}\right)_{j_s = 0,} = \frac{\hat{Q}_V k_0}{\eta_w \sigma} \qquad (3.212)$$

At high salinity, the conductivity can be approximated by

$$\sigma \approx \frac{1}{F} s_w^n \sigma_w \qquad (3.213)$$

We can now evaluate the effect of water saturation upon the streaming potential coupling coefficient by substituting into Equation (3.212) the volumetric charge density, the permeability, and the electrical conductivity with their expressions that include saturation. We obtain

$$C_0 = C_r C_S \qquad (3.214)$$

where C_r denotes the relative coupling coefficient (a concept first introduced by Revil & Cerepi, 2004) and where the streaming potential coupling coefficient at saturation is given by

$$C_S = \frac{\hat{Q}_V k_0}{\eta_w \sigma} \quad (3.215)$$

If Equation (3.215) is multiplied by $\rho_w g$ (g being the acceleration of the gravity in m s^{-2}), the coupling coefficient units transform to V m^{-1}. The relative streaming potential coupling coefficient is given by

$$C_r \approx s_e \quad \text{(Model A)}, \quad (3.216)$$

$$C_r = s_e^{2n-1} \quad \text{(Model B)}. \quad (3.217)$$

In Figure 3.10, we compare the two models to the existing data, replacing the water saturation by the irreducible saturation, to satisfy to the additional constrain that there is no flow at irreducible water saturation. Very clearly, Model B is unable to explain these data.

Note that the general form of the relative coupling coefficient with the Brooks and Corey model is

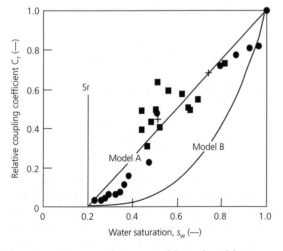

Figure 3.10 Comparison between Model A and Model B to predict the value of the relative streaming potential coupling coefficient as a function of the water saturation. We use an irreducible water saturation $s_r = 0.2$. Data from Revil et al. (2011), Vinogradov and Jackson (2011), and Guichet et al. (2003). The data seem to favor Model A over Model B.

$$C_r = s_w^{-(n+1)} s_e^{3+2/\lambda} \quad (3.218)$$

as proposed by Revil et al. (2007). In the next section, we will see how this equation needs to be modified if we use the van Genuchten model, instead of the Brooks and Corey (1964) model, to represent the capillary pressure curve.

3.5.4 Relative coupling coefficient with the Van Genuchten model

An alternative to the Brooks and Corey model is the van Genuchten model (1980) (see discussion in Linde et al., 2007; Revil et al., 2007). This model can be written as

$$s_e = \left(1 + \left|\frac{p_c}{p_e}\right|^{n_v}\right)^{-m_v} \text{ with } m_v \approx 1 - \frac{1}{n_v}, \quad p_c \geq p_e \quad (3.219)$$

$$k_r(s_w) \approx \sqrt{s_e}\left[1 - \left(1 - s_e^{1/m_v}\right)^{m_v}\right]^2 \quad (3.220)$$

where n_v and m_v are the van Genuchten exponents. Therefore, the relative coupling coefficient is given by

$$C_r = s_w^{-(n+1)}\sqrt{s_e}\left[1 - \left(1 - s_e^{1/m_v}\right)^{m_v}\right]^2 \quad (3.221)$$

Mboh et al. (2012) measured the relative streaming potential coupling coefficient of a clean sand characterized by the following properties: 99.3% silica, porosity $\phi = 0.41$, hydraulic conductivity $K = 8.25 \times 10^{-5}$ m s^{-1}, and mean grain diameter $d = 160$ μm (Table 3.2). Their

Table 3.2 Petrophysical properties of the samples discussed in the main text.

Property	Symbol	M	E3	E39
Saturation exponent	n (—)	1.87	2.7	3.5
Cementation exponent	m (—)	–	1.93	2.49
Permeability	k (m^2)	8.4×10^{-12}	48.4×10^{-15}	23.8×10^{-15}
Porosity	ϕ (—)	0.41	0.203	0.159
Residual saturation	s_r (—)	0.09	0.31	0.34
Grain size	d (μm)	160	–	–
Pore size	r (μm)	–	1.18	0.17

M corresponds to the sand sample investigated by Mboh et al. (2012), while Samples E3 and E39 are dolomitic samples investigated by Revil et al. (2007).

data are exceptionally good in terms of quality. It is indeed difficult to get very good data in unsaturated conditions because of the drift of the electrodes (see discussions in Revil & Linde, 2011, and Jougnot & Linde, 2013, for the drift associated with saturation effects and Petiau & Dupis, 1980, and Petiau, 2000, for other sources of noises). They measured a coupling coefficient at saturation of $C = -3.3\,\mathrm{mV\,m^{-1}}$ for a pore water conductivity of $\sigma_w = 0.044\,\mathrm{S\,m^{-1}}$. The second Archie's exponent (saturation exponent) was measured and found equal to $n = 1.87$. The van Genuchten exponent was measured and was found equal to $n_v = 3.88$ (by fitting the capillary pressure curve). This yields $m_v = 0.74$. A comparison between the data of Mboh et al. (2012) and Equation (3.220) is shown in Figure 3.11. The best fit of the data yields $m_v = 0.69 \pm 0.05$, very close to the value determined from the capillary pressure curve (see Figure 3.11). As mentioned by Mboh et al. (2012), this implies that the values of the relative coupling coefficient contains information regarding the van Genuchten parameters, as suggested by Linde et al. (2007) and Revil et al. (2007).

In Figures 3.12 and 3.13, we reanalyzed the data presented in Revil and Cerepi (2004) and Revil et al. (2007), correcting a few mistakes in the unit conversions found in these two papers. We analyzed the streaming potential coupling coefficient data of Samples E3 and E39 (dolomitic limestones; see properties in Table 3.2). In both cases, the second Archie's exponent (saturation exponent) was independently determined using resistivity measurements. The van Genuchten parameters were found to be roughly the same using the capillary pressure curves and the relative streaming potential coupling coefficient data.

3.6 Conclusions

We have extended the seismoelectric theory to unsaturated and two-phase flow conditions assuming that the rheology of the two immiscible fluids is viscous Newtonian. We developed the equations under three levels of models: (1) the full extension of the Biot theory to two-phase flow conditions, (2) a simple extension of the classical Biot theory assuming that the second (nonwetting) phase is infinitely compressible and connected to a reservoir at constant pressure (e.g., the atmosphere for vadose zone processes), and (3) a simple extension of the acoustic approximation discussed in Chapters 1 and 2. We have also showed that the theory agrees with

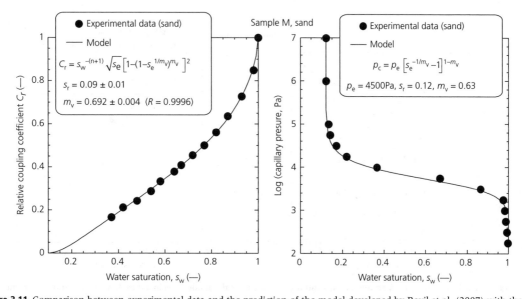

Figure 3.11 Comparison between experimental data and the prediction of the model developed by Revil et al. (2007) with the van Genuchten model (see also Linde et al., 2007). The experimental data are from Mboh et al. (2012) (Sample M, sand). Left panel: relative streaming potential coupling coefficient versus saturation. We used the measured value of the saturation exponent (second Archie's exponent) $n = 1.87$. Right panel: capillary pressure curve (nonwetting fluid: air).

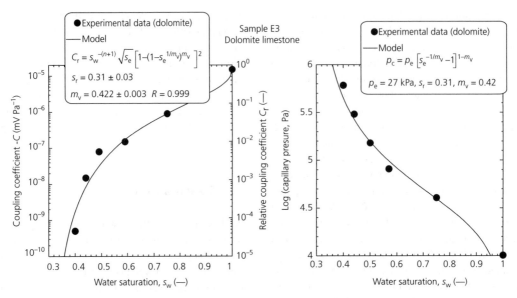

Figure 3.12 Comparison between experimental data and the prediction of the model developed by Revil et al. (2007) with the van Genuchten model (see also Linde et al., 2007). The experimental data are from Revil and Cerepi (2004) and Revil et al. (2007) (Sample E3). Left panel: absolute and relative streaming potential coupling coefficient versus saturation. We used the measured value of the saturation exponent (second Archie's exponent) $n = 2.7$. Right panel: fit of the capillary pressure curve with the same van Genuchten parameters that were obtained for the coupling coefficient (using a nonwetting fluid: mercury with water as the wetting fluid).

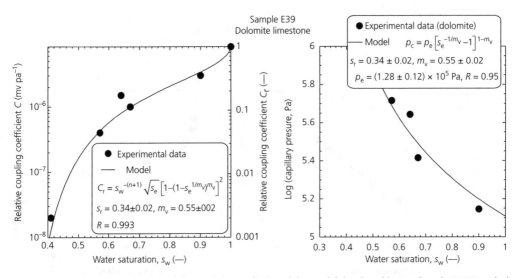

Figure 3.13 Comparison between experimental data and the prediction of the model developed by Revil et al. (2007) with the van Genuchten model (see also Linde et al., 2007). The experimental data are from Revil and Cerepi (2004) and Revil et al. (2007) (Sample E39). Left panel: absolute and relative streaming potential coupling coefficient versus saturation. We used the measured value of the saturation exponent (second Archie's exponent) $n = 3.5$. Right panel: fit of the capillary pressure curve with the same van Genuchten parameters that were obtained for the coupling coefficient (nonwetting fluid, mercury; wetting fluid, water).

the available experimental data. The relative permeability, the relative coupling coefficient, and the capillary pressure curve can be placed into a unified framework using the van Genuchten and Archie's law formulations. With this model and its experimental check, we can now simulate the propagation of seismic waves in unsaturated and two-phase fluid flow conditions and are now able to predict the characteristics of the seismoelectric conversions.

In Chapter 7, we will apply these equations to a vadose zone problem. In Chapter 4, the present models will be discussed in terms of their implementation in a finite-element package, and both forward and inverse modeling examples will be discussed using stochastic and deterministic approaches.

References

Archie, G.E. (1942) The electrical resistivity log as an aid in determining some reservoir characteristics. *Transactions of the American Institute of Mining, Metallurgical and Petroleum Engineers*, **146**, 54–62.

Bear, J., & Verruijt, A. (1987) *Modeling groundwater flow and pollution*. In: Martinus Nijhoff Publishers, Dordrecht, The Netherlands.

Biot, M.A. (1956a) Theory of propagation of elastic waves in a fluid-saturated porous solid, part I: low frequency range. *The Journal of the Acoustical Society of America*, **28**, 168–178.

Biot, M.A. (1956b) Theory of propagation of elastic waves in a fluid-saturated porous solid, part II: higher frequency range. *The Journal of the Acoustical Society of America*, **28**, 179–191.

Biot, M.A. (1962a) Generalized theory of acoustic propagation in porous dissipative media. *Journal of the Acoustical Society of America*, **34**(9), 1254–1264.

Biot, M.A. (1962b) Mechanics of deformation and acoustic propagation in porous media. *Journal of Applied Physics*, **33**(4), 1482–1498.

Brooks, R.H. & Corey, A.T. (1964) Hydraulic properties of porous media. *Hydrology Papers*, No. 3, Colorado State University, Ft. Collins, Colorado.

Detournay, E. & Cheng, A.H.-D. (1993) Fundamentals of poroelasticity. Chapter 5 in *Comprehensive Rock Engineering: Principles, Practice and Projects*, Vol. II, Analysis and Design Method, ed. C. Fairhurst, Pergamon Press, pp. 113–171.

Garambois, S. & Dietrich, M. (2001) Seismo-electric wave conversions in porous media: Field measurements and transfer function analysis. *Geophysics* **2001**(66), 1417–1430.

Guichet, X., Jouniaux, L. & Pozzi, J. P. (2003) Streaming potential of a sand column in partial saturation conditions. *Journal of Geophysical Research*, **108**, doi:10.1029/2001JB001517.

Huet C.C., Rushing, J.A., Newsham, K.E. & Blasingame, T.A. (2005) Modified Purcell/Burdine model for estimating absolute permeability from mercury-injection capillary pressure data. *International Petroleum Technology Conference*, Doha, Qatar, 21–23 2005, Paper IPTC 10994, 12 pp.

Jardani, A., Revil, A., Slob, E. & Sollner, W. (2010) Stochastic joint inversion of 2D seismic and seismoelectric signals in linear poroelastic materials. *Geophysics*, **75**(1), N19–N31, doi: 10.1190/1.3279833.

Johnson, D.L. (1986) Recent developments in the acoustic properties of porous media. In: Sette D, Editor. *Proceedings of the international school of physics « Enrico Fermi » course XCIII Frontiers in physical acoustics*. Amsterdam: North-Holland. 255–290.

Jougnot, D., Linde, N., Revil, A. & Doussan, C. (2012) Derivation of soil-specific streaming potential electrical parameters from hydrodynamic characteristics of partially saturated soils. *Vadose Zone Journal*, **11**(1), 1–15, doi:10.2136/vzj2011.0086.

Jougnot, D. & Linde, N. (2013) Self-Potentials in Partially Saturated Media: The Importance of Explicit Modeling of Electrode Effects. *Vadose Zone Journal*, **12**(2), 1–15, doi: 10.2136/vzj2012.0169.

Jun-Zhi, W. & Lile, O. B. (1990) Hysteresis of the resistivity index in Berea sandstone, Proceedings of the First European Core Analysis Symposium, London, UK, May 21–23, 1990, pp. 427–443.

Katz, A.J. & Thompson, A.H. (1987) Prediction of Rock Electrical Conductivity from Mercury Injection Measurements. *Journal of Geophysical Research*, **92**(**B**1), 599–607.

Leroy, P. & Revil A. (2004) A triple layer model of the surface electrochemical properties of clay minerals. *Journal of Colloid and Interface Science*, **270**(2), 371–380.

Li, K. & R.N. Horne, (2005) Inferring relative permeability from resistivity well logging, Proceedings, Thirtieth Workshop on Geothermal Reservoir Engineering Stanford University, Stanford, California, January 31–February 2, 2005 SGP-TR-176, 6 pp.

Linde, N., Jougnot, D., Revil, A., et al. (2007) Streaming current generation in two-phase flow conditions. *Geophysical Research Letters*, **34**(3), L03306, doi: 10.1029/2006GL028878.

Lo, W.-C., Sposito, G. & Mayer, E. (2002) Immiscible two-phase fluid flows in deformable porous media. *Advances in Water Resources*, **25**, 1105–1117.

Lo, W.-C., Sposito, G. & Majer, E. (2005) Wave propagation through elastic porous media containing two immiscible fluids. *Water Resources Research*, **41**, W02025, doi:10.1029/2004WR003162.

Mboh, C.M., Huisman, J.A., Zimmermann, E. & Vereecken, H. (2012) Coupled hydrogeophysical inversion of streaming potential signals for unsaturated soil hydraulic properties. *Vadose Zone Journal*, **11**(2), doi:10.2136/vzj2011.0115.

Morency, C. & Tromp, J. (2008) Spectral-element simulations of wave propagation in porous media. *Geophysical Journal International*, **175**, 301–345.

Petiau, G. & Dupis, A. (1980) Noise, temperature coefficient and long time stability of electrodes for telluric observations, *Geophysical Prospecting*, **28**(5), 792–804.

Petiau, G. (2000). Second generation of lead-lead chloride electrodes for geophysical applications. *Pure and Applied Geophysics*, **157**, 357–382.

Pride, S. (1994) Governing equations for the coupled electro-magnetics and acoustics of porous media. *Physical Review B*, **50**, 15678–15696.

Pride, S.R. & Haartsen, M.W. (1996) Electroseismic wave properties. *The Journal of the Acoustical Society of America*, **100**, 1301–1315.

Purcell, W.R. (1949) Capillary pressures-Their measurement using mercury and the calculation of permeability. *Transactions of the American Institute of Mining, Metallurgical and Petroleum Engineers*, **186**, 39–48.

Revil, A. (2012) Spectral induced polarization of shaly sands: Influence of the electrical double layer. *Water Resources Research*, **48**, W02517, doi:10.1029/2011WR011260.

Revil, A. (2013a) Effective conductivity and permittivity of unsaturated porous materials in the frequency range 1 mHz–1GHz. *Water Resources Research*, **49**, 306–327, doi:10.1029/2012WR012700.

Revil, A. (2013b) On charge accumulations in heterogeneous porous materials under the influence of an electrical field. *Geophysics*, **78**(4), D271–D291, doi: 10.1190/GEO2012-0503.1.

Revil, A. & Cerepi, A. (2004) Streaming potentials in two-phase flow conditions. *Geophysical Research Letters*, **31**, L11605, doi:10.1029/2004GL020140.

Revil, A. & Jardani, A. (2010) Seismoelectric response of heavy oil reservoirs. Theory and numerical modelling. *Geophysical Journal International*, **180**, 781–797, doi: 10.1111/j.1365-246X.2009.04439.x.

Revil, A., & Linde, N. (2006) Chemico-electromechanical coupling in microporous media. *Journal of Colloid and Interface Science*, **302**, 682–694.

Revil, A. & Linde, N. (2011) Comment on 'Streaming potential dependence on water-content in Fontainebleau sand' by V. Allegre, L. Jouniaux, F. Lehmann, and P. Sailhac. *Geophysical Journal International*, **186**, 113–114, doi: 10.1111/j.1365-246X.2010.04850.x.

Revil, A. & Mahardika, H. (2013) Coupled hydromechanical and electromagnetic disturbances in unsaturated clayey materials. *Water Resources Research*, **49**, doi:10.1002/wrcr.20092.

Revil, A., Schwaeger, H., Cathles, L.M. & Manhardt, P. (1999a) Streaming potential in porous media. 2. Theory and application to geothermal systems. *Journal of Geophysical Research*, **104**(B9), 20,033–20,048.

Revil, A., Pezard, P.A. & Glover P.W.J. (1999b) Streaming potential in porous media. 1. Theory of the zeta-potential. *Journal of Geophysical Research*, **104**(B9), 20,021–20,031.

Revil, A., Linde, N., Cerepi, A., Jougnot, D., Matthäi, S. & Finsterle, S. (2007), Electrokinetic coupling in unsaturated porous media, *Journal of Colloid and Interface Science*, **313**(1), 315–327, 10.1016/j.jcis.2007.03.037.

Revil, A., Woodruff, W.F. & Lu, N. (2011) Constitutive equations for coupled flows in clay materials. *Water Resources Research*, **47**, W05548, doi:10.1029/2010WR010002.

Richards, L.A. (1931) Capillary conduction of liquids through porous media. *Physics*, **1**, 318–333.

Rubino, J.G. & Holliger, K. (2012) Seismic attenuation and velocity dispersion in heterogeneous partially saturated porous rocks. *Geophysical Journal International* **188**(3), 1088–1102.

Santos, J.E., Douglas, J., Corbero, J. & Lovera, O.M. (1990) A model for wave propagation in a porous medium saturated by a two-phase fluid, *Journal of the Acoustical Society of America*, **87**(4), 1439–1448.

Santos, J.E., Ravazzoli, C.L. & Geiser, J. (2006) On the static and dynamic behavior of fluid saturated composite porous solids: A homogenization approach. *International Journal of Solids and Structures*, **43**(5), 1224–1238.

Strahser, M. (2006) Near Surface Seismoelectrics in Comparative Field Studies. PhD thesis, Christian-Albrechts-Universität, Kiel.

Teja, A.S. & Rice, P. (1981) Generalized corresponding states method for viscosities of liquid mixtures. *Industrial & Engineering Chemistry Fundamentals*, **20**, 77–81.

Thomas, L.K., Katz, D.L. & Tek, M.R. (1968) Threshold pressure phenomena in porous media. *Society of Petroleum Engineers Journal*, **243**, 174–184.

Tuncay, K. & Corapcioglu, M.Y. (1997) Wave propagation in poroelastic media saturated by two fluids. *Journal of Applied Mechanics*, **64**, 313–20.

Van Genuchten, M. Th. (1980) A closed-form equation for predicting the hydraulic conductivity of unsaturated soils. *Soil Science Society of America Journal*, **44**, 892–898.

Vasco, D.W. & Minkoff S.E. (2012) On the propagation of a disturbance in a heterogeneous, deformable, porous medium saturated with two fluid phases. *Geophysics*, **77**(3), L25–L44.

Vinegar, H.J. & Waxman, M.H. (1984) Induced polarization of shaly sands. *Geophysics*, **49**, 1267–1287.

Vinogradov, J. & Jackson M. D. (2011) Multiphase streaming potential in sandstones saturated with gas/brine and oil/brine during drainage and imbibition. *Geophysical Research Letters*, **38**, L01301, doi:10.1029/2010GL045726.

Warden, S., Garambois, S., Jouniaux, L., Brito, D., Sailhac, P. & Bordes, C. (2013) Seismoelectric wave propagation numerical modelling in partially saturated materials. *Geophysical Journal International*, **194**(3): 1498–1513, doi:10.1093/gji/ggt198

Waxman, M.H. & Smits, L.J.M. (1968) Electrical conductivities in oil bearing shaly sands. *Soc. Pet. Eng. J.*, **243**, 107–122.

Wood, A.W. (1955) *A textbook of sound*, MacMillan Publishing Company.

CHAPTER 4

Forward and inverse modeling

In Chapters 2 to 3, we developed all the field equations required to model the seismoelectric effect in saturated and unsaturated conditions (including two-phase flow conditions). We wish to use these field equations to numerically forward-model the occurrence of seismo-electric signals for various applications in earth sciences. Therefore, we need to discuss how to numerically implement the forward modeling of these equations with the finite-element method. Then, as geophysicists, we want to solve the inverse problem in order to determine the information content of these seismoelectric signals (e.g., White, 2005; White & Zhou, 2006). To accomplish the solution of the inverse problem, we present in this chapter both stochastic and deterministic algorithms to invert seismoelectric signals in terms of material properties or in terms of localizing boundaries between geological formations.

4.1 Finite-element implementation

4.1.1 Finite-element modeling

In water-saturated conditions, the poroelastic model described in Chapter 2 (dynamic Biot–Frenkel theory) can be expressed in the frequency domain under the following equations for the displacement of the solid phase, $\mathbf{u}$, and the pore fluid pressure, p (see Eqs. 2.201 and 2.203):

$$\omega^2 \rho_\omega^s \mathbf{u} - \theta_\omega \nabla p + \nabla \cdot \hat{\mathbf{T}}(\mathbf{u}) = 0 \qquad (4.1)$$

$$\frac{1}{M} p + \nabla \cdot (k_\omega \nabla p) + \theta_\omega \nabla \cdot \mathbf{u} = 0 \qquad (4.2)$$

where the bulk modulus, M, is given by Equation (2.189), k_ω by Equation (2.193), ρ_ω^s by Equation (2.195), and the coupling term θ_ω by Equation (2.202) and $\hat{\mathbf{T}}$ denotes the effective stress tensor (see Eqs. 2.199 and 2.200). The so-called weak formulation can be obtained by multiplying the governing Equations (4.1) and (4.2) by the admissible variations of displacement of the solid phase and pore water pressure fields, δu and δp. These equations are then integrated on the volume, Ω, of the poroelastic material

$$\int_\Omega \left(\omega^2 \rho_\omega^s \mathbf{u} - \theta_\omega \nabla p + \nabla \cdot \hat{\mathbf{T}}(\mathbf{u}) \right) \delta u d\Omega = 0 \qquad (4.3)$$

$$\int_\Omega \left(\frac{1}{M} p + \nabla \cdot (k_\omega \nabla p) + \theta_\omega \nabla \cdot \mathbf{u} \right) \partial p d\Omega = 0 \qquad (4.4)$$

The integral of the each of these equations is then rewritten using Green's formula (Attalla et al., 1998):

$$\int_\Omega \hat{\mathbf{T}}(u) \varepsilon^s(\delta u) d\Omega - \omega^2 \int_\Omega \rho_\omega^s \mathbf{u} \delta u d\Omega$$
$$+ \int_\Omega \theta_\omega \nabla p \delta u d\Omega - \int_\Gamma \hat{\mathbf{T}}(u) \cdot \mathbf{n} \delta u d\Gamma = 0 \qquad (4.5)$$

The Seismoelectric Method: Theory and Applications, First Edition. André Revil, Abderrahim Jardani, Paul Sava and Allan Haas.
© 2015 John Wiley & Sons, Ltd. Published 2015 by John Wiley & Sons, Ltd.

$$\int_{\Omega} \frac{1}{M} p \delta p d\Omega - \int_{\Omega} k_{\omega} \nabla p \nabla \delta p d\Omega + \int_{\Omega} \theta_{\omega} \nabla p u d\Omega$$

$$+ \int_{\Gamma} [k_{\omega} \nabla p \cdot \mathbf{n} \delta P + \theta_{\omega} u_n] d\Gamma = 0 \qquad (4.6)$$

where u_n denotes the normal component of $\mathbf{u}$, ε^s is the solid strain tensor, and $\mathbf{n}$ represents the unit outward normal vector to the surface Γ. In the next step, we discretize these integrals and each of their elements can be represented in the following matrix form:

$$u^e = [N_s]\{u_n\}^e, \quad p^e = [N_f]\{p_n\}^e, \qquad (4.7)$$

where $\{u_n\}^e$ and $\{p_n\}^e$ denote the solid displacement and fluid pressure at the finite-element grid nodes, respectively, and $[N_s]$ and $[N_f]$ correspond to the shape functions of the solid phase displacement and the fluid pressure within each element, noted by the superscript e.

In the next step, we rebuild each integral under the following discretized forms:

$$\int_{\Omega} \rho_{\omega}^s \mathbf{u} \delta u d\Omega = [\delta u]^T \mathbf{M} u, \qquad (4.8)$$

$$\int_{\Omega} \hat{\mathbf{T}}(u) \varepsilon^s(u) d\Omega = [\delta u]^T \mathbf{K} u, \qquad (4.9)$$

$$\int_{\Omega} \theta_{\omega} \nabla p \delta u d\Omega = [\delta u]^T \mathbf{C} p, \qquad (4.10)$$

$$\int_{\Omega} k_w \nabla p \nabla \delta p d\Omega = [\delta p]^T \mathbf{H} p, \qquad (4.11)$$

$$\int_{\Omega} \frac{1}{M} p \delta p d\Omega = [\delta p]^T \mathbf{Q} p, \qquad (4.12)$$

$$\int_{\Omega} \theta_{\omega} \nabla p u d\Omega = [\delta p]^T \mathbf{C} u, \qquad (4.13)$$

where $\mathbf{M}$ denotes a mass matrix and $\mathbf{K}$ a stiffness matrix. The matrices $\mathbf{H}$ and $\mathbf{Q}$ are kinetic and compression energy matrices for the fluid phase, and $\mathbf{C}$ is the volumetric hydromechanical coupling matrix. This leads to the following linear system of equations:

$$\begin{pmatrix} -\omega^2 \mathbf{M} + \mathbf{K} & -\mathbf{C} \\ -\mathbf{C}^T & (\mathbf{H} + \mathbf{Q}) \end{pmatrix} \begin{pmatrix} \mathbf{u} \\ p \end{pmatrix} = \begin{pmatrix} \mathbf{F}^s \\ \mathbf{F}^f \end{pmatrix}, \qquad (4.14)$$

where $\mathbf{F}^s$ is the skeleton's loading vector, while $\mathbf{F}^f$ is the kinematic coupling vector for the interstitial fluid.

With the poromechanical equations defined, we can now define the electrical equations. Computing the poroelastic problem is a necessary step prior to determining the volumetric electrical source current density $\Im = \nabla \cdot \mathbf{J}_S$ (expressed in $A\,m^{-3}$). The volumetric electrical source current density will be used as a source term when solving the following Poisson equation for the electrical potential ψ (see Eq. (2.215)):

$$\nabla \cdot (\sigma \nabla \psi) = \Im, \qquad (4.15)$$

$$\mathbf{J}_S = -i\omega \, \hat{Q}_V^0 \tilde{k}_w (\nabla p - \omega^2 \rho_f \mathbf{u}), \qquad (4.16)$$

where σ denotes the electrical conductivity of the material. We reformulate the electric part of the problem using variational discretized equations. This yields (e.g., Soueid et al., 2013)

$$\int_{\Omega} \sigma \nabla \psi_i^e \nabla \varphi_j = \int_{\Omega} \Im \varphi_j + \underbrace{\int_{\Gamma} \sigma \nabla \psi_i^e \mathbf{n} \cdot \varphi_j}_{= Jn}, \quad \text{with} \quad \psi^e = \sum_{j=1}^m \psi_i \varphi_i \qquad (4.17)$$

$$\sum_{j=1}^m \psi_i \int_{\Omega} \sigma \nabla \varphi_i \nabla \varphi_j = \int_{\Omega} \Im \varphi_j + \int_{\Gamma} J_n \varphi_j \qquad (4.18)$$

where ψ^e represents the discrete formulation of the electrical potential ψ and φ_i is an interpolation function. In terms of matrices, we write

$$\mathbf{K}^e \psi = \overline{\mathbf{Q}}^e, \qquad (4.19)$$

$$K_{ij}^e = \int_{\Omega} \sigma \nabla \varphi_i \nabla \varphi_j, \qquad (4.20)$$

$$\overline{\mathbf{Q}}^e = \int_{\Omega} \Im \varphi_j + \int_{\Gamma} J_n \varphi_j. \qquad (4.21)$$

Equation (4.19) can be solved using a Gaussian elimination approach with partial pivoting to determine the distribution of the electrical field (Kaw and Kalu, 2008).

4.1.2 Perfectly matched layer boundary conditions

Equations (4.1) and (4.2) describe the propagation of the seismic waves in a poroelastic framework within an infinite (unbounded) medium. However, when one performs numerical simulations, the domain investigated is always bounded. A finite domain will always produce reflections at the domain boundaries unless the boundary conditions are adjusted to prevent these reflections. A common approach to limit reflection at the boundaries of the domain is to use the one-way wave equation, based on the paraxial approximations of the seismic wave equations (Clayton & Engquist, 1977). The perfectly matched layer (PML) method is described later as an alternative way by Berenger (1994), first for electromagnetic problems and then for seismic problems (Chew & Weedon, 1994; Zeng & Liu, 2001; Zeng et al., 2001). With PML boundary layers, (almost) no reflection is expected to occur at the interface between the physical domain and the absorbing layer for any frequency and any angle of incidence of the seismic waves.

In this book, we use the convolutional perfectly matched layer (C-PML) approach for our numerical examples. The C-PML method, for first-order system of partial differential equations, has been developed for electromagnetic waves by Roden and Gedney (2000) and in simulation of elastic wave propagation by Bou Matar et al. (2005). This method can be extended for second-order systems written in terms of displacements. The main advantages of the C-PML approach over the classical PML approach concern its numerical stability and its high efficiency (Martin et al., 2008). Using the concept of complex coordinates (Chew & Weedon, 1994) in the frequency domain (with a time dependence of $e^{-i\omega t}$), the complex coordinate stretching variables are

$$\tilde{x}_i = \int_0^{x_i} s_{x_i}(x')dx', \quad i = 1, 2, \tag{4.22}$$

$$s_{x_i} = k_{x_i}(x_i) + \frac{\sigma_{x_i}(x_i)}{\alpha_{x_i} + j\omega}, \tag{4.23}$$

where α_{x_i}, σ_{x_i} are positive real damping coefficients and k_{x_i} are real and positive-definite numbers that are equal or larger than unity. In this paper, we consider $k_{x_i} = 1$ to keep the waves continuous (see Collino & Tsogka, 2001, for details). To determine the value of the two other damping coefficients, we use the following formula:

$$\sigma_i = \begin{cases} \dfrac{3c}{2L_0}\log\left(\dfrac{1}{R}\right)\left(\dfrac{x_{i\min} - x_i}{L_0}\right)^3, & \text{as } x_{i\min} \geq x_i, \\ 0, & \text{as } x_{i\min} \leq x_i \leq x_{i\max}, \\ \dfrac{3c}{2L_0}\log\left(\dfrac{1}{R}\right)\left(\dfrac{x_i - x_{i\max}}{L_0}\right)^3, & \text{as } x_i \geq x_{i\max}, \end{cases} \tag{4.24}$$

$$\alpha_i = \begin{cases} \pi f_0\left(\dfrac{x_{i\min} - x_i}{L_0} + 1\right), & \text{as } x_{i\min} \geq x_i, \\ \pi f_0, & \text{as } x_{i\min} \leq x_i \leq x_{i\max}, \\ \pi f_0\left(\dfrac{x_i - x_{i\max}}{L_0} + 1\right), & \text{as } x_i \geq x_{i\max}, \end{cases} \tag{4.25}$$

where c is the highest of all of the velocities in the domain, $R = 1/1000$ represents the amount of reflected energy at the outer boundary of the PML, L_0 is the thickness of the PML, and f_0 is the dominant frequency of the source (see Section 4.1.3).

In the succeeding text, the derivative $\partial(\cdot)/\partial\tilde{x}_i$ can be expressed in terms of the regular coordinate stretching variables, $\partial(\cdot)/\partial\tilde{x}_i = (1/s_{x_i})(\partial(\cdot)/\partial x_i)$. Finally, after replacing $\partial(\cdot)/\partial x_i$ by $\partial(\cdot)/\partial\tilde{x}_i$ and after some algebraic manipulations, the reduced set of equations for the modified poroelastic formulation is

$$-\omega^2 \tilde{\rho}_w^s \mathbf{u} + \tilde{\boldsymbol{\theta}}_w \nabla p = \nabla \cdot \tilde{\mathbf{T}}, \tag{4.26}$$

$$\tilde{\boldsymbol{\theta}}_w = \begin{pmatrix} \theta_w s_{x_2} & 0 \\ 0 & \theta_w s_{x_1} \end{pmatrix}, \tag{4.27}$$

$$\tilde{\rho}_w^s = \rho_w^s s_{x_1} s_{x_2}, \tag{4.28}$$

$$\tilde{\mathbf{T}} = \begin{pmatrix} \tilde{T}_{11} & \tilde{T}_{12} \\ \tilde{T}_{21} & \tilde{T}_{22} \end{pmatrix}, \tag{4.29}$$

$$\tilde{T}_{11} = (\lambda + 2G)\frac{s_{x_2}}{s_{x_1}}\frac{\partial u_1}{\partial x_1} + \lambda\frac{\partial u_2}{\partial x_2}, \tag{4.30}$$

$$\tilde{T}_{22} = \lambda\frac{\partial u_1}{\partial x_1} + (\lambda + 2G)\frac{s_{x_1}}{s_{x_2}}\frac{\partial u_2}{\partial x_2}, \tag{4.31}$$

$$\tilde{T}_{12} = G\left(\frac{s_{x_1}}{s_{x_2}}\frac{\partial u_1}{\partial x_2} + \frac{\partial u_2}{\partial x_1}\right), \tag{4.32}$$

$$\tilde{T}_{21} = G\left(\frac{\partial u_1}{\partial x_2} + \frac{s_{x_2}}{s_{x_1}}\frac{\partial u_2}{\partial x_1}\right), \tag{4.33}$$

$$\frac{s_{x_1} s_{x_2}}{M}p + \nabla \cdot \left[\tilde{\mathbf{k}}_{1\omega}\nabla p - \tilde{\mathbf{k}}_{2\omega}\omega^2 \rho_f \mathbf{u}\right] = \alpha s_{x_1} s_{x_2}\nabla \cdot \mathbf{u}, \tag{4.34}$$

$$\tilde{\mathbf{k}}_{1\omega} = \begin{pmatrix} k_\omega s_{x_2}/s_{x_1} & 0 \\ 0 & k_\omega s_{x_1}/s_{x_2} \end{pmatrix}, \tag{4.35}$$

$$\tilde{\mathbf{k}}_{2\omega} = \begin{pmatrix} k_\omega s_{x_2} & 0 \\ 0 & k_\omega s_{x_1} \end{pmatrix}. \tag{4.36}$$

For the electrical problem associated with the electrokinetic conversions, we solve the following modified equations:

$$\nabla \cdot (\tilde{\sigma} \nabla \psi) = \nabla \cdot \tilde{\mathbf{J}}_S, \tag{4.37}$$

$$\tilde{\mathbf{J}}_S = (j_{x_1} s_{x_2}; j_{x_1} s_{x_1}), \tag{4.38}$$

$$\tilde{\sigma} = \begin{pmatrix} \sigma \dfrac{s_{x_2}}{s_{x_1}} & 0 \\ 0 & \sigma \dfrac{s_{x_1}}{s_{x_2}} \end{pmatrix}. \tag{4.39}$$

This novel formulation of the electrical problem includes the perfect matching layers. For the boundary values, we applied the Neumann boundary condition at the insulating air–ground interface and $\psi = 0$ (Dirichlet boundary condition) at the other boundaries.

4.1.3 Boundary conditions at an interface

If we consider an interface between two poroelastic media 1 and 2, the boundary conditions at this interface are given by (Pride & Haartsen, 1996)

$$\mathbf{u}_1 = \mathbf{u}_2, \tag{4.40}$$

$$p_1 = p_2, \tag{4.41}$$

$$\hat{\mathbf{n}} \cdot (\mathbf{w}_1 - \mathbf{w}_2) = 0, \tag{4.42}$$

$$\hat{\mathbf{n}} \cdot (\mathbf{T}_1 - \mathbf{T}_2) = 0, \tag{4.43}$$

$$\hat{\mathbf{n}} \times (\mathbf{E}_1 - \mathbf{E}_2) = 0, \tag{4.44}$$

where $\hat{\mathbf{n}}$ denotes the unit vector normal to the interface between the media 1 and 2. These boundary conditions express the continuity in the solid displacement, the pore fluid pressure, the fluid displacement, the momentum flux, and the tangential components of the electrical field across the interface.

4.1.4 Description of the seismic source

In the following example, we use a source generating P-waves only. This force creates a net force on the solid phase of the porous rock. Because the source generates a displacement of pore water relative to the grain framework, it creates a source current density and therefore an electromagnetic disturbance. This disturbance diffuses nearly instantaneously to all receivers with an amplitude that can be pretty strong in the vicinity of the seismic source. Because this contribution can be easily removed from the electrograms (using its temporal and spatial characteristics), we will not model it in the succeeding text. We will focus on this contribution in Chapter 5.

Using the Fourier transform of the first time derivative of the Gaussian function for the source yields the following expression for the bulk force acting on the solid phase:

$$|\mathbf{F}(x, y, \omega)| = F(\omega) \nabla [\delta(x - x_0) \delta(y - y_0)], \tag{4.45}$$

$$F(\omega) = \mathrm{FT}\left[(t - t_0) \exp\left\{-[\pi f_0 (t - t_0)]^2\right\}\right], \tag{4.46}$$

where $\mathrm{FT}[f(t)]$ is the Fourier transform of the function $f(t)$, t_0 is the time delay of the source, and f_0 is its dominant frequency. In this chapter, we will neglect the pore fluid pressure source term, S, which is equivalent to neglecting the electromagnetic effects of the seismic source itself. A complete analysis of the electromagnetic effects generated by elementary sources can be found in Pride and Haartsen (1996) and will be developed in Chapter 5.

4.1.5 Lateral resolution of cross-hole seismoelectric data

In the seismic wave domain, the first Fresnel zone is defined as the area of the reflector that contributes energy constructively to the total reflection energy reaching an observation point P. The same definition can be used for the transmission part of problem as well. In our case, the seismoelectric Fresnel zones may be defined similarly as the area of an interface that contributes constructively to the total transmitted energy reaching an observation point P. If we consider a monochromatic seismic source, S, located above a horizontal interface between media of different electrical properties, the spherically spreading seismic wave intersects the interface and causes fluid flow across the interface. The resulting electrical field is due to the streaming current imbalance at the interface. This is equivalent to having electrical dipoles oscillating in phase with the seismic wave along the interface. As a consequence, the electromagnetic disturbances are radiated away from the dipole sources and are recorded remotely at the observation point P. Assuming that the shot point S and the observation point P are collocated, the first Fresnel zones correspond to two circles of radii r_S and r_{SE}, respectively. Fourie (2003) provided a complete analysis of the reflection problem and found that the seismic and seismoelectric first Fresnel zone radii are given by

$$r_S = \left[\left(d + \frac{\lambda_S}{4}\right)^2 - d^2\right]^{1/2} \approx \sqrt{\frac{d \lambda_S}{2}}, \tag{4.47}$$

$$r_{SE} = \left[\left(d + \frac{\lambda_S}{2}\right)^2 - d^2\right]^{1/2} \approx \sqrt{d \lambda_S}, \tag{4.48}$$

respectively, where d denotes the distance between S and the interface and λ_S corresponds to the wavelength of the seismic wave. The approximation in Equations (4.47) and (4.48) are obtained by assuming that $d \gg \lambda_S$, yielding $r_{SE} \approx \sqrt{2} r_S$. In Equation (4.48), it is further assumed that the diffusion of the electromagnetic disturbances are much faster than the propagation of the seismic energy, a very good approximation as discussed in Chapters 2 and 3. Therefore, the lateral resolution of surface seismoelectric data is poorer when compared with the lateral resolution of surface seismic data. Fourie (2003) also showed that for a horizontal interface, the first seismoelectric Fresnel zone is nearly circular and centered beneath the shot point S. For the seismic case, the first Fresnel zone is an ellipse centered halfway between the shot point and the observation point P.

4.1.6 Benchmark test of the code

We check the reliability of the finite-element formulation we use in this book through the use of a 2D benchmark test. In this test, we simulate the fast and slow P-waves associated with an explosive source in a homogeneous porous material filled with a Newtonian fluid (water). The dimensions of the 2D domain are 800 m × 800 m, where the reference of the Cartesian coordinate system, O(0, 0), is at the bottom left corner. To put seismic energy into the system, we use a time-dependent source function, $F(t)$, that is a Ricker wavelet with a dominant frequency of 10 Hz, located at the source point S(x, y) = (400 m, 400 m). The position of the receiver (observation point) is P(x, y) = (200 m, 300 m). The four edges are absorbing boundaries for which we use C-PML boundary conditions described in Section 4.1.2. The material properties used for this

Table 4.1 Material properties used for the numerical benchmark.

Parameter	Value	Units
ρ_s	2650	kg m^{-3}
ρ_f	1000	kg m^{-3}
K_s	35	GPa
K_{fr}	5	GPa
G	11	GPa
K_f	2.25	GPa
η_f	1×10^{-3}	Pa s
k_0	1×10^{-11}	m^2
ϕ	0.30	—
σ	1×10^{-2}	S m^{-1}

simulation are reported in Table 4.1. Figure 4.1 shows the two components (horizontal and vertical) of the displacement of the solid skeleton and the two components of the relative displacement **w**. The results are shown at 15 Hz. To benchmark the numerical code, we compare the numerical solution with the analytical solution given by Dai et al. (1995). As shown in Figure 4.2, both solutions are in excellent agreement.

4.2 Synthetic case study

Next, we show an application of the previously presented model to the detection of a NAPL(oil)/water encroachment front during the remediation of a NAPL-contaminated aquifer by flooding the aquifer with water. Such a remediation process can also be enhanced with the use of surfactants (e.g., Mercier & Cohen, 1990; Pope & Wade, 1995; Londergan et al., 2001). We will use the subscript "o" to characterize the properties of oil.

4.2.1 Simulation of waterflooding of a NAPL-contaminated aquifer

We use the following two steps to simulate the remediation of a NAPL-contaminated aquifer.

Step 1. We use the approach developed in Karaoulis et al. (2012) to generate a 2D heterogeneous aquifer in terms of porosity and permeability (Figure 4.3). A random field for the clay content was generated with the SGeMS library (Stanford University). We used an isotropic semivariogram to compute the clay content distribution (Karaoulis et al., 2012). The porosity and permeability were then computed according to the petrophysical model defined by Revil and Cathles (1999). This heterogeneous aquifer is assumed to be initially saturated with 75% of light motor oil resulting from an oil spill. The initial water saturation of the aquifer is therefore $s_w = 0.25$, which corresponds to the irreducible water saturation, s_r. The aquifer is located between two wells, Wells A and B. Well B is located 250 m away from Well A. The reference position, O(−80,30), for the coordinate system is located at the upper left corner of this domain.

Step 2. Waterflooding of this aquifer is simulated in 2.5D by injecting water in Well A (constant injection rate) while removing the NAPL with Well B (constant pressure condition). The computations are done in two-phase flow conditions following the same equations as in Karaoulis et al. (2012). The properties of the NAPL and water are

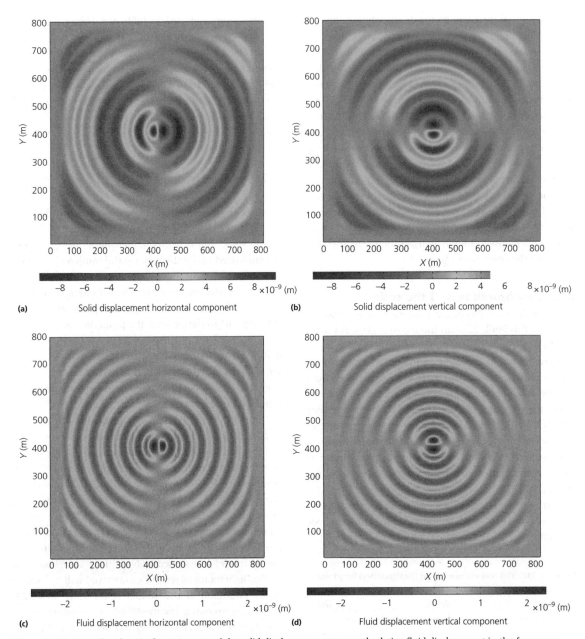

Figure 4.1 Horizontal and vertical components of the solid displacement vector and relative fluid displacement in the frequency domain (real components). **a)** Horizontal component of the solid displacement vector. **b)** Vertical component of the solid displacement vector. **c)** Horizontal component of the fluid displacement vector. **d)** Vertical component of the fluid displacement vector. Note the efficiency of the C-PML approach, at the boundaries, in attenuating the seismic waves. The use of the C-PML effectively eliminates reflections from the outer boundaries of the model. (*See insert for color representation of the figure.*)

reported in Table 4.2. We use a relatively low viscosity for the NAPL, and as is usually done, the injected water is heated to decrease the NAPL viscosity. In this analysis, we neglect all thermally induced effects other than the reduction of the viscosity of the NAPL. Also, the NAPL is assumed to be the nonwetting phase in this numerical experiment. This is realistic only a short time after an oil spill. Indeed, after a period of few years, the wettability

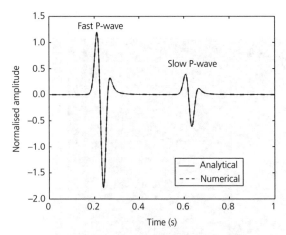

Figure 4.2 Comparison between analytical and numerical solutions of the seismic problem in a homogeneous medium (benchmark test) with properties summarized in Table 4.1. The figure shows a snapshot of the vertical component of the macroscopic solid displacement at t =0.58 s. The two P-waves can be observed. The synthetic numerical seismograms (solid black line) and the analytical solution calculated by Dai et al. (1995) are in close agreement as they cannot be distinguished from each other.

(surface tension) of the oil can change because bacteria produce biopolymers bridging the oil molecules to the surface of the grains. This effect is not accounted for here. We display six time step snapshots (T1–T6) of the oil and water saturations determined by the numerical simulation (Figure 4.4).

4.2.2 Simulation of the seismoelectric problem

For each of the snapshots shown in Figure 4.4, we simulated a seismoelectric acquisition between the two wells where the simulation of the seismoelectric problem is done in two steps as described in the succeeding text. Because this problem is formally different from the unsaturated case discussed earlier (two-phase flow problem vs. unsaturated problem), somehow we need to accommodate this issue as discussed in the following text.

Step 1. This step models the propagation of the seismic waves between the two wells. We use the material property values given in Table 4.2 to compute the seismic properties. The bulk modulus of the fluid is related to the NAPL saturation through the use of the Wood formula, as discussed in Section 4.3.1. From the Wood formula, we have

$$\frac{1}{K_f} = \frac{1 - s_w}{K_o} + \frac{s_w}{K_w}, \qquad (4.49)$$

where K_o and K_w denote the bulk moduli of NAPL and water, respectively. The shear modulus is independent of the saturation because neither of the fluids sustains shear stresses. The bulk modulus of the fluid is given by Equation (4.49), and the density and viscosity of the fluid are given by

$$\rho_f = (1 - s_w)\rho_o + s_w\rho_w, \qquad (4.50)$$

$$\eta_f = \eta_o \left(\frac{\eta_w}{\eta_o}\right)^{s_w}, \qquad (4.51)$$

where ρ_o and ρ_w denote the density of the NAPL and water, respectively, and η_o and η_w represent the dynamic viscosity of oil and water, respectively. The difference of fluid pressure between the two phases is controlled by the capillary pressure curve which is given by the same capillary pressure curve used to simulate the two-phase flow problem (e.g., Karaoulis et al., 2012).

The geometry of the model used for the computation of the seismic waves is shown in Figure 4.5. The seismic sources is an explosive-like source located at position So in Well A (Figure 4.5). The receivers comprise 28 pairs of seismic stations and electrodes (noted as E1–E28), which are located in Well B. The separation between these receivers is equal to 4 m.

First, we solve the poroelastodynamic wave equations in the frequency domain, taking into account the variable saturation of the water phase. We use the (**u**, p) formulation (see Jardani et al., 2010). The multiphysics modeling package COMSOL Multiphysics 4.2a and the stationary parametric solver PARDISO were used to solve the resulting partial differential equations (Schenk & Gärtner, 2004, 2006; Schenk et al., 2007, 2008). The problem is solved as follows: first (i), we compute for the poroelastic and electric properties distribution for the given porosity, fluid permeability, and saturation distribution of the NAPL and water phases, and then (ii) we solve for the displacement of the solid phase, **u**, and the pore fluid pressure, p, in the frequency domain. The solution in the time domain is computed by using an inverse Fourier transform of the solution in the frequency–wave number domain.

In the frequency domain, we use the frequency range 8–800 Hz since the appropriate seismic wave and associated electric field in this setting operate in this range. Then, using this frequency range, we compute

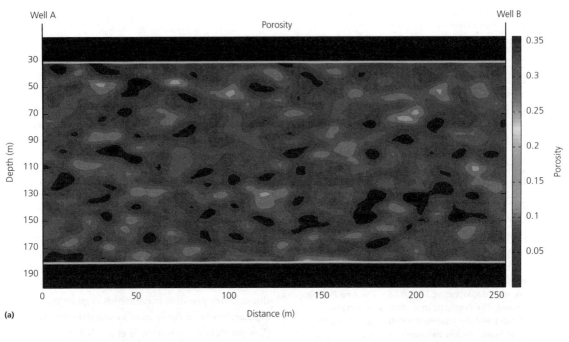

(a)

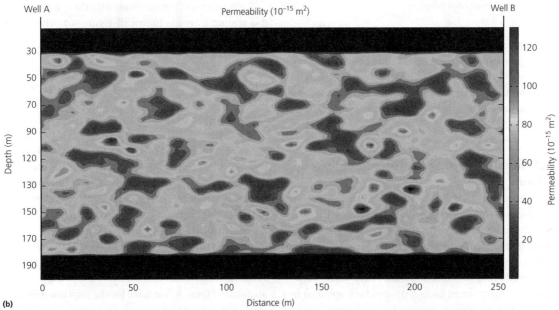

(b)

Figure 4.3 Porosity and permeability map of a NAPL contaminated aquifer between two wells for a water-flood simulation.
a) Porosity distribution. **b)** Permeability distribution. (*See insert for color representation of the figure.*)

the inverse fast Fourier transform (IFFT) (FFT^{-1}) to get the time series of the seismic displacements u_x and u_z, along with the time series of the electric potential response, ψ. We generated a rectangular mesh at least ten times smaller than the smallest wavelength of the

seismic waves. We have checked that this corresponds to the smallest mesh for which the solution of the partial differential equation is mesh independent. The seismic source is located at position So($x_s = 0$ m, $z_s = 116$ m) with a magnitude of 1.0×10^4 N m. For the time domain

Table 4.2 Material properties used in the saturation front detection. We use Model A described in Chapter 3 with $n = 2$.

Parameter	Value	Units
ρ_s	2650	kg m^{-3}
ρ_w	1000	kg m^{-3}
ρ_o	900	kg m^{-3}
K_s	36.5	GPa
K_{fr}	18.2	GPa
G	13.8	GPa
K_w	2.25	GPa
K_o	1.50	GPa
η_w	1×10^{-3}	Pa s
η_o	50×10^{-3}	Pa s

source function, we choose a Gaussian temporal function that starts with a 30 ms delay with a dominant frequency of 160 Hz. At the four external boundaries of the domain, we constructed a 20 m thick C-PML using the approach described previously. The sensors located at Well B are 30 m away from the right-side C-PML, and therefore, the solution is not influenced by the PML boundary condition. The receiver arrangement mimics the acquisition that would be obtained with triaxial geophones and dipole antennas. The C-PML boundary condition consists of a strip with a finite distance (or width, or thickness, or volume) that simulates the propagation of seismic waves out to an infinite distance without any reflections going back inside the domain of interest. PMLs are absorbing

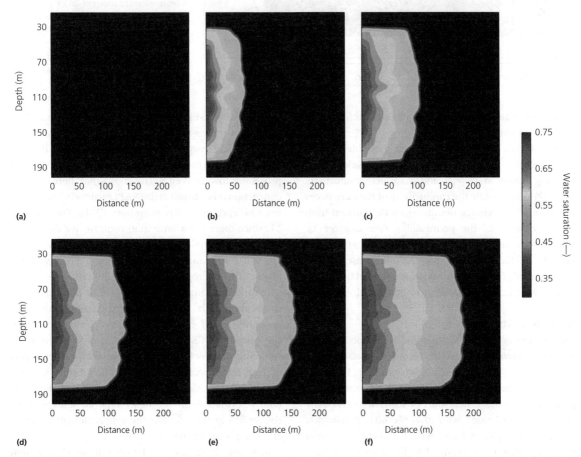

Figure 4.4 Six snapshots showing the evolution of the water saturation s_w over time in a 150-m-thick NAPL contaminated aquifer. The initial water saturation in the aquifer is equal to the irreducible water saturation $s_r = 0.25$ (which correspond to a NAPL saturation of 0.75). In this study the NAPL is considered to be the non-wetting phase. **a)** Reference snapshot T1. **b)** Snapshot T2 at 200 days. **c)** Snapshot T3 at 400 days. **d)** Snapshot T4 at 600 days. **e)** Snapshot T5 at 800 days. **f)** Snapshot T6 at 1000 days. (*See insert for color representation of the figure.*)

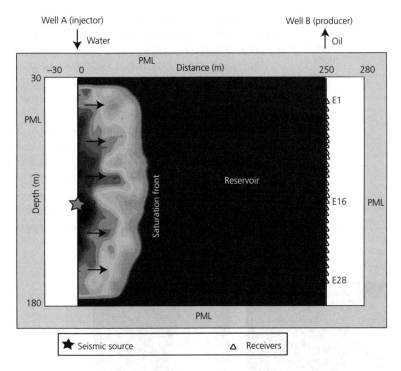

Figure 4.5 Sketch of the domain used for the modeling. The total modeling domain is a 410 m × 250 m rectangle. Injector Well A is used, located at position x =0 m, and is also used for the seismic source. Production and recording Well B is located at x =250 m. The discretization of the domain comprises a finite-element mesh of 205 × 125 rectangular cells. 28 receivers are located in Well B, approximately 30 m away from the nearest PML boundary (the PML boundary layers are shown in gray). (*See insert for color representation of the figure.*)

regions simulating therefore propagation to infinity (see previous equations for damping).

Step 2. We compute the electrical problem using the electrical conductivity equation developed in Chapter 1 with $m = n = 2$ and a surface conductivity $\sigma_S = 0.01$ S m^{-1} while the conductivity of the pore water is 0.1 S m^{-1}. The charge density $\hat{Q}_V^0$ is determined from the distribution of the permeability (see Chapter 1). The source current density is determined from the Darcy velocity and the displacement of the solid phase. Finally, the electrical potential distribution is obtained by solving a Poisson equation for the electrical potential.

4.2.3 Results

The evolution of the seismic displacement and the electrical potential time series recorded at station E12 (see Figure 4.5) for each saturation profile (T1–T6) are shown in Figures 4.6. The seismic displacements are generated from the seismic source which is a P-wave-only source. In this case, with the porosity distribution displayed in Figure 4.3 and with the water saturation variations shown in Figure 4.4, the average P-wave velocity of profiles, T1–T6, is about 4800 m s^{-1}. The P-wave arrivals in Well B are therefore roughly the same in the snapshots T1–T6 (Figure 4.6). Figure 4.6 shows that

seismoelectric (SE) conversions occur in each snapshot. These conversions always arrive earlier than the coseismic electrical field associated with the arrival of the P-wave. They also arrive later and later as the water front progresses toward Well B. This shows that there is a clear conversion mechanism at the NAPL/water encroachment front for each of the five snapshots, T2–T6. For snapshot T1, since there is no saturation contrast, we do not see any strong seismoelectric conversions. That said, there are still some small seismoelectric conversions taking place at the heterogeneities in the aquifer.

Putting the saturation profile T4 into the model, we show that both the seismoelectric conversion (SE) generated at the NAPL/water encroachment front and the coseismic (CS) electrical signals are shown at all receiving stations E1 to E28 (Figure 4.5 for the position of the electrodes). Figure 4.7 shows the seismic displacement and the electrical time series for station E12 with information on the delay time t_0, seismoelectric conversion at time t_1, and the similarities of coseismic and P-wave arrival times t_2.

In terms of amplitudes, the type of signal measured here is easily recordable in the laboratory and in the field through stacking. Dupuis, Butler, and colleagues have developed methods to improve the signal-to-noise ratio

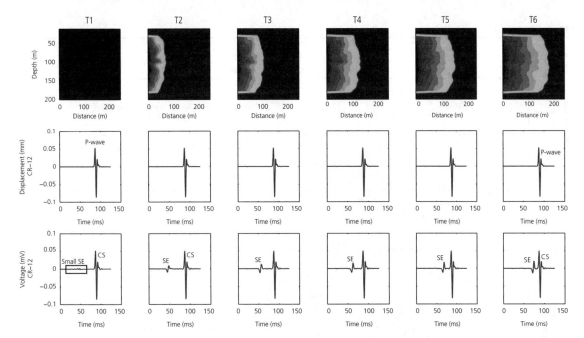

Figure 4.6 Evolution of the seismic displacement and the associated electric potential time series from receiver point E12 (see Figure 4.5) due to changes in the position of the oil/water encroachment front during snapshots T1 to T6. The arrival of the P-wave is identified in the seismic time series. SE denotes the seismoelectric conversions occurring at any electrical and mechanical heterogeneity in the aquifer (especially at the oil/water encroachment front), while CS denotes the coseismic electrical field associated with the P-wave. (*See insert for color representation of the figure.*)

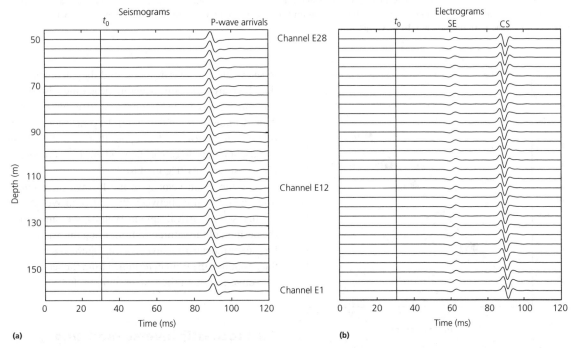

Figure 4.7 Seismogram and electrograms from implementing saturation profile T4. **a)** The seismograms reconstructed by the geophones show the *P*-wave propagation from the seismic source in Well A to recording Well B. **b)** The electrograms show the coseismic electrical potential field associated with the P-wave (CS) and the seismoelectric (SE) conversions with a smaller amplitude and the same time arrival. A very small moveout can be seen in the seismic and coseismic data that is due to the distance from the source. This moveout is not present in the seismoelectric response.

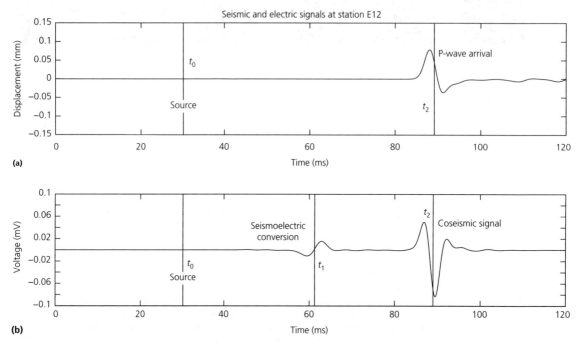

(a)

(b)

Figure 4.8 Seismogram and electrogram at receiver E12 (see Figure 4.5). Here, t_0 denoted the time of source ignition or the delay time (30 ms). The arrival times of the seismoelectric signals and the coseismic disturbance associated with the P-wave are denoted as t_1 and t_2, respectively. The strongest signal on the electrogram corresponds to the coseismic disturbance associated with the P-wave propagation. **a)** Seismogram. **b)** Electrogram.

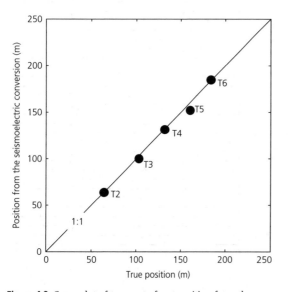

Figure 4.9 Cross-plot of true waterfront position from the two-phase flow simulator and position determined from the arrival time of the seismoelectric conversion and using the mean velocity of the P-wave between the two wells. It is clear that the seismoelectric conversion can be used to evaluate the position of the saturation front.

in seismoelectric investigations, and they demonstrated that there are no serious issues in measuring seismoelectric conversions in field conditions (Dupuis & Butler 2006; Dupuis et al., 2007, 2009). In conclusion, we see that our numerical model implies that the NAPL/water encroachment front can be detected through seismoelectric measurements (Figure 4.8).

Using the mean velocity of the P-wave (determined from the travel time between the source in Well A and the first seismic arrival time in Well B), we can localize the position of the saturation front using the arrival time for the seismoelectric conversion. In Figure 4.9, we show the position of the front determined from the seismoelectric conversion with the true position of the front.

4.3 Stochastic inverse modeling

4.3.1 Markov chain Monte Carlo solver

In this section, we develop a general algorithm to invert for the material properties of the formations (the intrinsic permeability k_0, the porosity ϕ, the electrical conductivity

σ, the bulk modulus of the dry porous frame K_{fr}, the bulk modulus of the fluid K_f, the bulk modulus of the solid phase K_s, and the shear modulus of the skeleton G) using a joint inversion of seismic and seismoelectric signals. We use a probabilistic framework to map the distribution of errors in the observed data into the model space as a complex error distribution. Additionally, this probabilistic framework can easily handle the nonuniqueness of the inverse problem, which is reflected in the probability distribution of the model parameters.

The first criterion for joint inversion requires that the predicted seismic and electroseismic data fit their observed counterparts. Therefore, we perform a Bayesian approach to estimating the following vector, $\mathbf{m}$, of material properties where $\mathbf{m} = [\log_{10}(k_0)$, $logit(\phi), \log_{10}(\sigma)$, $\log_{10}(K_{fr})$, $\log_{10}(K_s)$, $\log_{10}(K_f)$, $\log_{10}(G)]$ with $logit(\phi) = \log[\phi/(1-\phi)]$. Note that the connected porosity ϕ is a concentration of connected voids and is the only material property that is constrained to be between 0 and 1; this is why we use the logit of the function (rather than the log) to define the associated model parameter. The logit function is often used for linearizing sigmoid distributions of proportions.

Another point is that we consider the set of material properties (k_0, ϕ, σ, K_{fr}, K_f, K_s, and G) as corresponding to independent properties and k_0 and $\bar{Q}_V$ as two completely interrelated properties. We could also add additional petrophysical relationships between the porosity on one side and the permeability, the electrical conductivity, and the bulk modulus of the skeleton on the other side. These relationships could be defined in a probabilistic sense, which would improve the convergence of the inversion. However, we prefer to stay conservative in this book and maintain the aforementioned properties independent from each other.

The Bayesian solution to an inverse problem is based on combining the information coming from geophysical data with some a priori knowledge. In the Bayesian approach we use, we consider the acquisition of geophysical data as experiment E. The Bayesian analysis considers both data vector, $\mathbf{d}$, and model parameter vector, $\mathbf{m}$, of model M as random variables. Therefore, several geometrical or petrophysical models, M, are possible to explain the data. Random variables are characterized with distributions, and we assume that all distributions are characterized by probability density functions (Tarantola, 2005).

The objective of inverse modeling process is to update the information on $\mathbf{m}$, assuming a petrophysical model or a geometrical model M, given data $\mathbf{d}$ and a priori information regarding $\mathbf{m}$. The a priori information can come from independent observations and petrophysical relationships. In a probabilistic framework, the inverse problem maximizes the conditional probability of m in M occurring given the data vector $\mathbf{d}$. We note this probability with the statement $P_0(\mathbf{m}|M)$, the a priori probability density, or belief that the parameters of $\mathbf{m}$ in model M are such that the model, M, generates the probability density of likelihood $P(\mathbf{d}|\mathbf{m},M)$ corresponding to the data fit.

In a Bayesian approach, the posterior probability density $\pi(\mathbf{m}|\mathbf{d})$ of the model parameters $\mathbf{m}$ given the data $\mathbf{d}$ is obtained by using Bayes' formula:

$$\pi(\mathbf{m}|\mathbf{d},M) = \frac{P(\mathbf{d}|\mathbf{m},M)P_0(\mathbf{m}|M)}{P(\mathbf{d}|M)}, \quad (4.52)$$

where $P(\mathbf{d}|M)$, the evidence, is defined as

$$P(\mathbf{d}|M) = \int P_0(\mathbf{m}|M)P(\mathbf{d}|\mathbf{m},M)d\mathbf{m}. \quad (4.53)$$

This marginal likelihood is usually ignored because it is not a function of the parameters $\mathbf{m}$. For any given model M, the evidence will have a constant value. Consequently, the nature of the evidence is not needed if we are interested only in the nature of the variation of the a posteriori distribution rather than its exact value.

In the following, we assume that model M is certain (e.g., we know the position of the sedimentary layers and reservoir from seismic data and we know that the petrophysical model defined by Equations (4.26)–(4.39) is exact). Therefore, we drop the term M from Equations (4.52) and (4.53). The posterior probability density $\pi(\mathbf{m}|\mathbf{d})$ of model parameters $\mathbf{m}$ given data $\mathbf{d}$ is written as

$$\pi(\mathbf{m}|\mathbf{d}) \propto P(\mathbf{d}|\mathbf{m})P_0(\mathbf{m}). \quad (4.54)$$

The Bayesian solution of the inverse problem is the whole posterior probability distribution of the material properties. An estimate of the unknown parameters can be computed, for example, as the expectation value with respect to the posterior distribution (i.e., as the mean value) or as the maximum a posteriori value which can be understood as the most likely value.

As usually accepted, the likelihood function used to assess for the quality of model $\mathbf{m}$ is Gaussian distributed:

$$P(\mathbf{d}|\mathbf{m})$$
$$= \frac{1}{\left[(2\pi)^N \det C_{\mathbf{d}}\right]^{1/2}} \exp\left[-\frac{1}{2}(g(\mathbf{m})-\mathbf{d})^{\mathrm{T}} C_{\mathbf{d}}^{-1}(g(\mathbf{m})-\mathbf{d})\right] \tag{4.55}$$

$$\mathbf{d} = (\mathbf{d}_{\mathrm{S}}, \mathbf{d}_{\mathrm{E}})^{\mathrm{T}}, \tag{4.56}$$

where $g(\mathbf{m})$ is the forward modeling operator for the seismic and seismoelectric semicoupled problem. It connects nonlinearly the generation of seismograms and electrograms to the variation of the material properties of the ground, where $\mathbf{d}$ is an N-vector of the observed seismic data, $\mathbf{d}_{\mathrm{S}}$, and seismoelectric data, $\mathbf{d}_{\mathrm{E}}$. The $(N \times N)$ covariance matrix is given by

$$C_{\mathbf{d}} = \begin{bmatrix} \sigma_{\mathrm{S}}^2 & 0 \\ 0 & \sigma_{\mathrm{E}}^2 \end{bmatrix} \tag{4.57}$$

(where S stands for seismic and E for electrical data). This matrix comprises the measurement errors for both the seismic and electrical data, which are usually considered to be uncorrelated and are assumed to obey Gaussian distributions.

The a priori distribution on the model parameters, if available, is also assumed to be Gaussian:

$$P_0(\mathbf{m}) = \frac{1}{\left[(2\pi)^M \det C_{\mathbf{m}}\right]^{1/2}}$$
$$\times \exp\left[-\frac{1}{2}(\mathbf{m}-\mathbf{m}_{\mathrm{prior}})^{\mathrm{T}} C_{\mathbf{m}}^{-1}(\mathbf{m}-\mathbf{m}_{\mathrm{prior}})\right], \tag{4.58}$$

where $\mathbf{m}_{\mathrm{prior}}$ is the prior value of the distribution of the three petrophysical parameters in the ground and C_{m} denotes the model diagonal covariance matrix incorporating the uncertainties related to the a priori model of material properties. In the results, discussed in Section 4.3.3, we will use a null prior as information on the model parameters.

In the classical Bayesian approach, model parameters $\mathbf{m}$ that fit geophysical observations $\mathbf{d}$ maximizes the posterior probability density $\pi(\mathbf{m}|\mathbf{d})$. The problem is to explore the posterior probability density, $\pi(\mathbf{m}|\mathbf{d})$, expressed by Equation (4.54). The denominator of Equation (4.54) is the normalizing factor required for the integral of the probability density function to be

one. Normalizing is not required to perform the inversion except if we want to compute explicitly the probability for a given parameter to be in a given interval. In addition, the normalization can be done at the end of the computation of the probability density function (restricting to the numerator), just by normalizing it (dividing it by its integral) (Grandis et al., 1999).

The Markov chain Monte Carlo (MCMC) family of algorithms is well suited for Bayesian inference problems (Malinverno, 2002). MCMC algorithms consist of random walks where different states (i.e., different values of a model vector) are tested and where the choice of the next state depends only on the value of the current state. After an initial period in which the random walker moves toward the highest a posteriori probability regions, the chain returns a number of model vectors sampling the posterior probability density $\pi(\mathbf{m}|\mathbf{d})$. The characteristic of probability density $\pi(\mathbf{m}|\mathbf{d})$, like the mean and the standard deviation or the number of extrema in the probability density, can therefore be easily determined. Memory mechanisms of the MCMC algorithms (that keep the chain in the high a posteriori probability regions of the model space) are responsible for a greater efficiency of the algorithm in comparison to the Monte Carlo methods, for which the models are independently chosen and tested against the observations.

The basic Metropolis–Hastings algorithm is a two-step procedure. In the first step, the current model parameter vector, $\mathbf{m}$, in the Markov chain is modified randomly to obtain a candidate vector. This candidate is drawn from a proposal distribution $q(\mathbf{m}, \mathbf{m}')$ where the choice of $\mathbf{m}'$ depends on the current vector $\mathbf{m}$. The proposal distribution could be, for example, a multidimensional Gaussian distribution. In the second step, the candidate model is accepted with the acceptance probability (Malinverno 2002):

$$\alpha(\mathbf{m};\mathbf{m}')$$
$$= \min\left[1, \frac{\pi(\mathbf{m}'|\mathbf{d})}{\pi(\mathbf{m}|\mathbf{d})} \frac{q(\mathbf{m}|\mathbf{m}')}{q(\mathbf{m}'|\mathbf{m})}\right],$$
$$= \min\left[1, \frac{P_0(\mathbf{m}')}{P_0(\mathbf{m})} \frac{P(\mathbf{d}|\mathbf{m}')}{P(\mathbf{d}|\mathbf{m})} \frac{q(\mathbf{m}|\mathbf{m}')}{q(\mathbf{m}'|\mathbf{m})}\right],$$
$$= \min[1, (\text{prior ratio}).(\text{likelihood ratio}).(\text{proposal ratio})]. \tag{4.59}$$

If the candidate is accepted, the state of the chain is changed to $\mathbf{m}'$; otherwise, the chain stays at $\mathbf{m}$.

The acceptance probability depends only on the proposal, likelihood, and prior functions at the current and candidate models, all of which can be easily computed. Assuming that the proposal distribution is symmetric, $q(\mathbf{m}'|\mathbf{m}) = q(\mathbf{m}|\mathbf{m}')$ (e.g., a Gaussian distribution centered at the current point), the acceptance probability reduces to

$$\alpha(\mathbf{m};\mathbf{m}') = \min\left[1, \frac{P(\mathbf{m}'|\mathbf{d})}{P(\mathbf{m}|\mathbf{d})}\right]. \qquad (4.60)$$

This algorithm is known as the original Metropolis algorithm (Metropolis et al., 1953).

To improve the performance of the standard Metropolis–Hastings algorithm, Haario et al. (2001) introduced an algorithm called the adaptive Metropolis algorithm (AMA) to find the optimal proposal distribution. This algorithm is based on the traditional Metropolis algorithm with a symmetric Gaussian proposal distribution centered at the current model $\mathbf{m}^i$ and with the covariance $\mathbf{C}^i$ that changes during the sampling in such a way that the sampling efficiency increases over time (Haario et al., 2001, 2004). The AMA, though not Markovian, simulates correctly the posterior probability distributions of the model parameters. An important advantage of the AMA is that it starts by using the cumulating information right at the beginning of the simulation. The rapid start of the adaptation ensures that the search becomes more effective at an early stage of the simulation, which diminishes the number of iterations to reach the convergence of the chain.

The AMA is described as follows. Let us assume that we have sampled the states $(\mathbf{m}^0, ..., \mathbf{m}^{i-1})$ where $\mathbf{m}^0$ corresponds to the model vector of the initial state. Then a candidate point, $\mathbf{m}'$, is sampled from the Gaussian proposal distribution, q, with a mean point at the present point, $\mathbf{m}^{i-1}$, and a covariance of

$$\mathbf{C}^i = \begin{cases} \mathbf{C}^0, & \text{if } i \leq n_0, \\ s_n \mathbf{K}^i + s_n \varepsilon \mathbf{I}_n, & \text{if } i > n_0, \end{cases} \qquad (4.61)$$

where $\mathbf{I}_n$ denotes the n-dimensional identity matrix, $\mathbf{K}^i = \text{Cov}(\mathbf{m}^0, ..., \mathbf{m}^{i-1})$ is the regularization factor (a small positive number that prevents the covariance matrix from becoming singular), $\mathbf{C}^0$ is the initial covariance matrix that is strictly positive (note that the AMA is not too sensitive to the actual values of $\mathbf{C}^0$), and $s_n = (2.4)^2/n$ is a parameter that depends only on the dimension of the vector $\mathbf{m} \in \mathfrak{R}^n$ (Haario et al., 2001).

This choice of s_n yields an optimal acceptance in the case of a Gaussian target distribution and a Gaussian proposal distribution. The candidate model vector, $\mathbf{m}'$, is accepted with the acceptance probability:

$$\alpha(\mathbf{m}^{i-1};\mathbf{m}') = \min\left[1, \frac{\pi(\mathbf{m}'|\mathbf{d})}{\pi(\mathbf{m}^{i-1}|\mathbf{d})}\right]. \qquad (4.62)$$

If the candidate model vector is accepted, we consider that $\mathbf{m}^i = \mathbf{m}'$; otherwise, we choose $\mathbf{m}^i = \mathbf{m}^{i-1}$.

The AMA was written in a MATLAB routine that is coupled with a forward model in COMSOL Multiphysics 3.5 using the finite-element approach described at the beginning of this section. This algorithm is applied in the next section to the synthetic seismogram and electrogram data to invert the material properties, assuming that the position of the geological units is known.

4.3.2 Application

To check the usefulness of the joint inversion of seismic and seismoelectric data, we test this approach using a numerical case study. We consider two flat layers plus a rectangular reservoir embedded in the second layer (Figure 4.10). The geophones and the electrodes are located at the top surface of the system to simulate an onshore acquisition. The takeout for the electrodes and the geophones has a 10 m separation. The source wavelet is a first-order derivative of a Gaussian, as defined in Equation (4.46), with a dominant frequency, $f_0 = 30$ Hz, and a time delay factor, $t_0 = 0.1$ s (see Eq. (4.46)). The seismic source is located at a depth of 20 m (Figure 4.10). The true values of the material properties used for the simulation are reported in Table 4.3. With these values, the velocity of the fast P-wave is 1972 m s^{-1} in the first layer (labeled L1), 2188 m s^{-1} in the second layer (labeled L2), and 3118 m s^{-1} in the reservoir (labeled R). The receiver is located at an offset of $x = 150$ m from the source; the time required for the P-wave to reach this receiver is therefore 0.076 s, in agreement with the numerical simulations (see succeeding text). The finite-element modeling, based on COMSOL Multiphysics 3.5, is used to simulate both seismograms and electrograms at the ground surface.

In Figure 4.11d, the electrograms show the two types of seismoelectric signals described in the introduction. The first type of signal corresponds to the coseismic electrical signal associated with the propagation of the P-wave. The coseismic electric field related to the direct field wave from the source to the receiver is labeled

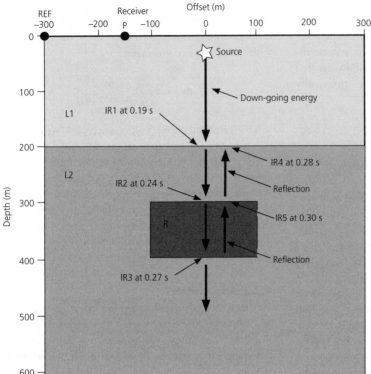

Figure 4.10 Sketch of the model used for the simulations. The geophones and the electrodes are collocated at the top surface of the system. All the electrodes are assumed to be connected to a reference electrode. REF corresponds to the position of the reference electrode. L1 and L2 correspond to the two layers of sediments, and R stands for the reservoir. Arrows indicate ray paths for seismic energy that creates the seismoelectric interface response labeled IRi. In addition to this signal, the direct field and reflected seismic arrivals are recorded as the coseismic electric field. The seismic source is located at a depth of 20 m below the top surface.

Table 4.3 True values of the material properties used for the synthetic model shown in Figure 4.10.

Parameter	Units	Unit L1	Unit L2	Unit R
k_0	m^2	10^{-12}	10^{-16}	10^{-11}
ϕ	—	0.25	0.10	0.33
K_s	GPa	36.50	6.90	37.00
K_f	GPa	0.25	0.25	2.40
K_{fr}	GPa	2.22	6.89	9.60
G	GPa	4.00	3.57	5.00
ρ_s	$kg\,m^{-3}$	2650	2650	2650
ρ_f	$kg\,m^{-3}$	1040	1040	983
η_f	Pa s	1×10^{-3}	1×10^{-3}	8×10^{-1}
σ	$S\,m^{-1}$	0.01	0.1	0.001
$\log \hat{Q}_V$	Log C m^{-3}	0.203	3.49	3.2

L1 and L2 stand for the two layers and R for the reservoir. Layer L1 corresponds to a clean sand, L2 to a clayey sand, and R to a sand reservoir partially filled with oil.

CS. Other coseismic signals are associated with the reflected P-waves at the various interfaces like the L1–L2 interface, the L2–reservoir interface, and the reservoir–L2 interface. These coseismic signals are labeled

RCS1, RCS2, and RCS3, respectively. A coseismic signal occurs when a seismic wave travels through a porous material, creating a relative displacement between the pore water and the solid phase. The associated current density is balanced by a conduction current density. It results in an electrical field traveling at the same speed as the seismic wave. Because shear waves are isochoric, they are not responsible for any source current density in a homogeneous medium, and therefore, they have no coseismic electric field associated with them (Haines & Pride, 2006).

The second type of seismoelectric signal corresponds to converted seismoelectric signals associated with the arrival of the P-waves at each interface (between the two layers and at the surface of the reservoir). These converted seismoelectric signals are labeled IR1, IR2, IR3, IR4, and IR5 (see Figures 4.10, 4.11, and 4.12). When crossing an interface between two domains characterized by different properties, a seismic wave generates a time-varying charge separation, which acts like a dipole that radiates electromagnetic energy. In our approach, we neglect the time used by this electromagnetic energy to diffuse from the geological interface, where it is

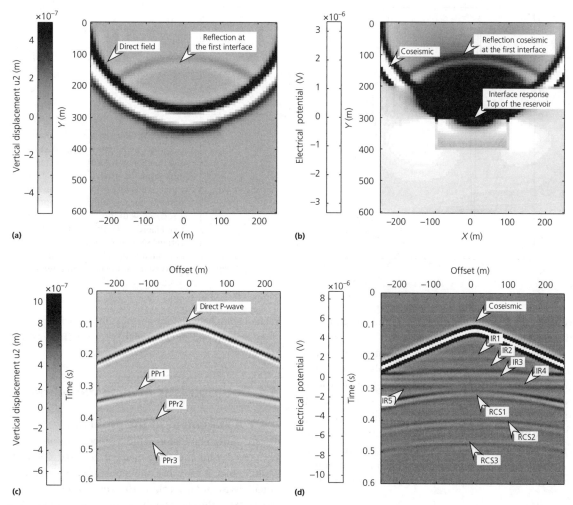

Figure 4.11 Snapshots of seismic and seismoelectric phenomena. **a), b)** Snapshots of the seismic and electrical fields at time t =0.24 s. This corresponds to the time when the P-wave reaches the top of the reservoir, which acts as a dipole radiating electromagnetic energy. **c)** The seismograms reconstructed by the geophones (with a takeout of 10 m) present direct field and different reflections of the P-wave: the reflection PPr1 on the interface L1–L2, the reflection PPr2 on the interface L2–top of the reservoir, and the reflection PPr3 on the interface L2–bottom of the reservoir. **d)** The electrograms show the coseismic electrical potential field associated with the direct wave and the reflections of the P-wave (labeled RCS1, RCS2, and RCS3) and the seismoelectric conversions with a smaller amplitude and a flat shape (labeled IR1, IR2, IR3, IR4, and IR5).

generated, to the receiver (quasistatic field approximation). This assumption is very good for investigations within the first kilometer below the ground surface. Due to constructive interferences, a significant portion of the first Fresnel zone acts like a disk of electric dipoles oriented normal to the interface. These dipoles oscillate with the waveform of the seismic wave (Figure 4.11b). Because the electromagnetic diffusion of the electrical disturbance is very fast, the seismoelectric conversions

are observed nearly at the same time by all the electrodes but with different amplitudes. Therefore, the seismoelectric conversions appear as flat lines in the electrograms. Note also that the polarity of the converted seismoelectric signals depends on the contrast of electrical material properties (volumetric charge density and electrical conductivity) at the interface where they are generated. On the contrary, the polarity of the coseismic electrical signals depends on the value of the streaming potential

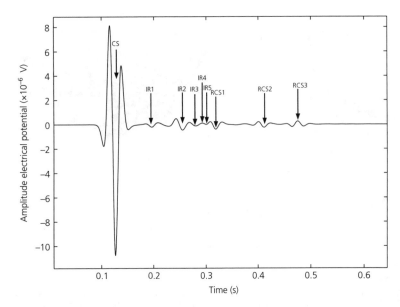

Figure 4.12 Electrogram at an electrode (receiver) with a horizontal offset of 150 m. CS stands for the coseismic disturbance associated with the direct wave (see Figure 4.11). RCS1 and RCS2 stand for the coseismic disturbances associated with the reflected P-waves (see Figure 4.11). IRi stand for the various seismoelectric disturbances associated with the seismoelectric conversions at the different interfaces of the system.

coupling coefficient at the position of the electrode and the polarity of the seismic waves.

Figure 4.12 shows the electric potential for a given electrode. In this figure, we can clearly discriminate the coseismic signals from the seismoelectric conversions. Note also that the amplitudes of the signals are small. However, they can easily be measured in the field using the type of ultrasensitive equipment discussed by Crespy et al. (2008). This equipment can be used to record the electrical potential with up to 256 simultaneous channels at several kHz with a sensitivity of 10 nV (see Section 4.3.3).

4.3.3 Result of the joint inversion

We use the AMA described in Section 4.4.3 to generate 25,000 realizations of the 21 parameters of the material properties of the different geological units using the data recorded in 60 geophones and 60 electrodes at only four frequencies (25, 30, 35, and 40 Hz). The position and the characteristic of the source are assumed to be perfectly known. The posterior probability distribution functions of the material properties of the three units (layers L1 and L2 and the reservoir R) are shown in Figure 4.13 using the last 5000 iterations. We observe that, except for the porosity, our algorithm does a very good job of properly inverting the seismic and seismoelectric data in terms of finding the mean value of the material properties. We believe that a better estimate of the porosity can be obtained through the use of additional petrophysical

relationships between the porosity, the electrical conductivity, and the bulk modulus of the skeleton (Figure 4.13).

4.4 Deterministic inverse modeling

4.4.1 A statement of the problem

The forward computation of the seismoelectric problem is performed with the finite-element package COMSOL Multiphysics 3.5a using the same partial differential equations as Jardani et al. (2010). The problem is defined in COMSOL through the following steps: (1) formulate the semicoupled field equations that describe the dynamic poroelastic phenomena with the associated electromagnetic disturbances (see Section 4.4.2), (2) define the geometry of the model (see Figure 4.15), (3) specify the model parameters (see Table 4.1), (4) design the finite-element mesh (we use triangular meshing in the present case), (5) select the boundary layer conditions (we use PML boundary conditions for the seismic part of the problem; see Figure 4.15), (6) solve the partial differential field equations, (7) run the inverse algorithms using the quasielectrostatic condition, and finally (8) postprocess the data to produce an image using a pixel-based approach. The flowchart for this process is shown in Figure 4.16. The PML boundary conditions consist of a strip simulating the propagation of the seismic waves out to infinity without any reflections going back inside

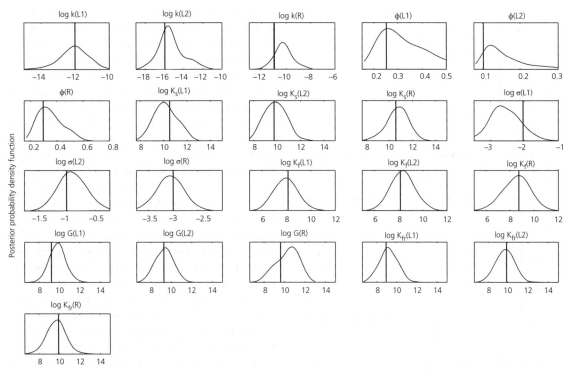

Figure 4.13 Posterior probability density functions of the material properties for the three geological units (the two layers L1 and L2 and the reservoir R). The vertical bars indicate the real value of the material properties (see Table 4.3).

the domain of interest (see Jardani et al., 2010, for further details on the implementation).

The first problem is to simulate the seismic wave propagation associated with the source in order to get the distribution of the two components of the displacement along the coordinates, x and z, and the mean pore fluid pressure, p, at each time, t. Since we are solving stationary partial differential equations in the frequency domain, we use the stationary parametric solver PARDISO (http://www.pardiso-project.org/). This solver is used to determine the distribution of the following parameters: first, the fields u_x, u_z, p are determined as a function of space and time using an IFFT of the solution to the spatial and temporal domain from the frequency domain. The quasistatic scalar potential ψ is computed by solving the Poisson equation coupled to the solution of u_x, u_z, p through its source term. In the frequency domain, we solve the partial differential equations from 1 to 100 Hz with a step of 1 Hz. The seismic forward modeling code we used was benchmarked by Jardani et al. (2010) using the analytical solution of Dai et al. (1995) for a

water-saturated poroelastic material. Then, we use the components of the solid displacement and the fluid pressure to determine the electrical potential distribution at each time step. In our modeling, we also neglect the Stoneley waves propagating along the boreholes in which the seismic sources are located. These Stoneley waves can generate seismoelectric signals (Mikhailov et al., 2000; Hunt & Worthington, 2000) and should be considered in the case of a real field experiment. However, we believe that they can be easily filtered out of the recorded signals by performing forward modeling of this contribution (see Ardjmandpour et al., 2011, for an example) using downhole measurements of the slowness and resistivity.

The geometry of the two synthetic case studies investigated in the succeeding text is shown in Figure 4.15. The domain consists of a 600 m × 600 m region and is infinite in the strike direction (2.5D assumption). In the first case study, the geometry is made of two half-spaces (U1 and U2) separated by a vertical interface located at $x = 300$ m. The seismic sources are located in borehole #1 at

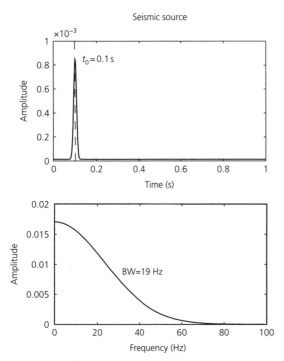

Figure 4.14 Description of the explosive seismic source. We use a sharp Gaussian pulse simulating an explosion at $t_0 = 0.1$ s. The standard deviation of the source in the frequency domain is 19 hertz.

$x = 100$ m and $z = 150$, 225, 300, 375, and 450 m for shots #1, #2, #3, #4, and #5, respectively. In borehole #2 (located at $x = 500$ m), we simulate an array of 50 sensors consisting of electrodes and geophones. The seismic sources are detonated sequentially in borehole #1. Meanwhile, seismic and electrical data are recorded for each individual seismic shot. Table 4.1 provides the material properties of media U1 and U2 used in the model.

Figure 4.17 shows eight snapshots of the seismic wave propagation, the associated electrical current density distribution, and the resulting voltage at electrode #25 in borehole #2. Figure 4.18 represents the signal at electrode #5 at location $z = 140$ m deep from the surface, which shows that seismoelectric signals are generated at the interface before the coseismic signal arrives at the receivers. While the coseismic signals show a characteristic hyperbolic shape, the seismoelectric signals arrive nearly instantaneously at all receivers. In the following, we call the "seismoelectric conversion (SC-)time window" the time between the shot of the seismic source and the time the seismic wave arrives at the receivers

located in borehole #2. The inverse problem involves locating the distribution of seismoelectric conversions at the interface. It will be described in the following sections using the data inside the SC-time windows for all the seismic shots.

Our main objective in this section is to estimate the spatial distribution of heterogeneities that generate the seismoelectric source current, regardless of the values of the material properties themselves associated with these heterogeneities. At each time step, the seismoelectric signals recorded in borehole #2 are like a self-potential profile (i.e., a distribution of voltages recorded at a set of electrodes with respect to a reference electrode due to a quasistatic source of current). At each time in the SC-time window and for each shot, we use the electrical potential distribution recorded on the array of electrodes located in the second well in order to find the position of the current source generated by the seismoelectric conversions between the two wells. At each time step, the inverse seismoelectric problem is therefore similar to a self-potential inverse problem for which several algorithms have been developed over the past few years using deterministic (Jardani et al., 2007, 2008) and stochastic (Jardani & Revil, 2009; Revil & Jardani, 2010) approaches. In general, we want to invert these voltages to recover the position, $\mathbf{r}$, and amplitude of volumetric current source, $\Im(\mathbf{r}, t)$. In our test case, at each time step, the seismic waves impinging on the interface are responsible for the source current density, which has compact support (i.e., the spatial distribution of sources at any given time is very sparse). Therefore, the algorithm we use in the succeeding text is based on compactness as a regularization tool, as developed for the self-potential source inversion problem. A mathematical description of the algorithm is given in Section 4.3.2 (Figure 4.16).

We now use a small test to analytically check the validity of our numerical computations and to compute the size of the first Fresnel zone for the seismic and seismoelectric problems. To estimate the time at which the seismoelectric conversion occurs, we can estimate the P-wave velocity based on the material properties reported in Table 4.4. We obtain $c_p = 1935.5$ m s^{-1} in unit U1 and 2163.5 m s^{-1} in U2. Taking the geometry of Figure 4.15 (case study #1), we obtain a travel time of 0.115 s from seismic source #5 to the interface and a travel time of 0.213 s from seismic source #5 to geophone #25. Because the source has a time delay of $t_0 = 0.1$ s, the seismoelectric conversion occurs at $t_1 = 0.215$, and

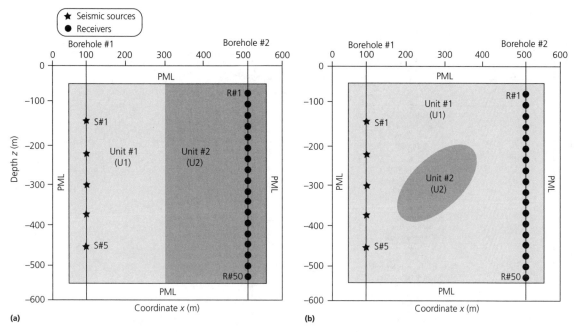

Figure 4.15 The model domain is a 600 m × 600 m square. Borehole #1, the shooting borehole, is located at position $x = 100$ m, and the measurement borehole #2 is located at $x = 500$ m. The discretization of the domain comprises a finite-element mesh of 60x60 rectangular cells. We consider 5 seismic sources (from S#1 at the top to S#5 at the bottom), equally spaced in borehole #1, and 50 receivers (R#1 to R#50), located in borehole #2. PML boundary conditions are used at the borders of the domain. **a)** Case study #1 concerns a vertical interface separating two homogeneous half-spaces. This interface is located at $x = 300$ m, an equal distance between the two sources. **b)** Case study #2 corresponds to an inclusion, U2, embedded into a homogeneous material, U1.

the coseismic signal occurs at $t_2 = 0.313$ s. This is in agreement with the numerical results of Figure 4.18. Using the relationship between the wavelength and the velocity of the P-waves, $\lambda_S = c_p/f$, where f is the dominant frequency, the first Fresnel zone radius for the seismic wave is $r_S = (dc_p/2f)^{1/2}$. Using $d = 200$ m (Figure 4.15), $f = 40$ Hz, and $c_p = 1935.5$ m s^{-1} in unit U1, we obtain $r_S = 69$ m. Also, using the relationship between the first seismic Fresnel zone and the seismoelectric Fresnel zone (see Section 4.2), we obtain $r_{SE} = 98$ m, which provides an idea of the lateral resolution of the seismoelectric method at this frequency.

4.4.2 5D electric forward modeling

The Poisson equation governing the electrostatic potential distribution corresponds to Equation 4.37. The source term of this Poisson equation is described as a Dirac (delta) function and a point current source, I (in A):

$$\Im(\mathbf{r}, t) = \nabla \cdot \mathbf{j}_s = I\delta(x - x_s)\delta(y - y_s)\delta(z - z_s), \quad (4.63)$$

where (x_s, y_s, z_s) denote the coordinates of the each point where the seismoelectric conversion takes place. Two assumptions are made in order to transform the 3D problem into a 2.5D problem. It is assumed that the model is homogeneous in the strike direction y, that is, $\partial\sigma(x, y, z)/\partial y = 0$, and the strike direction extends to infinity in both directions. Solving the Poisson equation in the wave number domain, where k_y is the wave number in the strike direction, and using the Fourier cosine transform,

$$\tilde{\psi}(x, z, k_y) = \int_0^\infty \psi(x, y, z)\cos(k_y y)\, dy. \quad (4.64)$$

Equation 4.64, in the wave number domain, takes the following form:

$$-\nabla \cdot \left(\sigma_{(x,z)}\nabla\tilde{\psi}(x, z, k_y)\right) + \sigma_{(x,z)}k_y^2\tilde{\psi}(x, z, k_y) = I\delta(x - x_s)(z - z_s). \quad (4.65)$$

Therefore, the initial Poisson equation is transformed to a Helmholtz-type differential equation in the wave

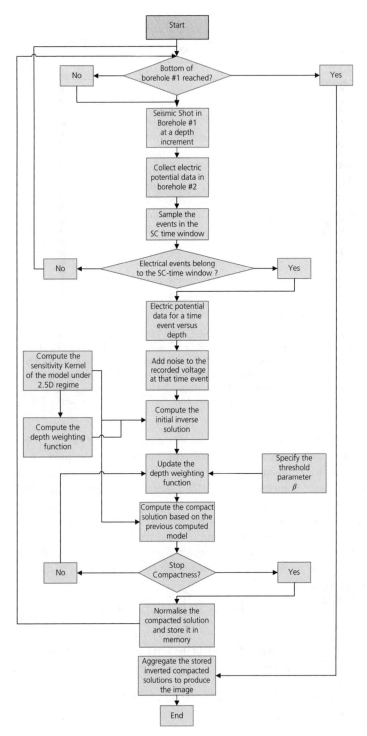

Figure 4.16 Flowchart of the seismoelectric forward and inverse modeling.

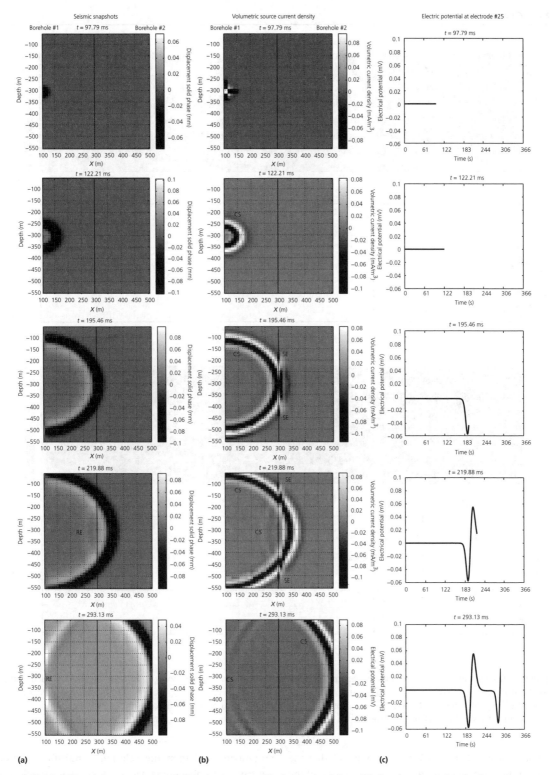

Figure 4.17 Modeling of the propagation of the seismic waves (displacement of the solid phase) and associated volumetric current density (case study #1, shot #3). CS, coseismic signals; SE, seismoelectric conversions; RE, reflected seismic wave.

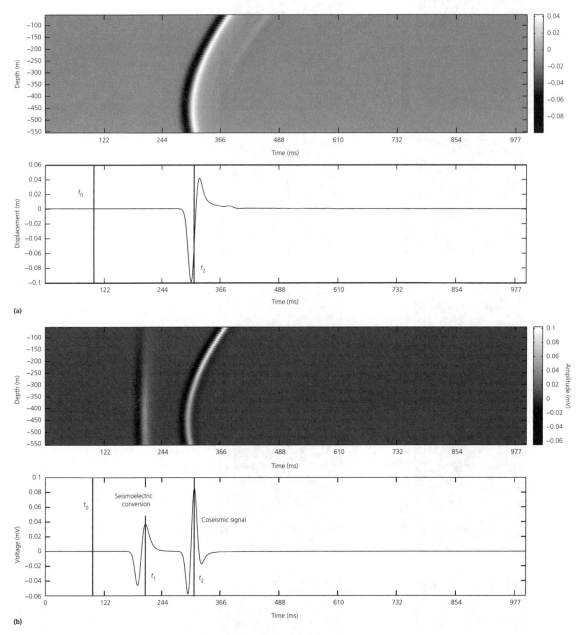

Figure 4.18 Example of seismograms and electrograms corresponding to shot #5 (deeper source). **a)** Seismograms recorded in the offset borehole with a typical parabolic shape. An example of seismogram is given for geophone #25. **b)** Electrograms recorded in the offset borehole showing the seismoelectric conversion at the vertical interface and the coseismic signal. An example of electrogram is given for electrode #25 (position $x = 500$ m, $z = 300$ m). The source occurs at $t = 110$ s.

number domain. It is important to point out that the wave number is related to the model geometry and the separation distances between source and receivers. Equation 4.65 is solved repeatedly for several wave numbers, and then the solution is transformed from the wave number domain to the spatial domain following the approach taken by Dey and Morrison (1979) using an inverse cosine transform.

Table 4.4 Material properties used for the numerical simulations corresponding to the case study #1.

Parameter	Description	Unit U1	Unit U2
σ	Conductivity of the medium	$0.01\,\mathrm{S\,m^{-1}}$	$0.1\,\mathrm{S\,m^{-1}}$
$\hat{Q}_V$	Excess of charge per unit pore volume	$0.203\,\mathrm{C\,m^{-3}}$	$3.49\,\mathrm{C\,m^{-3}}$
ρ_s	Bulk density of the solid phase	$2650\,\mathrm{kg\,m^{-3}}$	$2650\,\mathrm{kg\,m^{-3}}$
ρ_f	Bulk density of the fluid phase	$1000\,\mathrm{kg\,m^{-3}}$	$1000\,\mathrm{kg\,m^{-3}}$
ϕ	Porosity	0.25	0.10
K_s	Bulk modulus of the solid phase	$36.5 \times 10^9\,\mathrm{Pa}$	$6.9 \times 10^9\,\mathrm{Pa}$
K_f	Bulk modulus of the fluid phase	$0.25 \times 10^9\,\mathrm{Pa}$	$0.25 \times 10^9\,\mathrm{Pa}$
G	Shear modulus of the frame	$4.00 \times 10^9\,\mathrm{Pa}$	$3.57 \times 10^9\,\mathrm{Pa}$
K_{fr}	Bulk modulus of the frame	$2.22 \times 10^9\,\mathrm{Pa}$	$6.89 \times 10^9\,\mathrm{Pa}$
k_0	Low-frequency permeability	$10^{-12}\,\mathrm{m^2}$	$10^{-16}\,\mathrm{m^2}$
η_f	Dynamic viscosity of the pore fluid	$10^{-3}\,\mathrm{Pa\,s}$	$10^{-3}\,\mathrm{Pa\,s}$

Unit U1 simulates a sandstone, and unit U2 is used to simulate a clayey sandstone.

In the finite-element code COMSOL Multiphysics, the governing equation of the forward electrical model can be written at each time step as

$$\mathbf{K\psi} = \mathbf{s}, \qquad (4.66)$$

where $\mathbf{K}$ is the kernel matrix that contains the discrete form of the differential operator on the left hand of Equation (4.66), $\mathbf{s}$ is a vector containing the M source current density terms $\Im(\mathbf{x}, t)$, and $\mathbf{\psi}$ is the vector of electric potential observations at the N receiver locations.

At the boundaries of the domain, we used the following boundary conditions: A Neumann boundary condition is used at the interface between the model and the padding layer. At the outer edge of the padding layer, a Dirichlet boundary condition is used. The electric potential at the outer edge of the padding layer tends to zero, thus simulating an infinite domain (Dey & Morrison, 1979).

4.4.3 The initial inverse solution

Because the number of the measurement points is much less than the number of mesh elements of the model where the source can be located ($N \ll M$) (this is typical of potential field problems), the inverse problem is underdetermined. It is also ill posed and the solution is nonunique. The nonuniqueness of the inverse problem can be significantly reduced by using a regularization approach to select a solution that minimizes both the data misfit and, at the same time, carries a representation of the model structure that is consistent with some prior information (sparseness of the source distribution in this case).

The inverse problem involves reconstructing the spatial distribution of the volumetric source field (right-hand side of Eq. 4.66) at each time t through the optimization of the objective functions

$$C = \left\| \mathbf{W}_d \left(\mathbf{Gs} - \mathbf{\psi}^{obs} \right) \right\|_2^2 + \lambda \|\mathbf{\Lambda s}\|_2^2, \qquad (4.67)$$

$$\mathbf{\Lambda} = \mathrm{diag} \sqrt{\sum_{i-1}^{N} G_{kj}^{T\,2}}, \qquad (4.68)$$

where $\mathbf{G} = \mathbf{PK}^{-1}$ denotes the $N \times M$ matrix of the Green functions. This matrix is computed as the product of the inverse kernel matrix times a sparse selector operator matrix $\mathbf{P}_{(N \times M)}$ that contains a single 1 on each row in the column that corresponds to the location of that receiver. The rows of $\mathbf{G}$ can be computed effectively using reciprocity, which involves computing the forward response to a unit source located at each receiver. The vector $\mathbf{\Lambda}$ represents an inverse-sensitivity weighting function that accounts for distance from the receivers as well as the resistivity structure, $\mathbf{s}$ is the vector containing the discretized source current density terms $\Im(\mathbf{r}, t)$ with dimension M, $\mathbf{\psi}^{obs}$ is the observed electric potential vector at the N sensors, and $\mathbf{W}_d$ is a matrix that contains the information about the expected noise in the data. The parameter λ in Equation (4.67) corresponds to a trade-off parameter between the two contributions of the cost function. In other words, this term balances the relative influence of the data misfit term, $C_d = \left\| \mathbf{W}_d \left(\mathbf{Gs} - \mathbf{\psi}^{obs} \right) \right\|_2^2$, and the model misfit term, $C_m = \|\mathbf{\Lambda s}\|_2^2$. It is called the regularization term.

In the following, we consider $\mathbf{W}_d$ to be diagonal with each element on the diagonal being the inverse of the estimated variance of the measurement errors. In all the tests, we consider a Gaussian noise with a standard deviation equal to 10% of the computed data mean. This is a realistic noise level for this type of experiment accounting for the amplitude of the signals that are measured (see, for instance, Ardjmandpour et al., 2011). In addition, adding data weights helps to stabilize the inversion by eliminating artifacts that come from overfitting the data. The vector Λ represents the inverse-sensitivity weighting function. This function is needed because sensitivities decay quickly away from the receiver locations (typically as a power law function of the distance from the receiver in a homogeneous medium, but is also affected by heterogeneous resistivity distributions). The weighting function is therefore needed to recover sources that are distant from the receivers.

Applying the following transform $\mathbf{s_w} = \Lambda \mathbf{s}$ and minimizing Equation (4.67) gives the equation

$$\left[\left(\Lambda^{-1}\mathbf{G}^T\mathbf{W_d^T}\mathbf{W_d}\mathbf{G}\Lambda^{-1}\right) + \lambda \mathbf{I}\right]\mathbf{s_w} = \Lambda^{-1}\mathbf{G}^T\mathbf{W_d^T}\mathbf{W_d}\boldsymbol{\psi}^{obs} \tag{4.69}$$

The result of such an inversion is a smooth volumetric source current distribution. However, we know from the physics of the problem that the solution should be spatially compact. In the next section, we describe a modification to the model regularization term that promotes this compactness.

4.4.4 Getting compact volumetric current source distributions

Source compactness is a relatively classical technique that suits the nature of the electrical problem because the volumetric source current densities associated with the seismoelectric conversion tend to be spatially localized. The technique has been used in medical imaging and in geophysics (see, for instance, Last & Kubik, 1983, and Silva et al., 2001). Compactness is based on minimizing the spatial support of the source. The new global objective function is modified to include compactness as a regularization term,

$$C = \left\|\mathbf{W_d}\left(\mathbf{Gs} - \boldsymbol{\psi}^{obs}\right)\right\|_2^2 + \lambda \sum_{k=1}^{M} \frac{s_k^2}{s_k^2 + \beta^2}, \tag{4.70}$$

where β is the threshold term introduced to provide stability as $s_k \to 0$. This form of the objective function is

now nonlinear since the compactness portion of the objective function is nonquadratic. The compactness term is effectively a measure of the number of source parameters that are greater than β, regardless of their magnitude. Minimization of this objective function, Equation (4.70), results in the solution that uses the fewest number of source parameters that are still consistent with the measured data, which enforces sparseness of the source distribution. As model values fall below the threshold β, they no longer contribute to the sum in Equation (4.70) and will be effectively masked from the solution.

In order to make this compact source problem linear so that it can be solved in a least-squares framework and to incorporate the inverse-sensitivity scaling, the model weighting operator Λ in Equation (4.69) is modified as

$$\boldsymbol{\Omega} = \text{diag}\left\{\frac{\Lambda_{kk}^2}{s_{k_{(j-1)}}^2 + \beta^2}\right\}, \tag{4.71}$$

where diag(.) is an operator extracting the diagonal elements of the argument. Hence, the problem is transformed to a linear one by making the objective function quadratic in s_k by fixing the denominator of the model objective function $\boldsymbol{\Omega}$ with respect to the previous solution at step $(j-1)$ using an iteratively reweighted least-squares approach. The vector $\mathbf{s}_{j-1}$ is the initial model used to compute the first degree of compactness. A new vector $\boldsymbol{\Omega}$ is determined for every compactness degree based on the previous model generated from the immediate previous compactness degree. Using the renormalization with $\mathbf{s_w} = \boldsymbol{\Omega}\mathbf{s}$ and minimizing the global model objective function in Equation (4.70) give us the iterative solution which utilizes compactness

$$\left(\boldsymbol{\Omega}_{j-1}^{-1}\mathbf{G}^T\mathbf{W_d^T}\mathbf{W_d}\mathbf{G}\boldsymbol{\Omega}_{j-1}^{-1} + \lambda \mathbf{I}\right)\mathbf{s}_{w,j} = \boldsymbol{\Omega}_{j-1}^{-1}\mathbf{G}^T\mathbf{W_d^T}\mathbf{W_d}\boldsymbol{\psi}^{obs}. \tag{4.72}$$

The process is halted after several iterations. Focusing the image is a subjective choice. Nine iterations offer a good compactness level to localize the sources responsible for the observed self-potential data.

4.4.5 Benchmark tests

Before embarking on the process of inverting the seismoelectric signals generated by the seismoelectric forward model, it is essential to benchmark our inversion algorithm through a synthetic and well-defined 2.5D electrical model. The benchmark model is a $450 \times 500\,m$ section constructed with 45×50 cells. The conductivity distribution resembles that of the first case study of the vertical interface with two units U1 and U2 in the model. In the first test, we positioned a single point source at $x = 430\,m$ and $z = 310\,m$, and we recorded the electrical potentials at $x = 510\,m$ via an array comprised of 50 electrodes with $10\,m$ vertical separation between adjacent electrodes so that the array extends from $z = 50\,m$ to $z = 550\,m$. The initial inverse source solution is recovered using Equation (4.63), which is plotted in Figure 4.19a. This initial solution is diffusive and does not represent the spatial nature of the true compact source. Figure 4.19b shows the electric potential corresponding to the source from the initial inverse solution versus the observed data with added noise. It is clear that the algorithm is able to predict, with a great accuracy, the measured voltages without overfitting the noise. Nevertheless, the 2D representation of the model is overly smeared, thus the need to include compactness as a regularization tool as mentioned earlier. Upon performing only five iterations of Equation (4.72), the algorithm was able to locate the source precisely as shown in Figure 4.19b.

To test the ability of the algorithm to resolve two localized sources in a heterogeneous conductivity distribution, we conducted another benchmark test. Two volumetric current sources are located at position (x, z) = $(250\,m, 460\,m)$ and $(250\,m, 150\,m)$, respectively. The geometry is kept identical to the previous test. The resulting electrical voltage is aliased by the superstition of the two voltage distributions. The initial (diffusive) solution in terms of inverted current density is presented in Figure 4.20a using the observed data contaminated with added Gaussian noise (same as aforementioned), and in Figure 4.20b, we show the result of the inversion after five iterations in the focusing procedure. The algorithm is able to distinguish and localize two distinct current sources away from the receivers' location without losing sensitivity. Also, the predicted electrical voltages recorded in borehole #2 (see Figure 4.20b) are almost identical to the observed one. In other words, the algorithm is able to retrieve the true amplitude of the potentials without prior knowledge on the source.

These results highlight the ability of the compact source inversion to accurately recover the true spatial distribution of current sources. In these benchmark examples, the true sources are known to be highly compact. In other cases, the degree of compactness may not be known a priori. In other words, the electrical potential data can be accurately reproduced with diffuse or compact sources because of the nonunique nature of the problem; the user must decide how many compactness iterations are appropriate. The choice in focusing the tomogram is therefore the choice of the use, which is definitively a drawback of this approach. Our tests are showing that 5–10 iterations are usually a good iteration number to properly focus the tomogram. Future efforts to address this issue will include the use of auxiliary information from the recorded seismic response to help constrain the location of the heterogeneities.

4.4.6 Numerical case studies

Our goal is to invert the source current densities at heterogeneities regardless of their magnitudes. However, we note that our algorithm is able to retrieve the values of the divergence of the current source density as well. Hence, inverting for these sources using recorded voltages in the offset borehole will provide us with a tool to distinguish and localize heterogeneities from the background material. We start by sequentially triggering seismic sources in the source borehole (borehole #1) in order to produce electric potential time series recorded in borehole #2 for times belonging to the SC-time window. The process is repeated for five seismic shots separated by $75\,m$ in borehole #1 (see Figure 4.15). Electric potentials at each time event are recorded versus depth and processed by the inversion algorithm. As mentioned earlier, the decision to stop the compactness algorithm is arbitrary. It seems that focusing the tomograms up to the ninth iteration provides a suitable compact solution. All of the inversions for different time events and for different seismic shots are aggregated to produce the final image, which shows the position of the heterogeneity between the two boreholes.

Figure 4.21 shows eight voltage recordings as a function of depth for eight different time steps for shot #3 (localized in the middle of the borehole in which the seismic sources are located). This case corresponds to case study #1 in Figure 4.15a (vertical interface between two porous media). The eight times we chose belong to the SC-time window and sample the seismoelectric conversion signal such as that illustrated in Figure 4.21. At each time step, the vertical recordings of the potential field are processed using the source inversion algorithm to produce

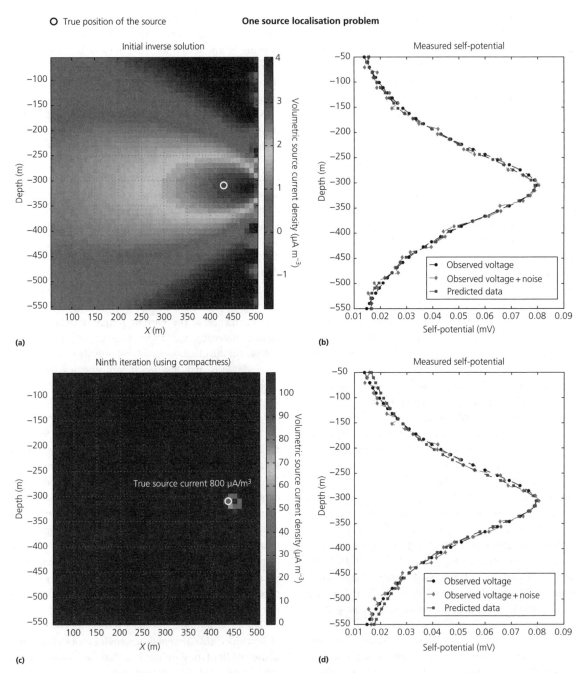

Figure 4.19 Test of the electrical source localization for a single source localized in an inhomogeneous medium. The resistivity distribution is given in Figure 4.15a (see Table 4.4). A Gaussian noise with a standard deviation equal to 10% of the computed data mean has been added to the data. The sources are analyzed in terms of volumetric source current distributions. **a)** Source current distribution at the first iteration. **b)** Comparison between the true self-potential distribution and the one determined from the inverted source current density distribution. **c)** Source current distribution using compactness. **d)** Comparison between the true self-potential distribution and the one determined from the inverted source current density distribution using compactness. (*See insert for color representation of the figure.*)

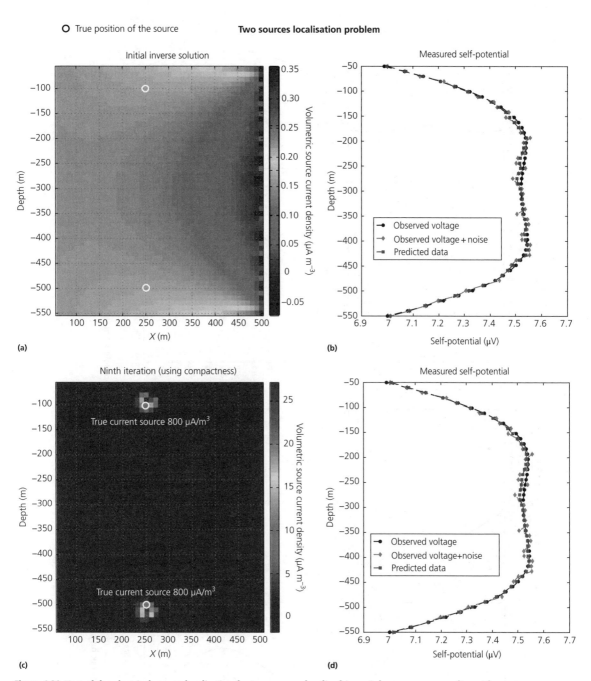

Figure 4.20 Test of the electrical source localization for two sources localized in an inhomogeneous medium. The sources are analyzed in terms of volumetric source current distributions. **a)** Source current distribution at the first iteration. **b)** Comparison between the true self-potential distribution and the one determined from the inverted source current density distribution. **c)** Source current distribution using compactness. **d)** Comparison between the true self-potential distribution and the one determined from the inverted source current density distribution using compactness. (*See insert for color representation of the figure.*)

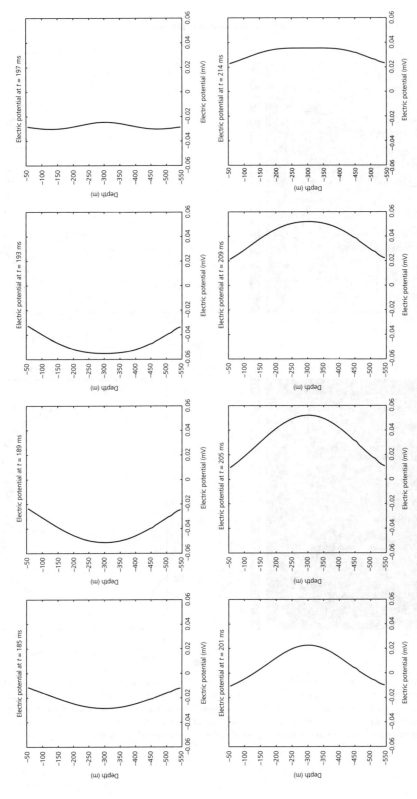

Figure 4.21 Vertical distribution of the electrical potential (in mV, reference at infinity) in borehole #2 at different times in the SC-time window in which the seismoelectric conversions taking place between the two boreholes occur (shot #3).

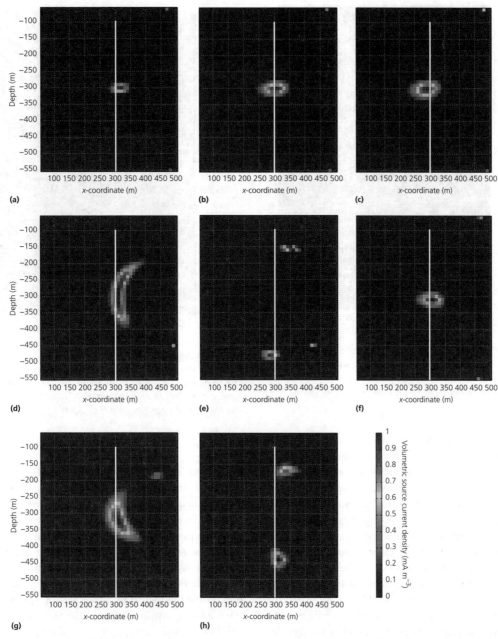

Figure 4.22 Source localization for the electrical potential distributions shown in Figure 4.21 (shot #3). The sources are analyzed in terms of volumetric source current distributions. The white line indicates the position of the discontinuity between the units U1 and U2. **a)** Source current distribution at time 184.96 ms. **b)** Source current distribution at time 187.77 ms. **c)** Source current distribution at time 190.58 ms. **d)** Source current distribution at time 193.38 ms. **e)** Source current distribution at time 196.19 ms. **f)** Source current distribution at time 210.23 ms. (*See insert for color representation of the figure.*)

a snapshot of the seismoelectric generated source on the interface, which is shown on a normalized scale in Figure 4.22. Afterward, the normalized source distributions at all time steps are aggregated to produce the

tomogram presented in Figure 4.23 that depicts the presence of a vertical contact at $x = 300$ m. The volumetric source current distributions are thresholded according to the global threshold level found by using the Otsu method

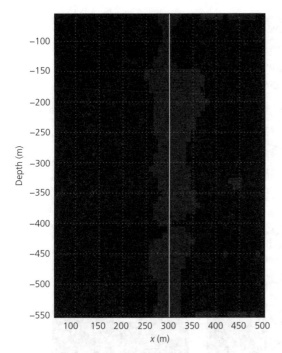

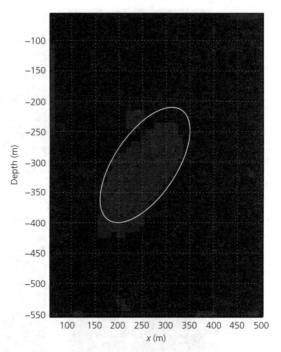

Figure 4.23 An aggregate of all the source distributions for the 5 shots using 6 characteristic times for each shot for case study #1. The white line indicates the true position of the discontinuity. (*See insert for color representation of the figure.*)

Figure 4.24 Aggregate of all the source current distributions for the 5 shots using 6 characteristic times for each shot for case study #2. The white line indicates the true position of the discontinuity between the domains U1 and U2. (*See insert for color representation of the figure.*)

(Otsu, 1979). This threshold is used to convert the distributions of the source current densities into a binary image by minimizing the interclass variance of the two types of pixels. Using multiple seismic shots enables us to "illuminate" certain parts of the interface, which is usually the part closest to the source. Thus, by utilizing several seismic shots along the source borehole, we were able to "illuminate" the interface using the seismoelectric signals.

Another synthetic case study is demonstrated to highlight the ability of the algorithm to image reservoirs or oil plumes with irregular geometries for more realistic scenarios. In this case study, we consider an inclusion of a poroelastic material embedded in another background poroelastic material (Figure 4.15b). The material properties of the two materials are those of Table 4.5. The same procedure is used as in the previous case study, and the tomogram produced is presented in Figure 4.24. We can distinguish the inclusion from the background using seismoelectric data only.

As explained in the introduction, this work focuses on the localization of the current sources corresponding to

the seismoelectric conversions, but not the amplitude of these sources. In order to obtain an image of the localization of the seismoelectric sources, we aggregated all the sources for the five shots and for eight different time steps that sample the seismoelectric voltage distribution at the receivers. The resulting source localization, shown in Figure 4.23, spreads around the position of the interface responsible for the seismoelectric conversion. While the tomogram is not perfect, we think that this is a fair comparison between the position of the true interface and the image produced by the algorithm. Note the presence of few ghosts in the tomogram corresponds to spurious source localizations. The second case study investigates an ellipsoidal anomaly embedded in a homogeneous material. The final result (following the exact same procedure as for the first case study) is shown in Figure 4.24. We see that the position of the anomaly is relatively well localized; however, there are also few ghosts that are present in the tomogram.

Table 4.5 Material properties for the numerical simulation corresponding to the case study #2 for which the inclusion U2 is used to simulate a porous formation with oil.

Parameter	Description	Unit U1	Unit U2
σ	Conductivity of the medium	$0.01\ \mathrm{S\,m^{-1}}$	$0.001\ \mathrm{S\,m^{-1}}$
$\hat{Q}_V^0$	Excess of charge per unit pore volume	$0.203\ \mathrm{C\,m^{-3}}$	$1585\ \mathrm{C\,m^{-3}}$
ρ_s	Bulk density of the solid phase	$2650\ \mathrm{kg\,m^{-3}}$	$2650\ \mathrm{kg\,m^{-3}}$
ρ_f	Bulk density of the fluid phase	$1000\ \mathrm{kg\,m^{-3}}$	$983\ \mathrm{kg\,m^{-3}}$
ϕ	Porosity	0.25	0.33
K_s	Bulk modulus of the solid phase	$36.5 \times 10^9\ \mathrm{Pa}$	$37 \times 10^9\ \mathrm{Pa}$
K_f	Bulk modulus of the fluid phase	$0.25 \times 10^9\ \mathrm{Pa}$	$2.40 \times 10^9\ \mathrm{Pa}$
G	Shear modulus of the frame	$4.00 \times 10^9\ \mathrm{Pa}$	$5 \times 10^9\ \mathrm{Pa}$
K_{fr}	Bulk modulus of the frame	$2.22 \times 10^9\ \mathrm{Pa}$	$9.60 \times 10^9\ \mathrm{Pa}$
k_0	Low-frequency permeability	$10^{-12}\ \mathrm{m^2}$	$10^{-11}\ \mathrm{m^2}$
η_f	Dynamic viscosity of the pore fluid	$10^{-3}\ \mathrm{Pa\,s}$	$10^{-1}\ \mathrm{Pa\,s}$

Following Linde et al. (2007) and Revil et al. (2007), the charge density of a partially water-saturated reservoir, $\hat{Q}_V^0$, should be replaced by $\hat{Q}_V^0/s_w$ where s_w (unitless) represents the partial saturation in water.

4.4.7 Discussion

In the present section, we address a few questions that may arise regarding the applicability of the present methods. The first is related to the sensitivity of the approach to the choice of the source parameters: What happens indeed to the model outputs if we change the source characteristics? Our choice of the source parameters was to make the source as impulsive as possible. That said, the choice of the source is totally arbitrary, and our approach is totally independent of the choice of the source. We can choose any type of source (e.g., a sweep), and we can always perform a deconvolution of the resulting seismoelectric signals with the (known) source in order to retrieve the impulse function of the system.

The second question we want to address is the noise level in the electrical data. Typically on land, it is easy to record the electrical field with a precision of at least one microvolt per meter, and the precision on the

measurements can be better than one nanovolt per meter for offshore applications (classically for CSEM) (Butler & Russell, 1993; Mikhailov et al., 1997; Dupuis et al., 2007). Therefore, the signal-to-noise ratio adopted in the present paper is rather pessimistic.

The third point is the use of compactness in the inversion of the electrical potential data. The source inversion problem identifies the position of the source between the two wells. This principle is different from the inverse problem corresponding to the imaging of the material properties using a smoothness regularization approach. A complete analogous application of the method we used previously has also been performed in the medical community using electroencephalographic (EEG) data (measuring time-varying electrical potentials on the scalp) and combining this information with resistivity information to spatially and temporally locate electrical sources within the brain. Portniaguine et al. (2001) proposed the compactness approach as a method to focus the position of the source responsible for the observed EEG anomalies. This approach has been very successful in EEG. We think that despite the drawback associated with the choice to let the user to stop the focusing of the tomogram, this method provides images that are more reliable than having only an unfocused image. The electrical signature collected at the boreholes due to seismoelectric conversion is due the occurrence of local electrical source current densities like in EEG. Also, we point out that the use of compactness does not remove the nonuniqueness of the inverse problem. As for other deterministic methods, this approach provides a set of solutions on the location of the sources. The choice of the best solution is dependent on the availability of independent information.

4.5 Conclusions

We have described the implementation of the seismic and seismoelectric equations in finite elements using the $\mathbf{u}$–p formulation. We used the softwares COMSOL and MATLAB to perform the forward modeling and MATLAB to implement routines to invert seismoelectric signals in terms of material properties and in terms of boundaries. We have introduced PML boundary conditions for the seismoelectric problem. Then, several synthetic case studies were discussed, and we demonstrated how the forward modeling is implemented and used to solve classical seismoelectric problems. Finally, we showed how inverse modeling can be used to take advantage of the information content of the seismoelectric signals, either to detect heterogeneities in the subsurface (imaging) or to determine petrophysical properties such as permeability. Inverse modeling approaches have been developed using deterministic and stochastic approaches.

References

Ardjmandpour, N., Pain, C., Singer, J., Saunders, J., Aristodemou, E. & Carter, J. (2011) Artificial neural network forward modelling and inversion of electrokinetic logging data. *Geophysical Prospecting*, **59**(4), 721–748, doi: 10.1111/j.1365-2478.2010.00935.x.

Atalla, N., Panneton, R. & Debergue, P. (1998) A mixed displacement-pressure formulation for poroelasticmaterials. *Journal of the Acoustical Society of America*, **104**(3), 1444–1452.

Berenger, J.P. (1994) A perfectly matched layer for the absorption of electromagnetic waves. *Journal of Computational Physics*, **114**, 185–200.

Bou Matar, O., Prebrazhensky, V. & Pernod, P. (2005) Two-dimensional axisymmetric numerical simulation of supercritical phase conjugation of ultrasound in active solid media. *Journal of the Acoustical Society of America*, **118**, 2880–2890.

Butler, K.E., & Russell, R.D. (1993) Subtraction of powerline harmonics from geophysical records. *Geophysics*, **58**, 898–903.

Chew, W.C. & Weedon, W.H. (1994) A 3-D perfectly matched medium from modified Maxwell's equations with stretched coordinates. *Microwave and Optical Technology Letters*, **7**, 599–604.

Clayton, R. & Engquist, B. (1977) Absorbing boundary conditions for acoustic and elastic wave equations. *Bulletin of the Seismological Society of America*, **67**, 1529–1540.

Collino, F, & Tsogka, C. (2001) Application of the PML absorbing layer model to the linear elastodynamic problem in anisotropic heterogeneous media. *Geophysics*, **66**, 294–307.

Crespy, A., Revil, A., Linde, N., et al. (2008) Detection and localization of hydromechanical disturbances in a sandbox using the self-potential method. *Journal of Geophysical Research*, **113**, B01205, doi:10.1029/2007JB005042

Dai, N., Vafidias, A. & Kanasewich, E.R. (1995) Wave propagation in heterogeneous porous media: a velocity-stress, finite-difference method. *Geophysics*, **60**, 327–340.

Dey, A., & Morrison, H.F. (1979) Resistivity modeling for arbitrarily shaped two-dimensional structures *Geophysical Prospecting*, **27**, 106–136.

Dupuis, J.C. & Butler, K.E. (2006) Vertical seismoelectric profiling in a borehole penetrating glaciofluvial sediments. *Geophysical Research Letters*, **33**, L16301, doi:10.1029/2006GL026385.

Dupuis, J.C., Butler, K.E., Kepic, A.W. & Harris, B.D. (2009) Anatomy of a seismoelectric conversion: Measurements and conceptual modeling in boreholes penetrating a sandy aquifer. *Journal of Geophysical Research*, **114**, B10306, doi:10.1029/2008JB005939.

Dupuis, J.C., Butler, K.E. & Kepic, A.W. (2007) Seismoelectric imaging of the vadose zone of a sand aquifer. *Geophysics*, **72**(6), A81–A85, doi: 10.1190/1.2773780.

Fourie, F.D. (2003) *Application of Electroseismic Techniques to Geohydrological Investigations in Karoo Rocks, Ph-D Thesis*, University of the Free State, Bloemfontein, South Africa, 195 pp.

Grandis, H., Menvielle, M., & Roussignol, M. (1999) Bayesian inversion with Markov chains, I. The magnetotelluric one-dimensional case. *Geophysical Journal International*, **138**, 757–768.

Haario, H., Saksman, E. & Tamminen, J. (2001) An adaptive Metropolis algorithm. *Bernoulli*, **7**, 223–242.

Haario, H., Laine, M., Lehtinen, M., Saksman, E. & Tamminen, J. (2004) MCMC methods for high dimensional inversion in remote sensing. *Journal of the Royal Statistical Society, Series B*, **66**, 591–607.

Haines, S. & Pride, S. (2006) Seismoelectric numerical modeling on a grid. *Geophysics*, **71**, N57–65.

Hunt, C.W. & Worthington, M.H. (2000) Borehole electrokinetic responses in fracture dominated hydraulically conductive zones. *Geophysical Research Letters*, **27**, 1315–1318.

Jardani, A. & Revil, A. (2009) Stochastic joint inversion of temperature and self-potential data. *Geophysical Journal International*, **179**(1), 640–654, doi:10.1111/j.1365-246X.2009.04295.x

Jardani, A., Revil, A., Bolève, A., Dupont, J.P., Barrash, W. & Malama, B. (2007) Tomography of groundwater flow from self-potential (SP) data. *Geophysical Research Letters*, **34**, L24403.

Jardani, A., Revil, A., Bolève, A. & Dupont, J.P. (2008) 3D inversion of self-potential data used to constrain the pattern of ground water flow in geothermal fields. *Journal of Geophysical Research*, **113**, B09204, doi: 10.1029/2007JB005302.

Jardani, A., Revil, A., Slob, E., & Sollner, W. (2010) Stochastic joint inversion of 2D seismic and seismoelectric signals in linear poroelastic materials. *Geophysics*, **75**(1), N19–N31, doi:10.1190/1.3279833.

Karaoulis, M., Revil, A., Zhang, J. & Werkema, D.D. (2012) Time-lapse cross-gradient joint inversion of cross-well DC resistivity and seismic data: A numerical investigation. *Geophysics*, **77**, D141–D157, doi: 10.1190/GBO2012-0011.1.

Kaw, A. & Kalu, E.E. (2008) Numerical Methods with Applications", 1st edition , http://www.autarkaw.com.

Last, B.J. & Kubik, K. (1983) Compact gravity inversion. *Geophysics*, **48**(6), 713–721.

Linde, N., Jougnot, D., Revil, A., et al. (2007) Streaming current generation in two-phase flow conditions. *Geophysical Research Letters*, **34**(3), L03306, doi:10.1029/2006GL028878

Londergan, J.T., Meinardus, H.W., Mariner, P.E., et al. (2001) DNAPL removal from a heterogeneous alluvial aquifer by surfactant-enhanced aquifer remediation. *Ground Water Monitoring and Remediation*, **21**(4), 57–67.

Malinverno, A. (2002) Parsimonious Bayesian Markov chain Monte Carlo inversion in a non-linear geophysical problem. *Geophysical Journal International*, **151**, 675–688.

Martin, R., Komatitsch, D. & Ezziani, A. (2008) An unsplit convolutional perfectly matched layer improved at grazing incidence for seismic wave equation in poroelastic media. *Geophysics*, **73**(4), T51–T61

Mercier, J.W. & Cohen, R.M. (1990) A review of immiscible fluids in the subsurface: Properties, models, characterization and remediation. *Journal of Contaminant Hydrology*, **6**, 107–163.

Metropolis, N., Rosenbluth, A.W., Rosenbluth, M. N., Teller, A. H. & Teller, E. (1953), Equations of state calculations by fast computing machine, *Journal of Chemistry Physics*, **21**, 1087–1091.

Mikhailov, O.V., Haartsen, M.W. & Toksoz, M.N. (1997) Electroseismic investigation of the shallow subsurface: Field measurements and numerical modeling. *Geophysics*, **62**(1), 97–105.

Mikhailov, O.V., Queen, J. & Toksöz, M.N. (2000) Using borehole electroseismic measurements to detect and characterize fractured (permeable) zones. *Geophysics*, **65**(4), 1098–1112.

Otsu, N. (1979) A threshold selection method from gray-level histograms. *IEEE transactions on Systems, Man, and Cybernetics*, **9**(1), 62–66.

Pope, G.A. & Wade, W.H. (1995) Lessons from Enhanced Oil Recovery Research for Surfactant-Enhanced Aquifer Remediation, in *Surfactant Enhanced Subsurface Remediation*, **11**, 142–160, doi: 10.1021/bk-1995-0594.ch011.

Portniaguine, O., Weinstein, D. & Johnson, C. (2001) Focusing inversion of electroencephalography and magnetoencephalography data. *Journal of Biomedical Technology*, **46**, 115–117.

Pride, S.R. & Haartsen, M.W. (1996) Electroseismic wave properties. *Journal of the Acoustical Society of America*, **100**(3), 1301–1315.

Revil, A. & Cathles, L.M. (1999) Permeability of shaly sands. *Water Resources Research*, **35**, 651–662.

Revil, A. & Jardani, A. (2010) Seismoelectric response of heavy oil reservoirs. Theory and numerical modelling. *Geophysical Journal International*, **180**, 781–797, doi: 10.1111/j.1365-246X.2009.04439.x.

Revil, A., Linde, N., Cerepi, A., Jougnot, D., Matthäi, S. & Finsterle, S. (2007) Electrokinetic coupling in unsaturated porous media. *Journal of Colloid and Interface Science*, **313**(1), 315–327, doi:10.1016/j.jcis.2007.03.037

Roden, J.A. & Gedney, S.D. (2000) Convolution PML, (CPML): An efficient FDTD implement of CFS-PML for arbitrary media. *Microwave & Optic, Technological Letters*, **27**, 334–339.

Schenk, O. & Gärtner, K. (2004) Solving unsymmetric sparse systems of linear equations with PARDISO. *Journal of Future Generation Computer Systems*, **20**(3), 475–487.

Schenk, O. & Gärtner, K. (2006) On fast factorization pivoting methods for symmetric indefinite systems. *Electronic Transactions on Numerical Analysis*, **23**, 158–179.

Schenk, O., Waechter, A. & Hagemann, M. (2007) Matching-based preprocessing algorithms to the solution of saddle-point problems in large-scale non-convex interior-point optimization, *Journal of Computational Optimization and Applications*, **36** (2–3), 321–341.

Schenk, O., Bollhoefer, M. & Roemer, R. (2008) On large-scale diagonalization techniques for the Anderson model of localization, *SIAM Review*, **50**, 91–112.

Silva, J.B.C., Medeiros, W.E. & Barbosa, V.C.F. (2001) Potential-field inversion: Choosing the appropriate technique to solve a geologic problem. *Geophysics*, **66**(2), 511–520.

Soueid, A.A., Jardani, A., Revil, A. & Dupont, J.P. (2013) SP2DINV: A 2D forward and inverse code for streaming potential problems. *Computers &Geosciences* **59**, 9–16.

Tarantola, A. (2005) *Inverse Problem Theory. Methods for model parameter estimation*. SIAM, Philadelphia.

White, B.S. (2005) Asymptotic Theory of Electroseismic Prospecting. *SIAM Journal of Applied Mathematics*, **65**(4), 1443–1462.

White, B.S. & Zhou, M. (2006) Electroseismic Prospecting in Layered Media. *SIAM Journal of Applied Mathematics*, **67**(1), 69–98.

Zeng, Y.Q. & Liu, Q.H. (2001) A staggered-grid finite-difference method with perfectly matched layers for poroelastic wave equations. *Journal of the Acoustical Society of America*, **109**(6), 2571–2580.

Zeng, Y.Q., He, J.Q. & Liu, Q.H. (2001) The application of the perfectly matched layer in numerical modeling of wave propagation in poroelastic media. *Geophysics*, **66**, 1258–1266.

CHAPTER 5

Electrical disturbances associated with seismic sources

In this chapter, we describe the electromagnetic disturbances directly associated with a seismic source. We first consider such a source in a water-saturated poroelastic material characterized by a moment tensor. We model the different types of electromagnetic disturbances associated with such a source. Then, based on the presentations in Chapters 1 to 4, we show how it is possible to observe and model electrical disturbances associated with various hydromechanical disturbances in saturated or partially saturated porous media. We present a laboratory experiment showing, at the scale of a cement block, the types of electrical disturbances that can be associated with the rupture of a seal during a hydraulic fracturing experiment. Next, we present an example of laboratory data generating bursts in the electrical field. These bursts are associated with the occurrence of Haines jumps, corresponding to jumps of the meniscus between two immiscible fluid phases, during drainage in a sandbox. Finally, we present a small-scale field experiment demonstrating how a brief burst of water injection in a small well produces a localizable transient electrical disturbance.

5.1 Theory

5.1.1 Position of the problem

Seismic event hypocenter localization and moment tensor solutions are integral components in the passive seismic monitoring of the hydraulic fracturing process

(Aki & Richard, 2002; McGarr et al., 2002; Stein & Wysession, 2003). Several methods have been developed in the past 50 years to simultaneously compute the point source mechanism and source time function of seismic events (Zhao & Helmberger, 1994; Ritsema & Lay, 1995; Sekiguchi et al., 2002). That said, in addition to the seismic signals that are routinely recorded, there have been several observations of direct electromagnetic disturbances associated with seismic sources (e.g., Kuznetsov et al., 2001; Byrdina et al., 2003). Gao and Hu (2010) demonstrated numerically that a seismic source is directly responsible for several seismoelectromagnetic signals that can also be recorded remotely. Their study opened the door to the possibility of extracting a seismic source location more accurately by using both seismic and electromagnetic signals. Moore and Glaser (2007) developed a comprehensive set of laboratory experiments and modeling to explain the occurrence of electrical fields associated with the fracturing of crystalline rocks. Wishart et al. (2008) performed a laboratory investigation of the use of passively recorded electrical disturbances associated with pneumatic fracturing (see also Ushijima et al., 1999, for a field experiment).

Our goal in this section is to show whether the joint inversion of related seismic and electrical signals is informative. Does the spatially distributed measurement of the electric field during seismic data acquisition bring useful information to the joint inversion problem? In such a case, the electrical information may help to reduce the uncertainty in the seismic source model parameters

The Seismoelectric Method: Theory and Applications, First Edition. André Revil, Abderrahim Jardani, Paul Sava and Allan Haas.
© 2015 John Wiley & Sons, Ltd. Published 2015 by John Wiley & Sons, Ltd.

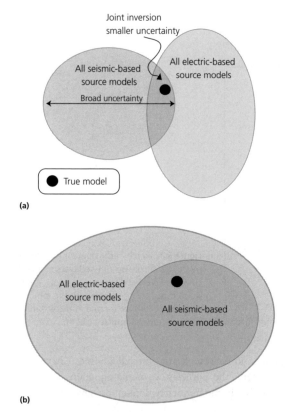

Joint inversion
smaller uncertainty

(a)

(b)

Figure 5.1 Information content of seismic and electromagnetic disturbances. **a)** Informative joint inversion case. The joint inversion of seismic and electrical data may help in reducing the uncertainty in the joint inversion problem by providing information that is not contained in the seismic data alone. **b)** Noninformative joint inversion case. All the seismic-based inverted models are contained in the set of electrical-based model. This can happen when the signal-to-noise ratio is much greater for the electrical data than for the seismic data.

relative to the uncertainty in the model parameters derived by using seismic data alone (Figure 5.1a). There is, however, the possibility (shown conceptually in Figure 5.1b) that the electrical information is noninformative. This is the case where the posterior probability densities for the model parameters determined from the seismic data alone are contained inside the posterior probability densities for the same model parameters derived from the electrical data alone. Such a case is possible since the level of noise in the seismic and electromagnetic data can be vastly different.

To determine the benefits of incorporating electrical data, we developed a simple finite-element model to compute the seismic and electrical fields associated with

a localized seismic source that is described only by a moment tensor. Because the electrokinetic coupling mechanism requires the computation of the relative water flow vector with respect to the solid phase, we use the Biot poroelastic theory to describe the elastodynamic part of the problem (see Chapter 2).

The second point that needs to be discussed is the choice of the inversion method. Whenever the information content of a data set is questioned, it is better to use Bayesian theory associated with a Markov chain Monte Carlo (MCMC) sampler techniques rather than a deterministic-based approach (Mosegaard, 2011 and references therein and Chapter 4). However, classical Monte Carlo samplers are not very efficient in retrieving the posterior probability density of model parameters, because of the slow convergence of the chain. In the present work, we will use the adaptive Metropolis algorithm (AMA; see Haario et al., 2001, 2004) for the inverse modeling, and we will discuss the advantages of this approach. We will also show how electrical data alone can be used to localize the seismic source and the heterogeneities that are present along the wave fronts between the source and the receivers.

5.1.2 Forward modeling

To test our forward and inverse modeling, we created a two-layer earth model comprised of layers L1 and L2 (L1 corresponds to the upper layer; see Figure 5.2). The material properties of L1 and L2 are summarized in Table 5.1. The mechanical property values reported in Table 5.1 yield a P-wave velocity $c_p = 4322 \, \mathrm{m \, s^{-1}}$ and an S-wave velocity $c_s = 2721 \, \mathrm{m \, s^{-1}}$ for L1 and a P-wave velocity $c_p = 1932 \, \mathrm{m \, s^{-1}}$ and an S-wave velocity $c_s = 1317 \, \mathrm{m \, s^{-1}}$ for L2. For the seismic source, we used a moment tensor source corresponding to a double couple on a normal fault (90° strike from the north direction and 45° dip angle from the horizontal plane). The seismic stations and electrodes are colocated close to the Earth's surface, at a depth of 5 m (as explained in the following text, we impose a perfectly matched layer (PML) just at the top surface of the system). The modeling domain itself corresponds to a two-dimensional slab (1.2 × 1.2) km², The reference position, O(0,0), is located at the upper left corner of this domain.

We will use the following assumptions in this section: (i) the porous material is isotropic, (ii) fully water-saturated with a connected porosity, (iii) the pore fluid (water) is a viscous Newtonian fluid and the skeleton is

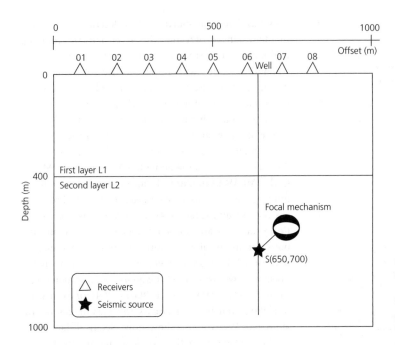

Figure 5.2 Sketch of the domain used for the modeling. It consists of two layers named L1 (shallow layer) and L2 (deeper layer). A set of seismic and electrical receivers (geophones and electrodes) are located at the ground surface (upper boundary of L1).

Table 5.1 Material properties of the two layers L1 and L2 used for the forward model.

Parameter	L1	L2	Units
σ	0.01	0.05	S m^{-1}
$\hat{Q}_V^0$	4.2	1234.8	C m^{-3}
ρ_s	2650	2650	kg m^{-3}
ρ_f	1040	1000	kg m^{-3}
ϕ	0.20	0.15	—
K_s	36.5	35.7	GPa
K_{fr}	2.22	17.9	GPa
K_f	0.25	2.25	GPa
G	4.0	17.7	GPa
k_0	1×10^{-12}	1×10^{-15}	m^2
η_f	1×10^{-3}	1×10^{-3}	Pa s
c_p	1925	4322	m s^{-1}
c_s	1310	2721	m s^{-1}

elastic, and (iv) other attenuation mechanisms like squirt-flow mechanisms are neglected. In the following, we assume an $e^{-i\omega t}$ time dependence for the mechanical disturbances (i being the pure imaginary number used in Chapter 1, $i^2 = -1$, ω is the angular frequency, and t is time). We use the displacement of the solid phase and the pore fluid pressure as unknowns instead of the solid and fluid phase displacement vectors like in most previous papers in seismoelectric modeling (see Chapter 2). Therefore, the poroelastodynamic equations of motion expressed in the frequency domain are given by

$$-\omega^2 \rho_\omega^S \mathbf{u} + \theta_\omega \nabla p = \nabla \cdot \hat{\mathbf{T}} + \boldsymbol{F}, \qquad (5.1)$$

$$\hat{\mathbf{T}} = (\lambda \nabla \cdot \mathbf{u})\mathbf{I} + G(\nabla \mathbf{u} + \nabla \mathbf{u}^{\mathrm{T}}), \qquad (5.2)$$

$$\frac{p}{M} + \nabla \cdot \left[k_\omega (\nabla p - \omega^2 \rho_f \mathbf{u} - f) \right] = -\alpha \nabla \cdot \mathbf{u}, \qquad (5.3)$$

where $\mathbf{I}$ is the identity matrix, $\boldsymbol{F}$ corresponds to the body (volumetric) force acting on the bulk porous medium, $\mathbf{u}$ denotes the (averaged) displacement of the solid phase, and f denotes the bulk force acting on the fluid phase. These three equations correspond to (i) a macroscopic momentum conservation equation for the solid phase, (ii) a constitutive equation for the effective stress tensor $\hat{\mathbf{T}}$, and (iii) a momentum conservation equation for the pore fluid. In these equations, p is the average pressure in the pore fluid phase, and the coefficient M is one of the Biot moduli defined by $M = K_f K_S / [K_f (1 - \phi - K_{fr}/K_S) + \phi K_S]$, where K_S and K_f denote the bulk moduli of the solid phase and fluid phase, respectively; K_{fr} symbolizes the bulk modulus of the solid frame (skeleton); and ϕ corresponds to the connected porosity.

The modulus λ corresponds to the drained Lamé coefficient defined by $\lambda = K_{fr} - (2/3)G$ where $G = G_{fr}$

denotes the shear modulus of the solid frame (skeleton). The density, ρ_ω^S, in Equation (5.1) signifies an apparent mass density of the solid phase. The coupling coefficient in Equation (5.1) symbolizes a hydromechanical coupling coefficient and in Equation (5.3) corresponds to a permeability-related constant. These three coefficients are given by (Chapter 2)

$$k_\omega = \frac{1}{\omega^2 \tilde{\rho}_f + i\omega b} \qquad (5.4)$$

$$\rho_\omega^S = \rho - \omega^2 \rho_f^2 k_\omega \qquad (5.5)$$

$$\theta_\omega = \alpha - \omega^2 \rho_f k_\omega \qquad (5.6)$$

and $b = \eta_f / k_0$ where η_f is the dynamic viscosity of the pore water and k_0 symbolizes the permeability for quasistatic flow. In the following, we will define the bulk density of material, ρ, as $\rho = (1-\phi)\rho_S + \phi\rho_f$, where ϕ denotes the connected porosity and ρ_S and ρ_f are the solid and pore solution mass densities, respectively. The Biot coefficient, α, is defined by

$$\alpha = 1 - \frac{K_{fr}}{K_S} \qquad (5.7)$$

The Biot coefficient also connects the other Biot poroelastic moduli, C and M, through $C = \alpha M$.

We now turn our attention to the body force, f, which corresponds to the averaged body force acting on the fluid phase. This source term is related to the bulk body force F. Indeed, the main mechanism for the pore fluid flow (relative to the solid phase) in these formulations is seismic disturbances (Pride & Haartsen, 1996). Therefore, we decompose the total bulk force generated by the seismic source as $F = F_S + f$, where $F_S = (1-\phi)F$. Here, F_S denotes the bulk force acting on the solid phase and f is the bulk force acting on the pore fluid phase, respectively (Huang, 2002; Huang & Liu, 2006). The relationship between F and the properties of the seismic source is described in the following text at the end of Section 5.1.2.

Dynamic electrokinetic theory describes the generation of an electrical current associated with the existence of an effective excess of electrical charge per unit pore volume in the pore space of a porous material. This excess of charge counterbalances the charge located on the mineral surface because of the chemical reactivity of the mineral surface in contact with water (see Revil et al., 1999a, b; Leroy & Revil, 2004 for electrical double

layer models for silica and aluminosilicates; see also Chapter 1). In our model, we note that $\hat{Q}_V^0$ is the net charge of the pore water that can be dragged along by the flow of the pore water (Revil & Linde, 2006; Bolève et al., 2007; Revil, 2007; Revil et al., 2011; Jougnot et al., 2012). The viscous drag of this net volumetric charge density is responsible for a net source current density (called the streaming current density) in a reference system attached to the solid phase. In this formulation, the excess of charge per unit pore volume, $\hat{Q}_V^0$, is related to the quasistatic permeability, k_0, by the following empirical formula (Jardani et al., 2007; see Chapter 1):

$$\log_{10} \hat{Q}_V^0 = -9.2349 - 0.8219\log_{10} k_0 \qquad (5.8)$$

This empirical relationship provides a pretty good fit to existing data for a wide variety of porous materials, as shown in Figure 1.9. The influence of salinity on $\hat{Q}_V^0$ is neglected. As shown recently by Jougnot et al. (2012), the empirical relationship of Equation (5.8) holds true in unsaturated conditions. This empirical relationship also implies that, for the same magnitude of hydromechanical disturbance, the generated source current density will be roughly the same in shales and in sands. Indeed, in the first case, the excess of charge per unit pore volume is very high, and the permeability controlling the Darcy velocity is very small, while in the second case, the charge per unit volume is very small, but the permeability is very high (Figure 1.9).

The dynamic streaming current generated by seismic signals is not a directly measurable quantity; however, the streaming potential is a directly measurable quantity. The streaming potential spreads out and away from the source current density that generated it through porous media with finite resistivities. This makes the streaming potential a remotely measurable quantity that can be measured with any appropriately configured voltage measurement system. Because of this property of the streaming potential, we seek out the solution for it using Maxwell's equations in their quasistatic limit and then invert for the source current density that generates the streaming potential. In the quasistatic limit formulation of these equations, frequency is low (<10 kHz), and diffusion is the transport process. A consequence of the quasistatic approximation is that there is no electromagnetic radiation of the fields, and therefore, all of the signals are contained entirely within media with finite and measurable conductivities. Manipulation of Maxwell equations under this formulation yields the electric field, $\mathbf{E} = -\nabla \psi$ (in V m^{-1}, where ψ denotes the electrostatic potential), and the magnetic field $\mathbf{H}$ (in A m^{-1}) is defined by the following scalar and vectorial Poisson equations:

$$\nabla \cdot (\sigma \nabla \psi) = \nabla \cdot \mathbf{J}_S, \tag{5.9}$$

$$\nabla^2 \mathbf{B} = -\mu \nabla \times \mathbf{J}_S, \tag{5.10}$$

where σ and μ are the electrical conductivity (in S m^{-1}) and the magnetic permeability (in H m^{-1}) of the medium, respectively, and where the dynamic streaming current density, $\mathbf{J}_S$, is given by (Jardani et al., 2010; Revil & Jardani, 2010)

$$\mathbf{J}_S = \hat{Q}_V^0 \dot{\mathbf{w}} = -i\omega \, \hat{Q}_V^0 \left(\nabla p - \omega^2 \rho_f \mathbf{u} - f \right) \tag{5.11}$$

The insulating boundary condition at the ground surface is written as $\hat{\mathbf{n}} \cdot \mathbf{E} = 0$ ($\hat{\mathbf{n}}$ denotes the normal unit vector to the ground surface). Therefore, the normal component of the electrical field at the ground surface is strictly equal to zero. This condition respects the non-radiative nature of the quasistatic limit of Maxwell equations. Note also that we will use, later on, a PML boundary condition at the ground surface to simplify the analysis (see Chapter 4). This assumption is, strictly speaking, incorrect for the Earth surface, but it is a good approximation if we consider just the first wave arrivals. In the following, we will compute the vertical component of the electrical field at a depth of 5 m.

When Equations (5.9) through (5.11) are coupled to the poroelastodynamic equations, three types of electromagnetic signals are generated by a seismic source. We use the following nomenclature with Roman numerals hereinafter:

i Type I corresponds to electromagnetic signals directly triggered by a seismic source, which moves the fluid with respect to the solid phase and therefore to the body force, f, as stated in Equation (5.1) (e.g., Gao & Hu, 2010). Type I signals can be described by a multipole expansion (the leading term is expected to be the dipolar or quadrupolar term). Therefore, the amplitude of type I electromagnetic disturbances quickly decreases with the distance between the seismic source and the receivers.

ii Type II denotes electromagnetic disturbances generated at heterogeneities in the electrical, mechanical, and transport properties (e.g., a geological interface) during the propagation of the seismic waves

themselves (e.g., Huang & Liu, 2006). Therefore, these types of conversions are sensitive to the level of heterogeneities of the subsurface, and they have been broadly recognized in the literature as seismo-electric conversions (e.g., Garambois & Dietrich, 2001). If the subsurface is described with piecewise constant material properties, these conversions are created at the interface between homogeneous blocks and are sometimes called "interfacial" conversions.

iii Type III corresponds to the coseismic electromagnetic signals. They are electromagnetic signals propagating at the same speed as the P- and S-seismic waves and are due to local fluid flow associated with the passage of the seismic wave. This flow generates a source current density that is locally compensated by the conduction current density in homogeneous materials. The type III electrical field is proportional to the acceleration of the seismic wave with a local transfer function that depends on the local electrical properties of the soil.

Because electrical source signals (type I in the earlier nomenclature) are observed at receivers at the moment when the seismic source occurs (the travel time of the information is quasi-instantaneous), they provide the time of occurrence of the seismic event (time zero). We will see that the electric field information is very useful to characterize a seismic point source location in addition to determining the seismic source mechanism information.

The bulk body force, F, described earlier, is related to the moment tensor of the seismic source by (Aki & Richards, 2002)

$$\mathbf{F}(r) = -\mathbf{M} \cdot \nabla \delta(r - r_S) S(\omega) \tag{5.12}$$

where $r_S(x_S, z_S)$ denotes the source position, r is the observation position or receiver position, $S(\omega)$ represents the source time function (written here in the frequency domain), and δ signifies the delta function. The moment tensor, $\mathbf{M}$, is a quantity that depends on the source strength and contains all the information about the source properties. Importantly, the distance between the source and the receiver needs to be much greater than the dimension of the fault plane (Aki & Richards, 2002). This forms the "localized source" assumption and is usually used to characterize seismic sources in the far field. In earthquake seismology, the seismic response is inverted to retrieve the position of the source; its seismic moment, M_0; the moment tensor components, M_{ij}; the time history of the source perturbation, $s(t)$; its dominant frequency, f_c; and to some extent the amount of stress change associated with the seismic source (Stein & Wysession, 2003). One example of the source time function used in the following is a Gaussian function:

$$s(t) = exp\left[-\left(\pi^2 f_c^2\right)(t - t_S)^2\right] \qquad (5.13)$$

$$S(\omega) = \int_{-\infty}^{+\infty} s(t) exp[i\omega t] dt \qquad (5.14)$$

where t_S denotes the initial time of the source.

In 3D, the seismic moment tensor, **M**, is composed of a set of nine force couples (Aki & Richards, 2002; Stein & Wysession, 2003). The tensor **M** is equivalent to $M_0 M_{ij}$ given by

$$M_0 M_{ij} = M_0 \begin{bmatrix} M_{xx} & M_{xy} & M_{xz} \\ M_{yx} & M_{yy} & M_{yz} \\ M_{zx} & M_{zy} & M_{zz} \end{bmatrix} \qquad (5.15)$$

5.1.3 Modeling noise-free and noisy synthetic data

We use a 2D finite-element-based numerical problem (in x and z directions) as a demonstration of the concepts. In this problem, we couple and solve the poroelastodynamic wave equations and the electrostatic Poisson equation for the electrical field in the frequency domain. To do this, we use the multiphysics modeling package COMSOL Multiphysics 3.5a and the stationary parametric solver PARDISO (http://www.computational.unibas.ch/cs/scicomp/software/pardiso/; see Schenk & Gärtner, 2004, 2006, Schenk et al., 2007, 2008). The problem is solved as follows: (i) first, we solve for the displacement of the solid phase and the pore fluid pressure using the poroelastodynamic equations, and (ii) then, we compute the electrical potential by solving the Poisson equation coupled to the solution of poroelastodynamic part of the problem. In both cases, the solution in the time domain is computed by using an inverse Fourier transform of the solution in the frequency (wave number) domain (see Jardani et al., 2010).

In the frequency domain, we use the frequency range 1–100 Hz since the seismic wave operates in this range and the associated electrical field occurs in the same frequency range (Garambois & Dietrich, 2001). Using this frequency range, we compute the inverse fast Fourier transform (FFT^{-1}) to get the time series of the seismic displacements, u_x and u_z, and the time series of the two components of the electrical field, E_x and E_z.

The finite-element computations require a mesh grid, and for this particular problem, we use a 10×10 m rectangular mesh grid on a 1000×1000 m domain for all of the computations, where the grid node spacing is small with respect to the smallest wavelength of the seismic wave. This dimension corresponds to the coarsest mesh for which the solution of the partial differential equation is still mesh geometry independent. The seismic source is located at $S(x_s = 650$ m, $z_s = 700$ m) with a magnitude of $M_w = 3.0$ equivalent to $M_0 = 3.98 \times 10^{13}$ N m, and a source time function is generated ($t_s = 0.15$ ms and $f_c = 30$ Hz). At the four external boundaries of the domain, we apply a 100 m-thick convolutional perfectly matched layer (C-PML) so the whole domain is actually 1200×1200 m so as to avoid edge effects. The sensors, located at 5 m of depth, are just below the C-PML, and therefore, the sampled fields are not influenced by the PML boundary condition. The receiver arrangement mimics the acquisition that would be obtained with triaxial geophones and dipole antennas. They are located at the 8 stations as shown in Figure 5.2.

In order to see the advantages of including the electrical field data versus the seismic data inversion alone, we generated a noisy data set by adding noise with respect to the maximum amplitude of the seismic and electrical signals. The construction of the added noise, as discussed in the previous sections, is performed as follows: given a wavelet function, $w(t)$, and reflectivity series, $r(t)$, the noise is computed using the following convolution product: $Noise(t) = w(t) \otimes r(t) = FT^{-1}[W(\omega) \cdot R(\omega)]$, where $W(\omega)$ and $R(\omega)$ denote the Fourier transforms of $w(t)$ and $r(t)$, respectively, and $\otimes$ represents the convolution product. The amplitude of the noise is scaled to 15% of the maximum amplitude of the seismic and electrical time series, with the average of the noise equal to zero. We chose a Gaussian wavelet to simulate the noisy signal with a characteristic frequency of $f_c = 35$ Hz. The reflectivity series are composed of randomly distributed numbers in the range [0, 1], as computed from a pseudo-random number generator. Note that the reflectivity noise series here are not related to reflections associated with the layering of the medium. The two distinctly generated noise time series are added to the noise-free seismic and electric data.

5.1.4 Results

Snapshots of the propagation of the seismic wave and the resulting electrical potential distribution are shown in Figures 5.3 and 5.4. Note that the electrical potential

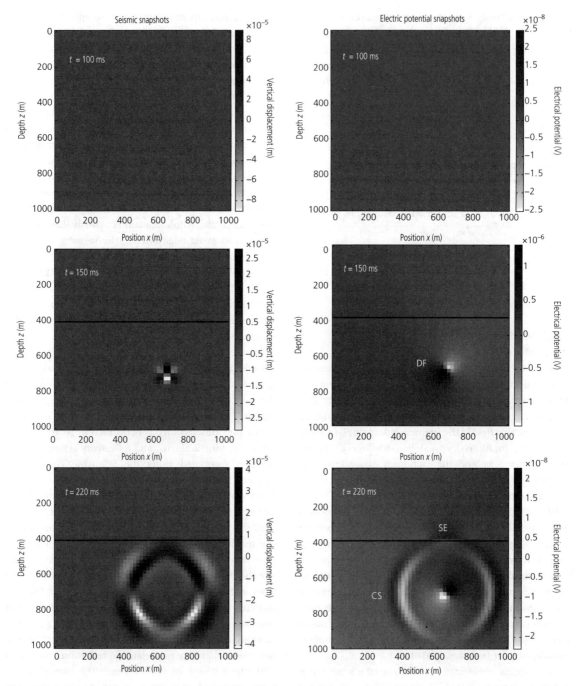

Figure 5.3 Snapshots of the vertical displacement of the solid phase during the propagation of the seismic waves and the associated electric potential distribution. DF, direct field associated with the seismic source itself (type I electrical signal); CS, coseismic P- and S-wave signals (type III electrical disturbance); SE, seismoelectric conversions at interfaces (type III electrical disturbance); and RE, reflected wave.

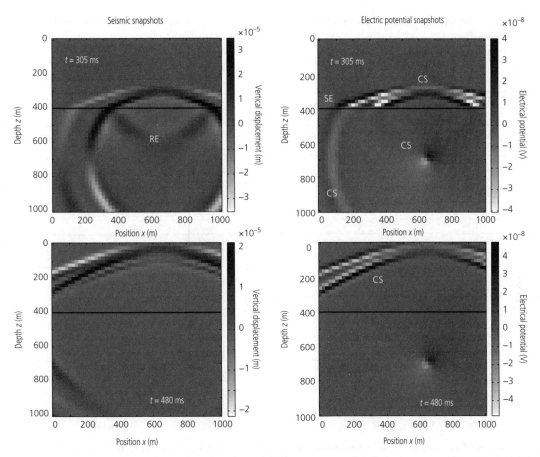

Figure 5.4 Same as Figure 5.3 for later times. Note that electrical field present around the seismic source and associated with the diffusion of the pore pressure around the source (type I electrical signal, which is dipolar in nature). CS, coseismic P- and S-wave signals (type III electrical signal); SE, seismoelectric conversions at interfaces (type II electrical signal); RE, reflected wave.

distribution associated with groundwater flow at the position of the seismic source diffuses slowly over time, depending on the value of the hydraulic diffusivity of the medium. The left-hand sides of Figures 5.3 and 5.4 show the direct seismic waves and the reflected waves. The right-hand sides of Figures 5.3 and 5.4 show the three types of electrical disturbances. The type I electrical disturbance corresponds to a dipolar field associated with the diffusion of the pore fluid pressure with time. After the occurrence of the seismic source, this field decreases exponentially with time. The type II electrical disturbances correspond to the electrical field generated by the arrival of the direct seismic wave at the interface between the layers L1 and L2. This field is also dipolar in nature. The type III electrical disturbances are seen to accompany the seismic waves.

The horizontal and vertical components of the seismic-solid displacements and two components of the electric field are shown in Figures 5.5 and 5.6, respectively, for each of the 8 stations located in the vicinity of the ground surface. The seismic displacements are created from the mechanical conversion mechanism at the interface between L1 and L2 such as the converted P–S-wave as well as the direct P–P- and S–S-waves. The converted S–P-wave is not visible on these time series because of the minimum radiation pattern of S–P-waves in the $(x–z)$ plane for the horizontal and vertical components of the seismic displacement (Aki & Richard, 2002; Gao & Hu, 2010).

The electrical field time series, due to the seismic disturbance, is shown in Figure 5.6. Note that the vertical component of the electrical field is computed at a depth of 5 m.

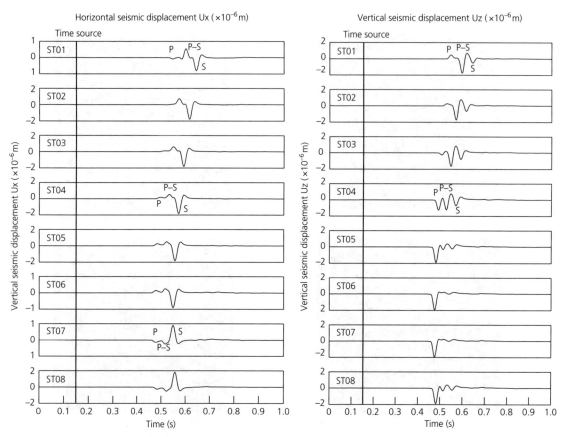

Figure 5.5 Horizontal and vertical seismic displacements responses. We can observe (i) some signals that are created from the mechanical P–S conversion mechanism at the interface (P–S conversion) and (ii) the P–P- and S–S-waves. The S–P interfacial signal is not visible on these time series because of the minimum radiation pattern of S–P-wave in (x–z) plane for horizontal and vertical components of the seismic displacement.

From these time series, we can observe the direct field (type I electrical disturbance), which occurs at the same time as the seismic source ($t_s = 0.15$ s). This direct electrical field is shown simultaneously at the eight stations. The seismoelectric conversion (type II) is observed at $t =$ 0.22 s when the wave arrives at the interface between the two layers. This seismoelectric disturbance is also observed simultaneously at the eight stations. For both type I and type II anomalies, the diffusion of the electromagnetic information is so fast that the signals are seen nearly instantaneously at all receivers. The coseismic (type III) electrical disturbance is observed at various times due to geometrical spreading of the seismic waves. This includes the P- and S-waves and also the converted P–S- and S–P-waves generated at the interface between L1

and L2. Following Snell's law and wave propagation theory, the S–P coseismic response is dominant at the region "near" the source midpoint because L2 has a higher velocity compared with L1. This contrast causes the S-wave to have less transmissivity at the interface and, in the process, switches all its incidence energy into the critically reflected wave along the L1–L2 interface (Stein & Wysession, 2003). In other words, the farther from the source that the conversion point is located along the interface, the smaller the S–P coseismic wave amplitude will be. From these synthetic data, we show that the electrograms contain much more information than the seismograms alone. The information content of the seismic and electrical data can be appraised by using a stochastic framework as shown in the next section of this paper.

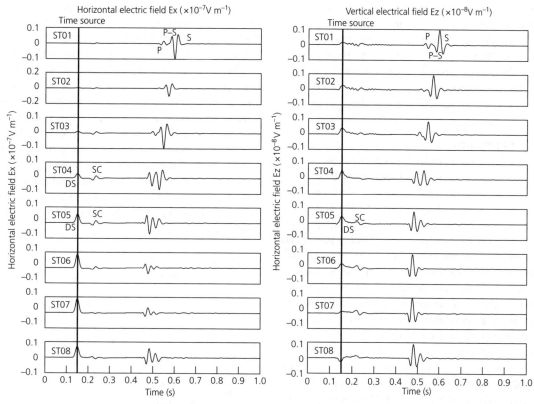

Figure 5.6 Horizontal and vertical electric field components (the vertical component is taken at a depth of 5 m below the ground surface). We can see the P–P and S–S coseismic electrical response and the coseismic responses associated with the P–S- and S–P-waves produced at the interface between the two materials. The occurrence of the direct field (DS) indicates the source time $t_s = 0.15$ s (see the horizontal component of the electric field). We can also see the seismoelectric conversion (SC) from the interface L1–L2 at $t = 0.22$ s.

5.2 Joint inversion of seismic and seismoelectric data

5.2.1 Problem statement

Next, the goal is to use seismic and electrical measurements to localize the position of a seismic source associated with hydraulic fracturing and to characterize its moment tensor through a full waveform inversion of the seismograms and electrograms. Because we are interested in the information content of the two types of geophysical signals, we use a stochastic, Bayesian approach to map the distribution of errors in the observed data into the model parameter space (e.g., see Jardani et al., 2010, for an example related to seismoelectric modeling). This stochastic approach is able to provide information that will allow the nonuniqueness of the inverse problem to be statistically characterized. The statistical characterization

of the inversion solution nonuniqueness is observed in the posterior probability distribution of the model parameters when dealing with noisy data.

We begin this inversion approach with the general equation used in the inversion process that relates model parameter estimations with data from measurements. This equation takes the form of

$$\mathbf{d} = \mathbf{Km} \qquad (5.16)$$

where $\mathbf{d}$ denotes the N-vector containing N separate observations (measurements), $\mathbf{m}$ is an M-vector containing M model parameters, and $\mathbf{K}$ symbolizes an $N \times M$ matrix that represents Green's functions of the system. The joint Bayesian solution to the inverse problem is based on combining information from seismograms, electrograms, and some prior knowledge of the problem.

Bayesian analysis considers both the data vector, $\mathbf{d}$, and the model parameter vector, $\mathbf{m}$, as random variables defined by probability density functions (Mosegaard & Tarantola, 1995; Gelman et al., 1996; Haario et al., 2001). The objective of the inverse modeling process is to update the information on $\mathbf{m}$ given $\mathbf{d}$; Green's functions of the system, $\mathbf{K}$; and prior knowledge of $\mathbf{m}$.

5.2.2 Algorithm

Now, we apply the Bayesian framework to estimate the posterior probability density of the model parameters. We chose the following vector of seven model parameters: $\mathbf{m} = [\log_{10}(x_S), \log_{10}(z_S), \log_{10}(h_d), M_w, M_{xx}, M_{xz}, M_{zz}]^T$ where x_s and z_s denote the coordinates of the source, h_d represents the thickness of the first layer, and the other parameters represent the magnitude and the independent components of the source moment tensor. Note that the variables involved in $\mathbf{m}$ are independent variables. Then, the Bayesian solution of the inverse problem is obtained by combining the information from the data and the prior information of the model parameters.

In a probabilistic framework, the objective of the inverse modeling is to look for the maximum posterior probability density, $\pi(\mathbf{m}|\mathbf{d}, M)$, values computed from the multiplication of a prior probability density of the model parameters, $\mathbf{m}$, in model M, $P_0(\mathbf{m}|M)$, with the probability density of likelihood, $P(\mathbf{d}|\mathbf{m}, M)$, corresponding to the data fit. In our test, the data corresponds to the seismic data alone or the seismic plus electrical data taken together. We will note that $\mathbf{m}_{Prior}$ is the prior vector of model parameters. Using this approach, the posterior probability density, $\pi(\mathbf{m}|\mathbf{d}, M)$, is obtained using Bayes' formula

$$\pi(\mathbf{m}|\mathbf{d}, M) = \frac{P(\mathbf{d}|\mathbf{m}, M)P_0(\mathbf{m}|M)}{P(\mathbf{d}|M)} \qquad (5.17)$$

where the denominator (the evidence) is defined as

$$P(\mathbf{d}|M) = \int P(\mathbf{d}|\mathbf{m}, M)P_0(\mathbf{m}|M)d\mathbf{m} \qquad (5.18)$$

Assuming that the model M is certain, we can drop the term M from Equations (5.17) and (5.18). If this is the case, the posterior probability density, $\pi(\mathbf{m}|\mathbf{d}, M)$, of the model parameters $\mathbf{m}$ given the data $\mathbf{d}$ is written as

$$\pi(\mathbf{m}|\mathbf{d}) = P(\mathbf{d}|\mathbf{m})P_0(\mathbf{m}) \qquad (5.19)$$

Another valid consideration for dropping the term M from the equations is that the division by the evidence factor is basically a normalization process which is not required to perform the inversion. The only exception for this is if we want to explicitly compute the probability for a given parameter in a particular interval. From the final expression of Equation (5.19), the inverse problem then can be solved by computing the expectation values of $\pi(\mathbf{m}|\mathbf{d})$, such as the mean, median, and the standard deviation of $\pi(\mathbf{m}|\mathbf{d})$ that can be considered as the final or most likely values of the reconstructed model.

The formulation of $\pi(\mathbf{m}|\mathbf{d})$, $P(\mathbf{d}|\mathbf{m})$, and $P_0(\mathbf{m})$ is described as follows. Assuming that the three quantities are Gaussian distributed, their likelihood functions are written as,

$$P(\mathbf{d}|\mathbf{m}) = \frac{1}{\left[(2\pi)^N \det(\mathbf{C}_d)\right]^{1/2}} \times$$

$$\exp\left[-\frac{1}{2}(\mathbf{d}-g(\mathbf{m}))^T \mathbf{C}_d^{-1}(\mathbf{d}-g(\mathbf{m}))\right] \qquad (5.20)$$

where N is the number of data, and the prior distribution on the M model parameters is written as

$$P_0(\mathbf{m}) = \frac{1}{\left[(2\pi)^M \det(\mathbf{C}_m)\right]^{1/2}} \times$$

$$\exp\left[-\frac{1}{2}(\mathbf{m}-\mathbf{m}_{prior})^T \mathbf{C}_m^{-1}(\mathbf{m}-\mathbf{m}_{prior})\right] \qquad (5.21)$$

To solve Equations (5.20) and (5.21), we need to set up the forward modeling operator for the seismic and electrical responses (based on the finite-element modeling of the field equations described in the main text), $g(\mathbf{m})$. We also need to establish the prior values of the probability distributions for the seven model parameters that we want to recover, $\mathbf{m}_{prior}$, and we need to define the covariance diagonal matrices for the data, $\mathbf{C}_d$, and the model, $\mathbf{C}_m$, incorporating the uncertainties related to the data and the prior model, respectively. The vector $\mathbf{d} = (\mathbf{d}_S, \mathbf{d}_E)^T$ is an N-vector consisting of the observed seismic data, $\mathbf{d}_S$, and electrical data, $\mathbf{d}_E$. The data vector, $\mathbf{d}$, is defined in a way that the $(N \times N)$ matrix, $\mathbf{C}_d$, is written as a diagonal matrix with the inverse of the covariance for the seismic and electrical data in the first and second set of diagonal elements, respectively. This allows for the weighting of the data according to the noise in the data. This is especially important if the electrical data are noisier than the seismic data.

In order to explore the "best" posterior probability density, we use the AMA to find the optimal proposal distribution of the model parameters (Haario et al., 2001). This algorithm is based on the Metropolis–Hastings algorithm with a symmetric Gaussian proposal distribution centered at the current model, $\mathbf{m}^i$, with a covariance, $\mathbf{C}^i$, that changes during the sampling process in such a way that the sampling efficiency increases over time (Haario et al., 2004).

The AMA is described as follows. First, we assume that the model vector, $\mathbf{m}$, is sampled by several states, ($\mathbf{m}^0$, ..., $\mathbf{m}^i$, ...,$\mathbf{m}^{RMax}$), where $\mathbf{m}^0$ corresponds to the initial state of $\mathbf{m}$ and R_{Max} is the maximum number of realizations. Each candidate point, $\mathbf{m}'$, is sampled from a pseudorandom Gaussian proposal distribution, $N(0.1)$, with a mean point at the present point, $\mathbf{m}^{i-1}$, and a covariance matrix given by

$$C^i = \begin{cases} C^0, & \text{if } i \le n_0 \\ s_n K^i + s_n \varepsilon_n I_n, & \text{if } i > n_0 \end{cases} \qquad (5.22)$$

where I_n denotes the n-dimensional identity matrix, $K^i =$ Cov($\mathbf{m}^0$, ..., $\mathbf{m}^{i-1}$) represents the covariance matrix, ε_n is a small positive number that prevents the covariance matrix from becoming singular, C^0 stands for the initial covariance matrix, and $s_n = (2.4)^2/n$ signifies a parameter that depends only on the dimension of the vector $\mathbf{m} \in \mathbb{R}^n$ (Haario et al., 2001). This yields an optimal acceptance factor for a Gaussian target distribution and a Gaussian proposal distribution (Gelman et al., 1996). The candidate model vector, $\mathbf{m}'$, is accepted with the acceptance probability

$$\alpha\left(\mathbf{m}^{i-1}; \mathbf{m}'\right) = \min\left[1; \frac{\pi(\mathbf{m}'|\mathbf{d})}{\pi(\mathbf{m}^{i-1}|\mathbf{d})}\right] \qquad (5.23)$$

If the candidate model vector is accepted, we consider $\mathbf{m}^i = \mathbf{m}'$; otherwise, $\mathbf{m}^i = \mathbf{m}^{i-1}$. This Bayesian framework is used to invert the model vector defined at the beginning of this section.

5.2.3 Results with noise-free data

A first-order solution is usually preferred to start the AMA sampler. There are many ways to find a first-order localization of the seismic source while ignoring the heterogeneity in the seismic velocity or in the electrical resistivity distribution. In Figure 5.7, we used a cross-correlation approach using only the electrical field distribution recorded at the 8 stations located at the ground surface. The cross-correlation is performed using the approach discussed by Revil et al. (2001).

To begin, using a first-order test solution, we applied the cross-correlation density algorithm developed by Revil et al. (2001) and Iuliano et al. (2002). The electrical field, $\mathbf{E(r)}$, due to a single dipole with moment $\mathbf{d}$, is written as $\mathbf{E(r)} = \mathbf{d}\nabla\mathbf{G}$, where $\mathbf{G}$ is Green's function of

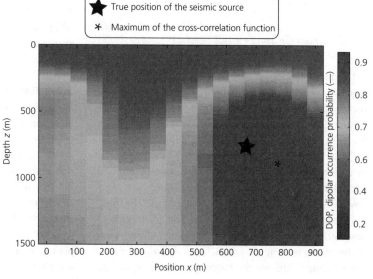

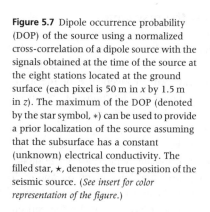

Figure 5.7 Dipole occurrence probability (DOP) of the source using a normalized cross-correlation of a dipole source with the signals obtained at the time of the source at the eight stations located at the ground surface (each pixel is 50 m in x by 1.5 m in z). The maximum of the DOP (denoted by the star symbol, ∗) can be used to provide a prior localization of the source assuming that the subsurface has a constant (unknown) electrical conductivity. The filled star, ★, denotes the true position of the seismic source. (*See insert for color representation of the figure.*)

the dipolar source. We note that E is the norm of the electrical field vector $\mathbf{E}(\mathbf{r})$. The power associated with the electrical field distribution is

$$\wp(E) = \int_S E^2(r)dS \tag{5.24}$$

$$\wp(E) = \sum_v d_v \int_S \mathbf{E}(\mathbf{r}) \frac{\partial \mathbf{G}}{\partial v_P} dS \tag{5.25}$$

where $v = x, y, z$ and $v_p = x_p, y_p, z_p$. The projection of S onto the (x, y) horizontal plane is adapted to a rectangle with sides of total length $2X$ and $2Y$ along the x and y-axis, respectively. The cross-correlation product is defined by

$$\eta_v(r_p) = C_v^p \int_{-X}^{X} \int_{-Y}^{Y} \mathbf{E}(\mathbf{r}) \frac{\partial \mathbf{G}(r_p - \mathbf{r})}{\partial v_p} dxdy \tag{5.26}$$

$$C_v^p = \left[\int_{-X}^{X} \int_{-Y}^{Y} E^2(\mathbf{r})dxdy \int_{-X}^{X} \int_{-Y}^{Y} \frac{\partial \mathbf{G}(r_p - \mathbf{r})}{\partial v_p} dxdy \right]^{-1/2} \tag{5.27}$$

where C_v^p denotes the normalization constant. The semblance function is the normalized scalar product between the form anomaly factor indicated by the electrical potential measurements at the ground surface and the form factor associated with the dipolar source located in the source volume. These cross-correlation densities have the following property: $-1 \leq \eta_v(r_p) \leq 1$. The norm of the cross-correlation vector $\mathbf{\eta}(\eta_x, \eta_y, \eta_z)$ is given by

$$\eta(r_p) = \sqrt{\eta_x(r_p)^2 + \eta_y(r_p)^2 + \eta_z(r_p)^2} \tag{5.28}$$

The finite-element model of the subsurface is generated first by the discretization of the 2D cross section into rectangular elements (50 m in x by 1.5 m in z). For each element, we compute the value of the cross-correlation function between all the electrical fields measured at the ground surface and those produced by a horizontal or a vertical dipole. The values of the cross-correlation coefficients for horizontal and vertical dipoles are combined into a dipole occurrence probability (DOP) function (comprised between 0 and 1). The maximum of this distribution provides a rough estimate of the position of the source (see Figure 5.7). This result is used to provide starting values for the model parameters x_s and z_s. We start with $M_w = 1.0$ as a starting value for the sampler.

The other model parameters are chosen randomly inside their interval of occurrence of $0 \leq h_d \leq 1$ km, and M_{xx}, M_{xz}, and M_{zz} are comprised between -2 and 2. Furthermore, we add $z_s \geq h_d$ as a constraint to the inversion (the seismic source is located in the deeper layer). Without this constraint, the solution of the MCMC sampler is unstable.

We generated 1000 realizations of the seven model parameters for the source characteristics and location using the data recorded in 8 geophones and 8 electrodes at only four frequencies (25, 30, 35, and 40 Hz). Figure 5.8 presents the behavior of the AMA sampler for the seven model parameters. The true values of the model parameters are represented by the solid horizontal bars. For the case corresponding to the noise-free data and the joint inversion of the seismograms and electrograms, the MCMC sampler converged after 521 realizations. Convergence success was determined by the variance of the distribution of the realizations for each of the seven model parameters (not shown here). The additional realizations (inside the gray boxes of Figure 5.8) are used to compute the posterior probability densities for each model parameter. These posterior probability distribution functions are displayed in Figure 5.9 for both the inversion of the seismograms alone and for the inversion of the seismograms and electrograms together. The following conclusions can be reached: (i) in the absence of noise on the data, our algorithm converges to the right solution, and correct estimates are obtained for the source location, (x_s, z_s); the thickness of the first layer, h_d; the magnitude of the seismic event; and the moment tensor components, M_{ij}. (ii) The posterior probability distributions for the two problems are the same. In other words, the information content of the seismic data and the electrical data are the same.

5.2.4 Results with noisy data

We performed the same exercise with noisy data. The first inversion was done with the seismic data alone, and the second was performed with the seismograms and electrograms simultaneously. After convergence of the algorithm, the posterior probability functions for the seven model parameters are shown in Figure 5.10. Here, we clearly see that the electrograms provide valuable information while inverting for the source properties. The peaks of the posterior probability densities of the model parameter resulting from the joint inversion approach are always closer to the true value of the model parameters (the vertical bars in Figure 5.10) by comparison

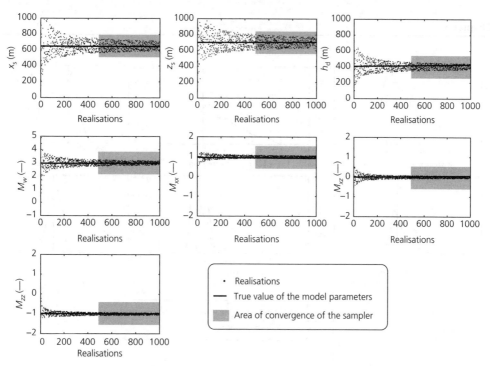

Figure 5.8 Joint inversion of the seismic and electrical waveforms. Application of the MCMC samplers to determine the posterior probability density of the seven model parameters (including the source location, the thickness of the first layer, the seismic magnitude, and the three components of the moment tensor). Each dot corresponds to one realization of the MCMC sampler. The solid horizontal lines correspond to the true values of the model parameters.

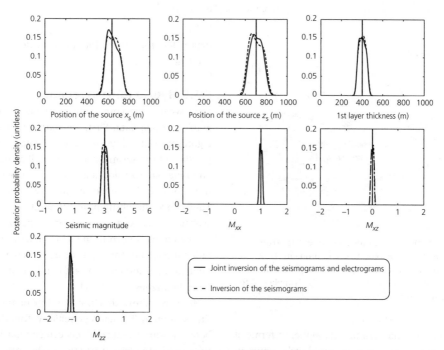

Figure 5.9 Posterior probability density function of the seven model parameters computed from the noise-free seismic and electrical data. The model parameters include the source location (x_s, z_s), the thickness of the first layer h_d, the seismic magnitude M_w, and the three moment tensor components (M_{xx}, M_{xz}, M_{zz}). The solid vertical bars indicate the true value of the model parameters. The information content of the seismic and electrical data is the same.

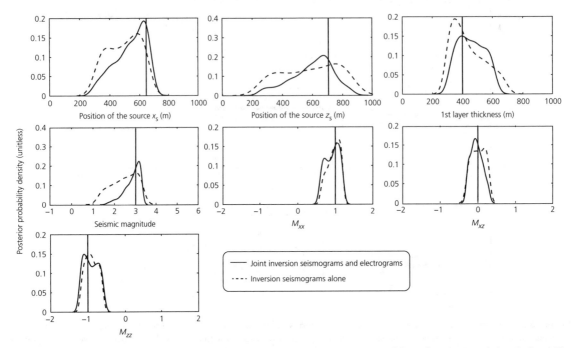

Figure 5.10 Posterior probability density function of the seven model parameters computed from the seismic and electric data with 15% noise level (2000 realizations in total, convergence reached after iteration 1356). The solid vertical bars indicate the true value of the model parameters. The joint inversion of the seismograms and electrograms does a better job than the inversion of the seismograms alone, in getting closer to the true values of the model parameters.

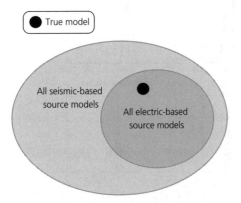

Figure 5.11 Information content for noise-free data. True information content for the joint inversion problem. The electrical field contains more information than the seismic field under the condition that the same level of noise is applied to the seismograms and electrograms.

with the inversion of the seismic data alone. In terms of information content, we can draw the sketch shown in Figure 5.11. For the same level of noise in the seismic and electrical data, the information content of the seismic data is less than the information content of the electrical data.

5.2.5 Hybrid joint inversion

Next, we provide a more efficient algorithm that could be applied to real data. First, we start with a deterministic algorithm using only the electrical data to invert for the position of the seismic source as well as the position of any heterogeneity located along the wave fronts using the approach developed recently by Araji et al. (2012). Then, once the position and time of occurrence of the seismic source and the position of heterogeneities are estimated, we switch to the AMA to determine the posterior probability functions of the independent components of the moment tensor to find the moment tensor with the highest likelihood.

Here, we show that the position of the source and the position of the heterogeneities within the wave field between source and receivers can be imaged using only the electrical potential signals. At the scale of this exercise, the seismic source geometry and associated seismic wavelength, along with the seismic related electrical source current density, can be considered to be small with respect to the system geometry. Therefore, each of the occurrences may be considered as a compact or point source. Additionally, as a seismic wave propagates through geological heterogeneities, the heterogeneities are considered to be contacted at single points and time, and as such, each point of contact can also be considered to be a compact point of contact, or a point source. Therefore, at each time step, the seismic source or the resulting seismic waves impinging on heterogeneities are responsible for a point source current density and, by the rationalization stated earlier, by definition can only have compact support (i.e., the spatial distribution of sources at any given time is very sparse). Compact electrical point current densities are vector quantities and therefore can be represented as electrical dipoles with quantifiable dipole moments. Based on these source current density compactness concepts, we recently developed an algorithm, presented in Araji et al. (2012), for the seismoelectric effect to find the location of these compact volumetric source current densities. The governing equation of the forward electrical model (Eq. 5.9) can be rearranged and written in a matrix form:

$$\mathbf{d} = \mathbf{Km} \qquad (5.29)$$

In Equation (5.29), $\mathbf{K}$ denotes the kernel matrix, which depends on the distance between the source of current and the receivers and the distribution of the electrical resistivity. This matrix corresponds to the discrete form of the leading field operator and accounts for the boundary conditions applied to the system. In this problem, we used Neumann boundary condition at the interface between the model and the boundary padding layer and Dirichlet boundary condition in the outer edge of the padding layer. The vector $\mathbf{m}$ is a vector containing the M source current density terms (right-hand side of Equation (5.29)), and $\mathbf{d}$ is the vector of the observed electric potentials at each time step, at N receiver locations. Araji et al. (2012) developed a method to invert the source current density from Equation (5.29) using compactness as a regularization tool.

If we assume that the resistivity structure and boundary conditions for the electrical field are known (via downhole measurement and/or resistivity tomography), a stable solution of $\mathbf{m}$ is obtained by first solving

$$\left(\mathbf{\Lambda}^{-1}\mathbf{G}^{-1}\mathbf{W}_d^{\mathrm{T}}\mathbf{G}\mathbf{\Lambda}^{-1} + \alpha\mathbf{I}\right)\mathbf{m}_{\mathrm{w}} = \mathbf{\Lambda}^{-1}\mathbf{G}^{-1}\mathbf{P}^{\mathrm{T}}\mathbf{P}\mathbf{d} \qquad (5.30)$$

where $\mathbf{G} = \mathbf{PK}^{-1}$ denotes Green's matrix $(N \times M)$ computed as a product of the inverse kernel matrix times the sparse selector operator matrix that contains a single 1 on each row in the column that corresponds to the location of that receiver. The rows of G can be computed using reciprocity, which involves computing the forward response to a unit source located at each receiver. Other variables, such as $\mathbf{\Lambda}$, denote the inverse-sensitivity weighting function that accounts for the distance from the receivers as well as the resistivity structure and is formulated as $\mathbf{\Lambda} = \mathrm{diag}\left(\sum_{i-1}^{N} G_{kj}^{\mathrm{T}^2}\right)^{1/2}$, matrix $\mathbf{W}_d$ corresponds to the selection operator related to the expected noise level in the data, α denotes the regularization parameter, $\mathbf{I}$ represents an identity matrix, and $\mathbf{m}_{\mathrm{w}}$ is a modified model vector, which is computed from $\mathbf{m}_{\mathrm{w}} = \mathbf{\Lambda}\mathbf{m}$. The calculation of $\mathbf{m}$ in the first stage of inversion is important since we need to have an initial model to feed the source compactness calculation.

The compact source distribution method is a relatively classical technique that suits the electrical part of the problem, because the source current densities associated with it tend to be spatially localized, as rationalized earlier. This method is basically computed by minimizing the spatial support for the source current density. In order to get a compact source solution, the model weighting parameter $\mathbf{\Lambda}$ is modified by

$$\mathbf{\Omega} = \mathrm{diag}\left(\frac{\Lambda_{kk}^2}{m_{k_{(i-1)}}^2 + \beta^2}\right)^{1/2} \qquad (5.31)$$

Using this modification, the problem now becomes linear by making the denominator of $\mathbf{\Omega}$ a function of the previous solution at step $(i-1)$ and using the new $\mathbf{\Omega}$ in an iteratively reweighted least-squares algorithm. The vector $\mathbf{m}_{i-1}$, where $i = 1$, denotes the initial model used to compute the first solution of the compact source method. Then, an updated $\mathbf{\Omega}$ is determined at each new iteration based on the previous solution. Using the transformed parameters $\mathbf{m}_{\mathrm{w}} = \mathbf{\Omega}\mathbf{J}$, we minimize the new weighted objective function $P^\alpha(\mathbf{m})$ defined by

$$P^\alpha(\mathbf{m}) = \|\mathbf{W}_d(\mathbf{G}\mathbf{m} - \mathbf{d})\|^2 + \alpha\sum_{k=1}^{M}\frac{m_k^2}{m_k^2 + \beta^2} \qquad (5.32)$$

to get the solution for $\mathbf{m}_{\mathrm{w}}$ from

$$\left(\mathbf{\Omega}_{i-1}^{-1}\mathbf{G}^{\mathrm{T}}\mathbf{W}_d^{\mathrm{T}}\mathbf{W}_d\mathbf{G}\mathbf{\Omega}_{i-1}^{-1} + \alpha\mathbf{I}\right)\mathbf{m}_{w,i} = \mathbf{\Omega}_{i-1}^{-1}\mathbf{G}^{\mathrm{T}}\mathbf{W}_d^{\mathrm{T}}\mathbf{W}_d\mathbf{d} \qquad (5.33)$$

After a certain number of iterations, the final vector, m_w, is transformed back to m using the original relationship $m = \Omega^{-1} m_w$. We found that eight iterations in the compactness algorithm are usually a good strategy to focus the inverted source current distribution.

We apply this algorithm to the electrical potential recorded at the 8 receiver stations three times. We

assume that the resistivity model is perfectly known, and 5% noise was added to the electrical potential data. The first time corresponds to the time of occurrence of the source, and therefore, we locate a type I anomaly. In Figure 5.12a, we observe that the source current density responsible for the electrical potential distribution is located in the vicinity of the true location of the

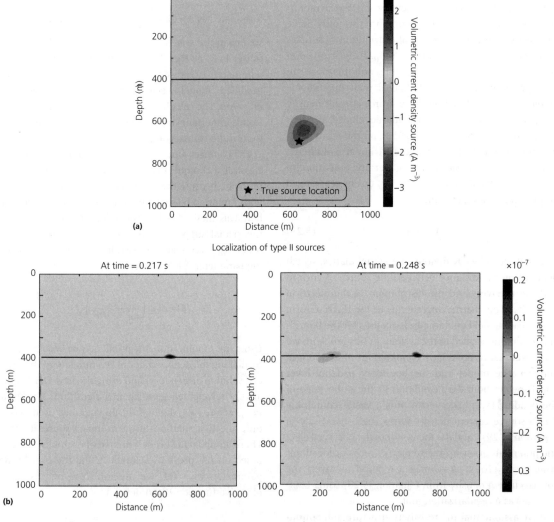

Figure 5.12 Localization of the source current density distribution responsible for the electrical potential distribution recorded at the surface stations (5% noise). **a)** Type I anomaly associated with the seismic source. **b)** Type II anomalies (seismoelectric conversions) (iteration #8, data root mean square error: 3%). The horizontal line denotes the interface between L1 and L2 where seismoelectric conversions take place. (*See insert for color representation of the figure.*)

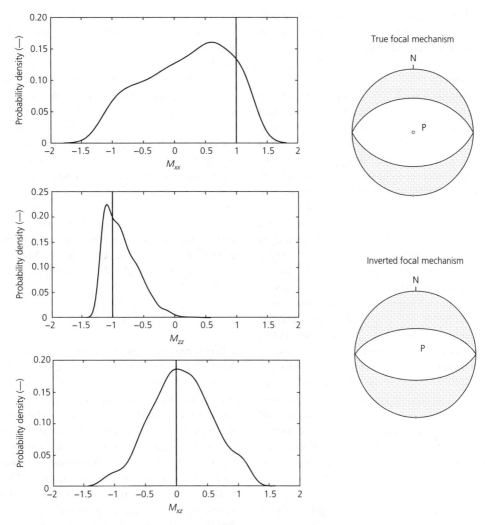

Figure 5.13 Probability density functions of the three components of the moment tensor determined from the AMA where the thickness of the first layer and the position of the source was determined from the tomographic approach of Figure 5.12. We also show the true focal mechanism and inverted focal mechanism with the highest likelihood.

seismic source. We then applied the same algorithms to the potential distributions for two subsequent times for which the source current densities are generated in the first Fresnel zone of the interface separating L1 and L2. Again, the algorithm is able to locate the source of current at the correct position along the interface. Therefore, this type of analysis can be successful in locating the source and the position of the heterogeneities in the system.

Once the thickness of the first layer and the position of the source have been determined from the electrical data using the deterministic, gradient-based algorithm presented earlier, we can switch to the AMA to determine the probability densities of the different components of the seismic source tensor. Again, we used the AMA with 1000 realizations. The posterior probability densities for the three components of the moment tensor are shown in Figure 5.13: we see that the peaks of the probability density functions are in reasonable agreement with the true model parameters represented by the solid vertical bars (Figure 5.13).

5.2.6 Discussion

The previous sections we presented a proof-of-concept exercise. However, the model used earlier is not intended to be directly applicable to real earth problems. The first limitation of our approach lies in the use of the Biot theory itself, which needs to be modified to account for the true quality factor of earth materials, especially for shales (Carcione, 2007). Modeling the quality factor of porous media is presently an area of fertile research (e.g., Gelinsky et al., 1998; Müller & Rothert, 2006), and the inversion of the intrinsic attenuation coefficient can be set up as an additional inverse problem in the full waveform inversion of the seismic data (e.g., Ravaut et al., 2004). Regarding 2D/3D effects, they can be incorporated in the present approach (actually our code is 3D + time), but the forward problem has rapidly become computationally expensive in 3D. This prohibits, at this point, the use of an MCMC sampler to solve the inverse problem in real situations. That said, the code can be parallelized, and the MCMC sampler strategy can be combined with the deterministic approach discussed earlier. Regarding the signal-to-noise ratio (SNR), the level of electrical noise is usually higher at the ground surface than at depth, and therefore, the electrical field should be preferentially monitored in shallow wells. The optimization of the SNR for an electrical monitoring sensor array is a subject that has been broadly discussed in electroencephalography (EEG) (Grech et al., 2008) but not in geophysics.

Additionally, electromagnetic disturbances have been observed to be associated with earthquakes, seismovolcanic activity, steam injection in hot dry rock reservoirs, detonation of explosive charges in boreholes, and underground nuclear explosions (e.g., Ushijima et al., 1999; Yoshida, 2001; Gaffet et al., 2003; Byrdina et al., 2003; Yoshida & Ogawa, 2004; Soloviev & Sweeney, 2005; Honkura et al., 2009). Multiphase flow in porous media, and more precisely the occurrence of Haines jumps at pore throats, is also responsible for tiny seismic and electrical field bursts as observed by Haas and Revil (2009). Therefore, the monitoring of shallow earthquakes along tectonic faults and active volcanoes, the hydraulic fracturing in geothermal fields and hydrocarbon reservoirs, and the secondary and enhanced

recoveries of oil reservoirs could benefit from the use of electromagnetic sensors coupled to three-component geophones to monitor this type of seismoelectric activity.

An impressive example of the occurrence of electromagnetic signals associated with a hydraulic fracturing experiment in a well has been documented by Ushijima et al. (1999) (see Figure 5.14). The electrical potential changes recorded at the ground surface were analyzed by Ushijima et al. (1999) in terms of a dipolar source associated with the plane of fracture and consistent with the fracture localization using seismic activity (see Figure 5.14d). In the case where the injected fluid has a pore water conductivity that is different from the conductivity of the groundwater, there are two contributions to the source current density given by (e.g., Martínez-Pagán et al., 2010; Revil et al., 2011; Bolève et al., 2011; Ikard et al., 2012)

$$\mathbf{J}_S = \hat{Q}_V^0 \dot{\mathbf{w}} - \frac{k_b T}{Fe}\left(2t_{(+)} - 1\right)\nabla\sigma_f \qquad (5.34)$$

where k_b represents the Boltzmann constant, T is the absolute temperature, F denotes the formation factor, e stands for the elementary charge, $t_{(+)}$ symbolizes the microscopic Hittorf number of the cations (0.38 for NaCl), and σ_f signifies the conductivity of the water. The first term on the right in Equation (5.34) is the electrokinetic contribution investigated earlier, and the second term in Equation (5.34) represents the diffusion current associated with the gradient of the conductivity of the water. In the case study reported by Ushijima et al. (1999), the self-potential was inverted assuming that the source is dipolar and the location of the source was found to be coincident with the fracture plane, illuminated by the localized acoustic emissions (AE). Therefore, the experiment reported by Ushijima et al. (1999) shows that recordable electrical signals can be measured remotely in association with hydraulic fracturing experiments.

The computed signals can be compared also to the level of the background electrical noise at the seafloor and onshore. At 1 Hz, the level of background noise on the seafloor is on the order of $10^{-12}\,\mathrm{V\,m^{-1}}$ (Chave et al., 1991). Instrumental noise caused by the nonpolarizable Ag–AgCl electrodes in contact with seawater is about $10^{-24}\,\mathrm{V^2\,m^{-2}}$ $\mathrm{Hz^{-1}}$ at frequencies above 1 Hz (Webbs et al., 1985). Therefore, the background noise level at the seafloor is well below the magnitudes required to detect the signals described in this chapter.

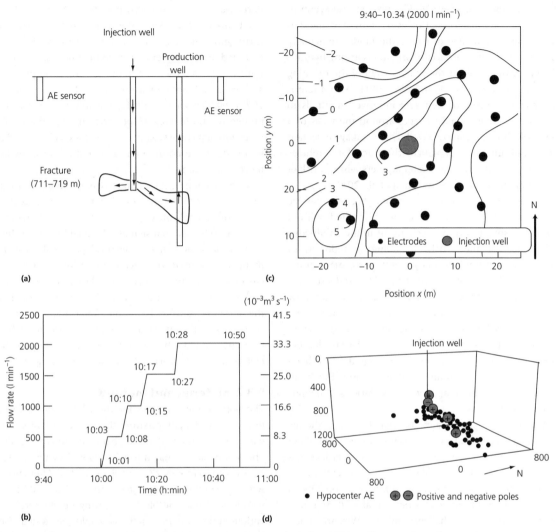

Figure 5.14 Hydraulic fracturing test and resulting electrical potential (SP stands for "self-potential") and acoustic emissions (AE) data from Ushijima et al. (1999). **a)** Field operations. Sketch of the position of the hydraulic fracturing. **b)** Water injection flow used during the hydraulic fracturing experiment showing the flow rate history. **c)** SP changes (in mV) with respect to the SP distribution at time 9:40, prior to the fracturing tests. **d)** Hypocenters determined from the AE and positions of the positive and negative electrical potential poles inverted from the distribution of the electrical potential distribution at the ground surface.

5.3 Hydraulic fracturing laboratory experiment

5.3.1 Background

Hydraulic fracturing has become a very important method to increase the permeability of shales and tight sandstones (Agarwal et al., 1979) for hydrocarbon recovery and to increase the heat coupling to thermal transport fluids to improve the energy production of geothermal fields (Kohl et al., 1979). Hydraulic fracturing can also be used for the nonexempt solid waste disposal (Keck & Withers, 1979) and in situ stress measurements (Kuriyagawa et al., 1989) and can occur during the grouting of the foundations of dams (Lee et al., 1999). The traditional method used for monitoring the progress of the hydraulic fracturing process remains AE (microseismic). Significant progress has been made in the last decade with passive seismic monitoring methods used to detect hydraulic fracturing events and localizing these events in heterogeneous materials. However, with AE, there is insufficient knowledge of where the fluids are actually moving within the subsurface formations during hydraulic fracturing as well as the actual extent of the resulting fracture network (e.g., Warpinski, 1991). Also, there is a recognized need for new and improved methods to detect and localize fluid leakages around the walls of a borehole during the life of a well. This is leakage detection and localization need is particularly critical for boreholes that traverse through freshwater aquifers where a risk of contamination is a concern (Cihan et al., 2011).

The self-potential method corresponds to the passive measurement of electrical signals associated with a variety of source current mechanisms in the conductive subsurface of the Earth, including a redox-based contribution (Sato & Mooney, 1960; Castermant et al., 2008; Mendonça, 2008; Revil et al., 2010) and a streaming current contribution related to groundwater flow (Revil & Linde, 2006; Revil et al., 2011; Ikard et al., 2012). The self-potential inverse problem is similar, in essence, to EEG in medical imaging. In the last decade, the recording and inversion of EEG signals have been instrumental in our understanding of how the brain works and in the mapping of its various functions (Grech et al., 2008).

The flow of pore water associated with hydraulic fracturing, waste and other injection wells, multiphase oil and gas production, and borehole and other leakages into the subsurface results in the generation of measurable voltages. This occurs during field operations in reservoir environments (Chen et al., 2005; Entov et al., 2010) and in shallow aquifers (Wishart et al., 2008) and has been associated with artificial seismic sources (Kuznetsov et al., 2001). Similar conclusions have been reached in volcanic environments (e.g., Byrdina et al., 2003), and there is a relatively broad base of literature on laboratory observations of electromagnetic fields associated with hydromechanical disturbances (Moore & Glaser, 2007; Nie et al., 2009; Jia et al., 2009; Chen & Wang, 2011; Wang et al., 2011; He et al., 2011, 2012; Onuma et al., 2011).

For instance, Moore and Glaser (2007) investigated unconfined and confined samples of granite subjected to hydraulic fracturing in the laboratory. Their results indicate that the principal mechanism for the self-potential response is due to the generation of a streaming source current density associated with the flow of the pore water (see also Wurmstich & Morgan, 1994, and Ushijima et al., 1999). We believe that these voltages carry information about the fracture network that is expected to be complementary to the information determined from microseismic, tiltmeter, wellhead pressure, and wellhead flow measurements (Keck & Withers, 1979). Mahardika et al. (2012) recently provided a comprehensive framework to perform full waveform joint inversion of passive seismic and associated electrical data to invert for the position and moment tensor of hydromechanical events.

In this section of this chapter, we are concerned with the time-lapse monitoring of a laboratory-based hydraulic fracturing/leak-off experiment using the time-lapse record of self-potential signals. The goal here is to show that the inversion of the electrical potential measurements can be used to detect and localize the streaming current disturbances caused by a borehole seal failure and the resulting undesirable fluid migration along the borehole during hydraulic fracturing operations. A leak-off resulted from an attempt to hydraulically fracture a porous block while our initial goal was to study the electrical signals associated with such fracturing events.

5.3.2 Material and method

The porous material used for the laboratory fracturing tests was a cement mixture (FastSet Grout Mix) and was cured for about 10 months before the tests proceeded. The porous sample had a roughly cubical shape ($x = 30.5$ cm $\times y = 30.5$ cm $\times z = 27.5$ cm; Figure 5.15). After curing, several 10 mm diameter holes (named #1 to #10 below) were drilled into the block to varying depths such that various tube sealing methods could be tested (see Figure 5.15). Stainless steel tubing with 9.5 mm outside diameter was placed into a few holes using Loctite Instant Mix 5 Minute epoxy as the tube sealing agent. The voltage measurement electrodes were attached to the top and one side of the block (16 electrodes on each face) using a plastic template for precise positioning. AE sensors were also mounted to three faces of the block for fracture event localization. The electrodes were solid sintered Ag grains with a solid AgCl coating and an active diameter of about 1 mm. Each electrode had a voltage amplifier built into the electrode casing, and each of these electrodes was electrically connected to the block surface through a drop of conductive gel that was normally used for EEG.

The electrical response during the experiment was measured using a very sensitive multichannel voltmeter

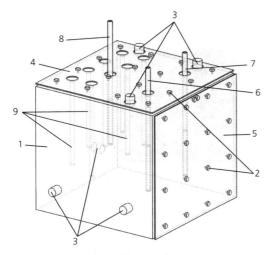

Figure 5.15 Unconfined cement block sensor configuration. (1) Cement block; (2) 34 Ag–AgCl electrodes (BioSemi); (3) six acoustic emission sensors (Mistras WSa); (4) plastic plate with top array of 16 channels of Ag–AgCl electrodes (BioSemi); (5) plastic plate with back array of 16 channels of Ag–AgCl electrodes (BioSemi); (6) Hole #9, high-pressure fluid injection tube; (7) Hole 10, high-pressure fluid injection tube; (8) Hole 6, high-pressure fluid injection tube (not used); and (9) other holes (no injection).

manufactured by BioSemi, Inc., designed for EEG research (http://www.biosemi.com/). During the experiments, the electrical potential measurements were acquired with 32 amplified nonpolarizing silver–silver chloride (Ag–AgCl) electrodes as described earlier. The electrode potentials were measured using the BioSemi ActiveTwo data acquisition system that is completely self-contained, battery powered, galvanically isolated, and digitally multiplexed with a single high-sensitivity analog-to-digital converter (ADC) per measurement channel. Thirty-two 24 bit ADC, employing the sigma-delta architecture, were used in the ActiveTwo data acquisition system. The ActiveTwo system has a typical sampling rate of 2048 Hz with an overall response of DC to about 400 Hz. This measurement system has a scaled quantization level of 31.25 nV (LSB) with 0.8 µV rms noise at a full bandwidth of 400 Hz and a specified $1/f$ noise of 1 µV pk/pk from 0.1 to 10 Hz. The common-mode rejection ratio was higher than 100 dB at 50 Hz, and the amplified nonpolarizing electrode input impedance was 300 MΩ at 50 Hz (10^{12} Ohm // 11 pF). Because there is a single ADC per channel, the digitization part of the system had very small sampling synchronization errors. The specifications for the system

state that the sampling skew is <10 ps among all of the channels and has a 200 ps overall sample rate jitter. This represents a very small amount of timing error causing all of the measurements to be synchronized to a very small fraction of a sample period, making the data among all of the channels temporally highly synchronous. The digitally multiplexed signals are subsequently serialized into a bit wide data stream and sent through a fiber-optic cable (to achieve galvanic isolation) to a USB-based computer interface. The entire system, including the computer, was operated on batteries to minimize conductive coupling with the electrical power system.

The voltage reference for the system is contained within the measurement area and was designed into the data acquisition system to be a part of the common-mode sense (CMS) and common-mode range control (DRL) electrodes (see Haas et al., 2013, and reference therein for further information on the common-mode control used in the BioSemi system and the effect of electrode impedance on the measured response). The CMS and DRL electrodes are used in combination to make a feedback control system that keeps the CMS electrode as close as possible to the reference voltage at the ADC. In this system, the CMS electrode becomes a dynamic reference potential. All of the digitized data are saved in the raw data form and are referenced to the CMS electrode. The data are recorded with all of the common-mode signals, and as a result, any channel can be used as the reference channel. In fact, the maximum common-mode rejection in the system only fully occurs when a single channel of choice is subtracted from all of the other channels (see http://www.biosemi.com/; see Crespy et al., 2008, and Ikard et al., 2012, for further explanations). In these measurements, we selected the least active signal in the data (channel 4) as the reference. This type of measurement system allows us to change the reference electrode as needed to correct for dynamic voltages occurring at the CMS electrode. Therefore, this type of system is best suited for dynamic self-potential signals. The flowchart used to analyze the raw electrical data is shown in Figure 5.16.

Any AE that would occur during the experiment were measured using 6 WSα sensors manufactured by Physical Acoustics Corporation (PAC; see position in Figure 5.15). All six sensors had an operational frequency of 100–900 kHz and a resonant frequency of 125 kHz. PAC's Micro-II PCI-2-8 digital AE system chassis was used to run the AEwin data collection and posttest data analysis software.

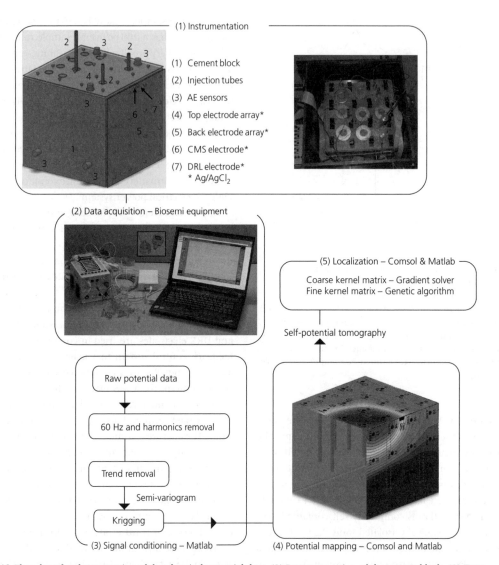

Figure 5.16 Flowchart for the processing of the electrical potential data. (1) Instrumentation of the porous block. (2) Data acquisition showing the BioSemi EEG system and the laptop computer. (3) Signal conditioning of the raw data. (4) Mapping the voltage response using ordinary kriging. (5) Localization of the causative sources in the block. (*See insert for color representation of the figure.*)

The Micro-II chassis acquires data at a 40 MHz acquisition rate with sample averaging and automatic offset control. Waveform streaming enables data acquisition to hard disk continuously up to 10 MHz. PAC's 2/4/6 (20/40/60 dB gain) single-ended preamplifiers were used on each channel throughout all testing. A 60 dB gain setting was preferred in order to amplify microfracture signals and increase SNRs. The AE data were also inverted to localize the position of the acoustic sources (see Hampton, 2012, for additional details).

The experiments were conducted on the porous block in equilibrium with the atmosphere of the laboratory (~30% relative humidity). Saline water was used as the injection or fracturing fluid (no proppant) containing 10 g of NaCl dissolved into 1000 ml of deionized water (conductivity of $1.76\,\mathrm{S\,m^{-1}}$ at 25°C). Note that because lower salinities imply higher electrokinetic signals, we position our experiment in the most difficult conditions we could meet in the field. We demonstrate below that even in such high salinity conditions, the self-potential signals can be easily observed. The fluid control system injects fluid through stainless steel tubes (Figure 5.15) using a computer-controlled Teledyne Isco 100DX syringe pump that is able to control flow rate or pressure during the various phases of the experiment. The injection tube was designed to have an open end at the bottom; there were no side ports for fluid to flow through. The system has a total fluid capacity of 103 ml and is capable of achieving pressures up to 68.9 MPa while maintaining constant flow rates of 0.001–60 ml min^{-1}. In this experiment, the injection tubes were initially pressurized to a PID controlled 1.17 kPa with the fracturing fluid and maintained at that pressure for a period of time to be sure that the system was maintaining pressure and to measure the static fluid leak-off rate. If fluid was measured to be moving, then there was some sort of leak-off into the block or the seal was failing to hold enough pressure to commence hydraulic fracturing. A constant fluid flow rate of 1 ml min^{-1} was then imposed on the system with the intent of inducing hydraulic fracturing. Under constant flow, either the cement block or the tubing seal would eventually fail.

The test procedure began by preparing the cement block for high-pressure injection (see Frash & Gutierrez, 2012, for details). The injector was filled with the saline solution described earlier and coupled to the injection tube that was also filled with the saline solution. The injection system was purged of air and then subjected to constant pressure of 1.17 kPa for about 30 min to monitor leak-off to be sure that there was no pressure loss. For the experiment associated with Hole #9, a 60 s preinjection (termed phase 0) electrical potential measurement period was acquired (Figure 5.17a). The goal of this phase was to establish individual channel offsets and drift trends for use during postacquisition signal processing. Constant pressure fluid injection at 1.17 kPa (termed phase I) was initiated at T0 = 60 s and terminated at T1 = 1632 s. This phase was followed by phase II, a 1 ml min^{-1} constant flow rate initiated at T2 = 1795 s (note that fluid pressure was maintained, but not actively controlled between T1 and T2.). Fluid injection was terminated well after the end of the electrical data acquisition, when seal failure was confirmed through the appearance of water on the surface of the block near the injection hole. For this experiment, self-potential data acquisition terminated at 2086 s, prior to completion of phase II injection due to an accidental interruption of the streaming data because of a poor USB 2.0 connection.

This did not affect the raw data saved or any subsequent use of the data.

5.3.3 Observations

Figure 5.17 shows the temporal evolution of the electrical potential for all of the electrodes, including the occurrence of bursts in the electrical potential that are similar in shape (but much larger in amplitude) to the electrical field bursts observed by Haas and Revil (2009) for Haines jumps during the drainage of an initially water-saturated sandbox. Figure 5.17a shows the entire 2086 s record, while Figure 5.17b, c, and d zoom in on specific areas of interest. There are seven major events, of which three are highlighted (Events E1 through E3), and two will be used in the following text to test our localization procedure. These events are shown in the time series of Figure 5.17b, c, and d. All major electrical potential events occurred during the phase II constant flow injection period. During phase I, the measured electrical potential gradually increases as fluid is being injected into the cement block. No bursts in the electrical field were observed during the constant pressure phase (phase I). Other types of information are contained within this part of the experiment, but are not analyzed here.

Each major event is characterized by a rapid change in the electrical potential time series followed by a slower exponential-type relaxation of the potential with a characteristic time comprised between several seconds to several tens of seconds. This relaxation is believed to be associated with the relaxation of the fluid flow under pressure as shown later. Because the relaxation of the potential distribution is relatively slow after each event, a sequence of overlapping events causes a superposition of the potentials from each event in the sequence to varying degrees (see Figure 5.17b and c). We term the superposition of a past event decay response with a new event a residual potential superposition. It can be clearly seen from Figure 5.17 that the degree of residual potential superposition is dependent on event physics (hydroelectric coupling), event magnitudes, event spatial distribution, time of occurrence, and event decay rate. Each of these factors is variable, and to localize and characterize individual impulsive events, the influence of residual potential superposition must be accounted for and removed to complete a comprehensive analysis of the data.

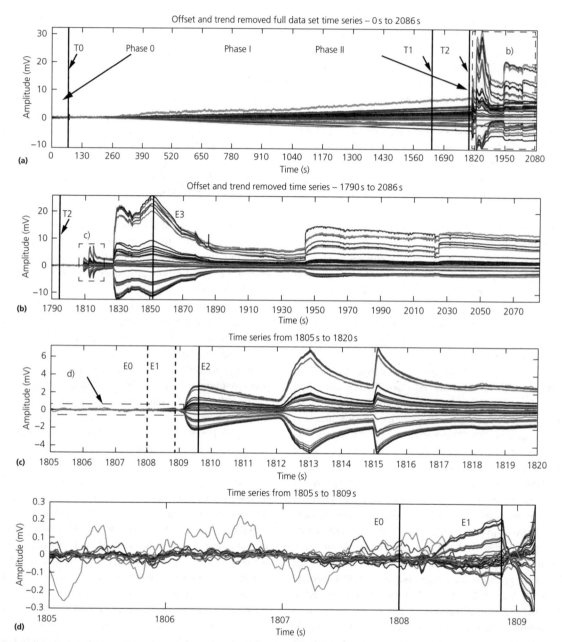

Figure 5.17 Self-potential time series related to Hole #9 saline water injection. **a)** Full time series data set showing the different fluid injection time periods during data acquisition (T0, T1, and T2). Note the significant change in electrical response after T2 that is bounded by region **b)**. **b)** Zoom highlighting the background normalized electrical response showing distinct electrical impulses related to the start of constant flow injection at T2 with selected peak events. **c)** Zoom in region **c)** highlighting the first series of impulsive signals with selected peak Events E2, E3, and E4 with temporal reference to E0 and E1. **d)** Zoom in region **d)** showing the temporal noise leading up to Event E1, with a voltage background time slice at E0. Note the change in potential after E1. (*See insert for color representation of the figure.*)

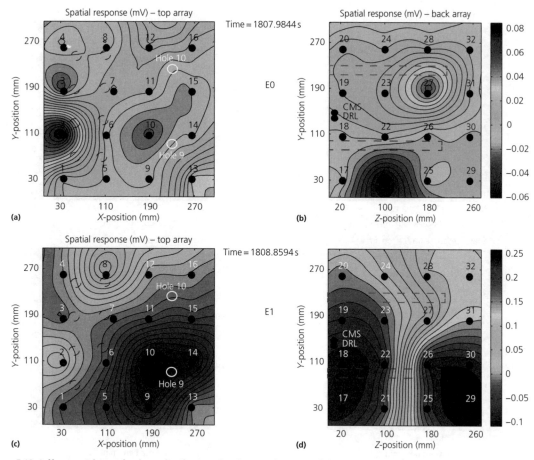

Figure 5.18 Self-potential spatial voltage distributions for the snapshot E0 and the Event E1. Each panel is a kriged contoured distribution of the electrical potential on the top and back panels (ordinary kriging). The black dots denote the position of the electrode positions. **a)** and **b)** Spatial electrical potential distribution for snapshot E0 showing the spatial variations associated with the background noise. Note the very small color bar voltage scale (+0.08 to −0.06 mV). **c)** and **d)** Voltage distribution for Event E1 showing the burst of the electrical field associated with the first hydraulic pulse taking place during constant flow injection. Note the voltage polarities in the spatial distributions and the much larger color bar voltage scale (+0.25 to −0.1 mV). (*See insert for color representation of the figure.*)

Figures 5.18 and 5.19 show the spatial evolution of the electrical potential on the monitored faces of the test block, starting with snapshot #E0 taken prior the occurrence of Events E1 and E3. For these snapshots, ordinary spatial kriging was performed on each face separately. It can be seen in Figure 5.17d, 4a, and 4b that the snapshot E0 shows random spatial electrical potential fluctuations associated with the temporal noise that can be seen in Figure 5.17d. In these figures, channel 13 is noisier with respect to the rest of the channels possibly from poor contact between the electrode and the cement block.

Event E1 in Figure 5.18c and d shows an initial voltage distribution with a small negative potential on the top surface of the block and a bipolar signal on the side of the block. This voltage distribution implies that there is a current source density possibly near Hole 9 (see position in Figure 5.15) that is pointing mostly downward into the block. The time series in Figure 5.17d shows the onset of this small peak (Event E1), followed by a quick decay and reversal of the polarity of the current source density as indicated by Event E2 as shown in Figures 5.17c and 5.19a and b. We consider that the polarity reversal may be described by a sequence of

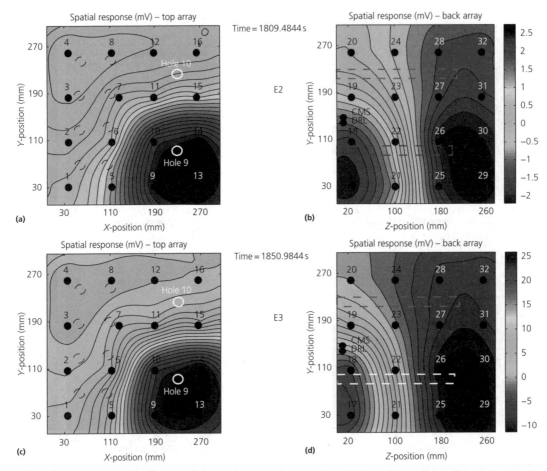

Figure 5.19 Self-potential spatial voltage distributions of Events E2 and E3. **a)** and **b)** Event E2 voltage distribution showing the peak voltage associated with the second hydraulic pulse during constant flow injection. Note the voltage polarities in the spatial distributions and the color bar voltage scale (+2.5 to −2.0 mV). Event E2 represents the first of a series of electrical field bursts. Panels **a)** and **b)** show a reversal in polarity and increase in peak magnitude relative to figure panels **c)** and **d)**. **c)** and **d)** Event E3 spatial voltage distribution showing the peak voltage associated with the highest magnitude pulse during constant flow injection. (*See insert for color representation of the figure.*)

events. First, a brief pressure drop (E1p), seen in Figure 5.20 just before the E1 peak, indicates some sort of pulse flow of fluid occurred, which may have led up to the E1 peak. The following reversal of polarity that peaks at E2 is correlated with another pressure drop (Figure 5.20; E2p) just prior to the peak at E2. This indicates that the initial fluid flow direction at E1 was in a downward direction, possibly an indication of the initial downward direction of a plastic failure in the epoxy seal before the reversal of flow direction due to other seal failures with higher volumes and mostly vertical flow directions. It is possible that the impulsive nature of these

progressive seal failure events was unique to this particular epoxy seal technique that caused a plastic seal failure. Additionally, gas pockets inside the epoxy interface with the hole wall could be an explanation of the burst nature of the seal failure. This could have been caused by unequal distribution of the epoxy along the hole wall. The rupture of each gas pocket would produce a drop in pressure followed by an increase in the fluid flow along the hole wall, resulting in an electrokinetic response. As we will show later, the direction of the current density corresponding to Event E2 is mostly pointing upward and grows in magnitude in an impulsive manner

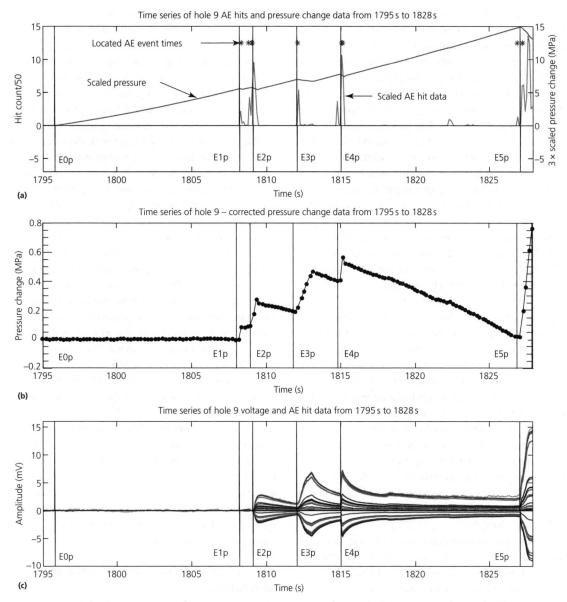

Figure 5.20 Fluid pressure, acoustic emissions, and electrical potential changes during a given time window. **a)** The acoustic emission (AE) data and pressure change correlation. The red asterisks denote the located AE events. **b)** The trend removed pressure changes. **c)** The voltage responses at all 32 measurement points. The events labeled E0p–E5p represent the same events in each panel of the figure. The AE hits, the pressure changes, and the voltage data all align, indicating that there is a relationship between these phenomena. (*See insert for color representation of the figure.*)

as the fluid injection proceeds. The magnitude of the electrical potential grows from Event E2 onward and maintains the spatial voltage distribution/polarity throughout the remainder of the data acquisition. This implies that fluid is moving upward in a persistent manner,

somewhere in the vicinity of Hole 9 during and after Event E2.

The Hole 9 fluid pressure change (sampled at 5 Hz) response during constant flow injection (phase II) is shown in Figure 5.20 along with the AE hit count versus time histogram shown in Figure 5.20a. The AE hit counts peak very close to and during the pressure changes. This indicates some sort of breakage is occurring during these times. The sequence is highly temporally correlated with pressure and voltage changes and indicates some sort of breakage followed by periods of low AE activity. The AE hit counts close to Event E1p peak at 108 hits; the E2p count peaks at 477 hits; the E3p event peaks at 270 hits; the E4p hit count peaks at 532 hits; and the hit count at E5p is complex having three peaks, with a maximum count above 680 hits. The hits are based on exceeding an AE threshold level on each channel in the AE detection system. Only a few of the hits contain enough SNR and channel-to-channel correlation without overlap to allow hit localization. If hits are localized, then they turn into AE events. Figure 5.20a also shows the temporal correlation of the located AE events. Figure 5.20b shows the trend removed pressure change data along with the event correlations, and Figure 5.20c shows the voltage response and pressure change event correlations.

Figure 5.20 shows that the observed bursts in the electrical field are directly related to pressure changes that were measured in the injection system and AE hits. The pressure data indicate that there were some sharp changes in the flow regime inside the system connected to Hole 9 and that the detected fluid flows were only occurring inside the block (the occurrence of electrical data shows that the fluid that moved was in contact with porous media; no electrokinetic phenomena would occur outside the block and directly in the hole). The large number of temporally correlated AE hits indicates that something was breaking at the times of the pressure and voltage changes.

Note that the pressure data shows only small pressure fluctuations in the early phases of the seal degradation around Hole 9 while the pressure continued to build. The drops in pressure and correlated increases in voltage indicate that fluid is moving in the system. The drops in pressure indicate that the fluid flow rate through the failing seal was momentarily higher than the controlled, constant fluid flow rate that the system was programmed to deliver. This higher fluid flow rate is thought to be caused by a small increase in the system fluid volume due to an increase in annulus volume, and as a result, the pressure momentarily decreases in the entire fluid injection system until the pathway associated with the momentarily higher flow rate closes. This flow rate decrease allows the pressure buildup, from the constant

flow rate injection into the block, to resume until another portion of the seal fails, generating another increase in system volume with an associated pressure drop and electrical impulse. This repeats until the seal completely fails. If pressure measurements were the only observations of these events, then these fluctuations could not be directly attributed to a seal failure mechanism, and an explanation for the pressure changes could not be established. However, the existence of electrical and AE data shows that there is a mechanism other than induced block fracturing going on. The electrical data provides information related to the fluid flow processes during the series of events. In general, the electrical data implies progressive seal failure, and the pressure data confirms fluid movement. In fact, the electrical data actually provides more detail of the early development of the seal failure process. We will show that the electrical field can be used to localize these events, showing a progressive seal failure was occurring during constant flow rate injection. Each of the pressure drops shown in Figure 5.20 indicates that fluid is moving in the system, resulting in the burst-like electrical behavior described in the previous section. However, only when the pressure decreases precipitously (E5p in Figure 5.20) and is sustained can full seal failure be identified from the pressure data alone.

This set of observations shows a strong correlation between mechanical effects and electrical responses, indicating the breakage of material along with the movement of fluid in the system. Each observation by itself is insufficient to explain the physical processes occurring within the block; however, the combination of observations strengthens the understanding of the physical changes within the block.

5.3.4 Electrical potential evidence of seal failure

The persistent spatial voltage distribution shown in Figures 5.18 and 5.19 indicates the effects of upward fluid migration somewhere near Hole 9. We believe that this set of observations provides a leading indicator of progressive borehole seal failure. Ultimately, the existence of a failed injection tube seal was further confirmed through large drops in the fluid pressure measurements (see previous subsection), and subsequently, fluid leakage external to the borehole was later visually confirmed through the observation of water at the top surface of the block in the vicinity of Hole 9. The temporal electrical signature shown in Figure 5.17 indicates numerous impulsive events that grow as seal failure progresses. The hypothesis for the explanation of this data considers that the epoxy seal failed in a plastic manner beginning with the onset of failure followed by subsequent repeated blockage and breakthrough events having a valve-like behavior until the end of the data acquisition. The seal failure occurs in the partially epoxy-filled annulus of the borehole between the steel tube and the cement; as more fluid contacts the cement walls of the borehole with higher and higher velocities, the magnitude of the electrical response grows accordingly. The approximate position of the fluid contact with the borehole wall can be determined from the data and will be demonstrated for Events E2 and E3. The position of the positive anomaly recorded by the top array is not centered on Hole 9, but is displaced, from the center of the hole, possibly because of the position of the electrodes, the electrical boundary conditions around the borehole, and the spatial interpolation process.

The data from the side face electrical potential array also contains source location and orientation information, indicating that the fluid flow encountered porous media somewhere well above the bottom of the borehole, also an indication of possible borehole seal failure. The observations imply that the fluid flow occurred along a pathway following the borehole and close to the lower right corner of the top array. The electrical boundary condition in Hole 9 is insulating between the borehole wall and the stainless steel tubing, causing the reflection of the electrical potential away from the borehole center. Generally, these electrical potential measurements are consistent with the subsequent observations of fluid leakage at the test block surface near Hole 9 due to complete borehole seal failure. These electrical observations occurred several minutes before surface fluid leakage was visually observed on the top surface.

5.3.5 Source localization algorithms

In this particular situation, an observed electrical signal is potentially caused by a point fluid leak source. Here, we consider that the electrical source current density generated by this type of leak will have a point characteristic. This point-like current density characteristic can be simply modeled with a single, ideal electrical dipole. However, even if the fluid leak has a finite flow dimension with respect to the measurement system geometry, generating an electrical current density with a finite

dimension (but still relatively small with respect to the system geometry), we consider that the centroid of the electrical current density can still be suitably approximated by a single electrical dipole. Overall, this argument establishes the expectation that a relatively compact fluid leak can be represented by a compact electrical source current density. Therefore, we can use ideal electrical dipoles in the numerical inversion process to model the electrical sources that are postulated to have generated the signals that were measured during the seal failure.

Thusly, our goal in this section is to apply a dipole-based inversion process to localize the centroid of the source of the electrical disturbances in the block for two of the events discussed earlier. We start by specifying the physical problem for the occurrence of the quasistatic electrical potential distribution. Then we introduce and apply a two-step inversion process as follows: (i) we apply a gradient-based inversion with reduction of the volume of support for the source (compactness), followed by (ii) a genetic algorithm (GA) localization to refine the position of the source. We apply a two-step inversion process that is designed to reduce the computational resource requirements and computational time primarily for the kernel matrix computations and ultimately develop an approach that can be used in a real-time computational environment.

In the first step, we use a coarse cubical dipole matrix grid that encompasses the majority of the sample volume, in an unbiased localization technique that uses a deterministic algorithm with compactness to locate the rough area of each of the event dipole sources. The coarse nature of the dipole grid greatly reduces the time it takes to compute the kernel matrix relative to a fine grid that would be needed to locate the source in an equivalently unbiased and detailed way. This approach is designed to allow the inversion computations to find the approximate location for the source within the definition of the dipole grid points.

In the second step, we use the results of the first step to bias the localization process by generating a cylindrical dipole grid point that more than encompasses the dipole sources located in step one. This dipole matrix has a much finer spatial resolution, but with about half the points used in step one. This reduction in points also reduces the size of the kernel matrix and thereby reduces the computational resources for that element of the inversion computation. We use a GA to search through each of the dipole grid points using a single dipole-based forward model computation with the new kernel matrix. For each dipole position, a series of statistically generated dipole moments are used to compute the forward model simulation of the data which is then used in a cost function that calculates the error between the data and the model. The GA computes numerous model realizations and converges to a minimum of the cost function for each dipole position. After all dipoles in the matrix have had their cost functions minimized, the resulting error function is searched for the minimum of all model realizations, the dipole solution for each event. A GA was selected for its computational parallelization potential, thereby generating a systematic approach that can approach real-time speeds with parallel computing resources.

5.3.5.1 Electrical and hydromechanical coupling

The coupling between the hydromechanical equations and the electromagnetic equations is described in Mahardika et al. (2012) including dynamic terms. The governing equation for the occurrence of self-potential signals is obtained by combining a constitutive equation with a continuity equation. The constitutive equation corresponds to a generalized Ohm's law for the total current density $\mathbf{J}$ (in A m^{-2}):

$$\mathbf{J} = \sigma \mathbf{E} + \mathbf{J}_S \tag{5.35}$$

where σ denotes the low-frequency electrical conductivity of the porous material (in S m^{-1}), $\mathbf{E} = -\nabla\varphi$ the electrical field in the quasistatic limit of the Maxwell equations (in V m^{-1}), and φ the electrical potential (in V). The source current density is given by $\mathbf{J}_S = \hat{Q}_V^0 \dot{\mathbf{w}}$ where $\dot{\mathbf{w}}$ denotes the Darcy velocity and $\hat{Q}_V^0$ the excess of charge (of the diffuse layer) per unit pore volume of the porous or fractured material (in C m^{-3}) that can be dragged by the flow of the pore water. As seen in Chapter 1, the Poisson equation for the electrical disturbance ψ (expressed in V) is written as

$$\nabla \cdot (\sigma \nabla \psi) = \Im \tag{5.36}$$

where $\Im$ denotes the volumetric current density (in A m^{-3}). This volumetric current density is given by

$$\Im \equiv \nabla \cdot \mathbf{J}_S = \hat{Q}_V^0 \nabla \cdot \dot{\mathbf{w}} + \nabla \hat{Q}_V^0 \cdot \dot{\mathbf{w}} \tag{5.37}$$

The electrical potential distribution at an observation point P is given by

$$\psi(P) = \frac{1}{2\pi}\int_\Omega \rho \frac{\nabla \cdot \mathbf{J}_S(M)}{x(P,M)} dV + \frac{1}{2\pi}\int_\Omega \nabla\ln\rho(M) \cdot \frac{\mathbf{E}(M)}{x(P,M)} dV \tag{5.38}$$

where x denotes the distance from the source at position M to the electrode located at position P where the electrical potential signal is recorded. In Equation (5.38), the two contributions associated with the primary field (first term of the right-hand side of Eq. 5.38) and the secondary potential (the second term of the right-hand side of Eq. 5.38) are separated. The primary source term is due to the hydromechanical disturbances, while the second term is due to the heterogeneities in the resistivity distribution of the resistive block.

The solution can be written in a more compact form using the following convolution integral:

$$\boldsymbol{\psi}(P) = \int_\Omega \mathbf{K}(P,M)\mathbf{J}_S(M)dV \tag{5.39}$$

where $\mathbf{K}(P, M)$ is called the kernel or the leading field and dV is a small volume around the source point M and $\boldsymbol{\psi}(P)$ denotes the vector of self-potential measurement at a set of stations P. We use this equation in the following computations in our attempt to localize the causative source of the electrical bursts shown in Figure 5.17. Generally, the elements of the kernel are Green's functions that mathematically relate voltage measurements at an array of observation stations, P, located at the variously defined measurement locations to each of the individual sources of current density defined by a set of source points, M, located in the conducting volume. In this system, the kernel computation accounts for the electrical resistivity distribution and the boundary conditions that are present within the test block. In the following computations, a uniform resistivity distribution is used within the forward model domain volume, excluding the holes that were drilled into the block. Indeed, these holes represent infinite impedance zones within the volumetric resistivity distribution of the modeled block and are explicitly accounted for in the computation of the kernel. In fact, accounting for the presence of the holes within the model volume is crucial to properly compute the kernel and localize the source current density. The kernel matrix used in Equation (5.39) takes the form of Green's functions of the resistivity in the block and has units of resistivity, conforming to Ohm's law. Figure 5.21 shows the

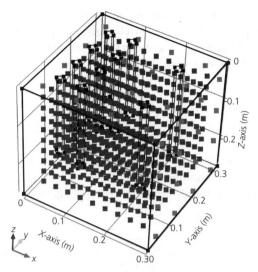

Figure 5.21 COMSOL model showing the distribution of points used to compute the coarse kernel matrix (blue points, 729 positions) and the voltage measurement points (red points, 32 positions). The model geometry includes each of the 10 holes that were drilled into the block (black cylinders) considered to be perfect insulators. The arrows labeled x, y, and z indicate the positive directions of the corresponding axes in the figure for the Cartesian coordinate system. (*See insert for color representation of the figure.*)

729 dipole positions within the block volume that are used to compute the kernel matrix that is used in the inversion process.

5.3.5.2 Inversion phase 1: gradient-based deterministic approach

Equation (5.39) can also be written in the following matrix form:

$$\mathbf{d} = \mathbf{Km} \qquad (5.40)$$

where $\mathbf{d}$ denotes the N-vector of the observed electrical potentials, at each time step, at N electrodes; the vector $\mathbf{m}$ is a $(3\,M)$-vector containing the M source current density terms (times three because each of the dipoles representing the source current densities in the model has a 3-component vector that represents the dipole moment in a 3D Cartesian coordinate system); and $\mathbf{K}$ represents the kernel matrix which has an $N \times 3\,M$-dimensional structure (the kernel matrix has a $3\,M$ dimension because each dipole point in the forward model has a three-component vector that represents a dipole moment in a 3D Cartesian coordinate system).

The inverse problem is solved in the 7 steps detailed below.

Step 1. Kernel computation. First, we place M elementary dipoles (numbered from 1 to M), in a cubical arrangement inside the three-dimensional COMSOL Multiphysics model of the block, with coordinates (x, y, z) for each dipole. Next, we position N voltage monitoring points on the two orthogonal surfaces of the block model. These voltage monitoring points were placed with the same positions and electrode numbers as the actual measurement electrodes. Next, we computed the voltage at each of the N monitoring points due to each of the individual current densities at each of the M elementary dipoles, separately. In each case, we subtracted the voltage at the position of the selected reference point. Indeed, as explained in detail in Jardani et al. (2008), the computed kernel must respect the voltage at the selected reference electrode. In the resulting kernel matrix configuration, the kernel is composed of three submatrices $\mathbf{K} = \left[\mathbf{K}_x, \mathbf{K}_y, \mathbf{K}_z\right]$, and each of these submatrices $\mathbf{K}_i(i = x, y, z)$ is a $N \times M$ matrix so that $\mathbf{K}$ corresponds to a $N \times 3\,M$ matrix. Each of these submatrices represents an orthogonal component of the leading field. In this representation, each of the elementary dipoles used in the computation is described by the electrical current dipole moment vector, $\mathbf{m_d} = I\mathbf{D}$, where I is the electrical current magnitude, $\mathbf{D}$ represents the displacement vector that points in the direction of the flow of the current (in the direction of the electrical field), and $\mathbf{m_d}$ stands for the electrical current dipole moment vector (this is equivalent to the current density vector divided by the volume of the cell). The current dipole moment is therefore expressed in units of A m.

Step 2. In our use of the dipole moment in this application, we break the moment vector down into three orthogonal components that formulate the three components of the electrical current source density model vector. Thusly, the model vector, $\mathbf{m}$, is composed of three subvectors $\mathbf{m} = [\mathbf{m}_x, \mathbf{m}_y, \mathbf{m}_z]^{\mathrm{T}}$ (the superscript T means transpose), where each subvector, $\mathbf{m}_i(i = x, y, z)$, has an $M \times 1$ dimension leading to a full model vector dimension of $3\,M \times 1$. The inversion process determines the magnitude of each of these model vector components and thereby approximates the direction of electrical current flow that was a result of the physical interactions that generated the measured spatial voltage distribution. This means that the number of model vector unknowns is $3\,M$ (three components for the current density times the number of point dipoles). We have $3M \gg N$ (model unknowns $\gg$ actual measurements); therefore, the inverse problem is strongly underdetermined. In this deterministic inversion process, the roles of data misfit and model objective functions are balanced using Tikhonov regularization (Tikhonov & Arsenin, 1977; Jardani et al., 2008) through a global objective function, $P_\lambda(\mathbf{m})$, defined as

$$P_\lambda(\mathbf{m}) = \|\mathbf{d} - \mathbf{Km}\|_2 + \lambda s(\mathbf{m}) \qquad (5.41)$$

where λ represents a Lagrange regularization parameter $(0 < \lambda < \infty)$ and $s(\mathbf{m})$ denotes the (stabilizing) regularizer. The vector $\mathbf{m}$ is the vector described earlier.

Step 3. The inversion process needs to be scaled with respect to the distance between the sources and the receivers to remove depth-based bias in the solution. This gives each dipole in the model an equal chance to participate in the numerical solution. This will be reversed later in the process to reflect the actual depth effects on the solution. Depth weighting is accomplished by using a $(3M \times 3M)$ diagonal matrix, $\mathbf{J}$, that can be computed from the kernel as follows:

$$\mathbf{J} = \begin{bmatrix} \mathbf{J}_x & 0 & 0 \\ 0 & \mathbf{J}_y & 0 \\ 0 & 0 & \mathbf{J}_z \end{bmatrix} \qquad (5.42)$$

where each of the $\mathbf{J}_i(i=x,y,z)$ components represents one orthogonal direction and is a $(M \times M)$ diagonal submatrix computed by

$$J_{ii}^{x,y,z} = \sum_{j=1}^{N} K_{ij}^{x,y,z} \qquad (5.43)$$

$$J_{ij}^{x,y,z}(i \neq j) = 0 \qquad (5.44)$$

This weighting matrix structure matches the kernel matrix and model vector organizations in terms of separating out the three dimensions of space in three separate subsections.

Given that we are looking for a compact source consisting of as few dipoles as possible that would model the electrical mechanism that generated the voltage measurements, compactness of the source must be assured. Therefore, we apply a compact source inversion process, where we seek to find one model (model vector) with the minimum volume of the source current density

(the fewest dipoles with moments that mathematically support the data). As shown by Last and Kubik (1983), the depth-weighted and compactness-based stabilizing functional is expressed as

$$\boldsymbol{\Omega} = \begin{bmatrix} \boldsymbol{\Omega}_x & 0 & 0 \\ 0 & \boldsymbol{\Omega}_y & 0 \\ 0 & 0 & \boldsymbol{\Omega}_z \end{bmatrix} \qquad (5.45)$$

where each of the $\boldsymbol{\Omega}_i(i=x,y,z)$ components represents one orthogonal direction and is a $(M \times M)$ diagonal submatrix. The elements of these three submatrices are computed by

$$\Omega_{ii}^{x,y,z} = \sqrt{\frac{J_{ii}^{x,y,z}}{m_i^{x,y,z} + \beta^2}} \qquad (5.46)$$

$$\Omega_{ij}^{x,y,z}(i \neq j) = 0 \qquad (5.47)$$

where β is a support parameter (our choice of β is explained further in the following text). This compactness matrix structure is also required to match the kernel matrix, weighting matrix, and model vector organizations in terms of separating out the three dimensions of space in three separate subsections.

Step 4. Normalization of the kernel matrix is required to prevent undesirable depth-based bias while invoking compactness constraints in the inversion process. Therefore, we form a new normalized kernel matrix that is used during compactness computations as follows:

$$\mathbf{K}^* = \begin{bmatrix} \mathbf{K}_x^*, \mathbf{K}_y^*, \mathbf{K}_z^* \end{bmatrix} \qquad (5.48)$$

$$\mathbf{K}_{x,y,z}^* = \mathbf{K}_{x,y,z} \boldsymbol{\Omega}_{x,y,z}^{-1} \qquad (5.49)$$

where $\mathbf{K}^*$ is a $(N \times 3M)$ matrix that is formed through the scaling of the original kernel with the compactness functional as shown in Equation (5.45).

Step 5. We solve the following system of equations for $\mathbf{m}^*$ using Tikhonov regularization (Tikhonov & Arsenin, 1977; Jardani et al., 2008) through the global objective function, $P_\lambda(\mathbf{m})$:

$$\begin{bmatrix} \mathbf{K}^* \\ \lambda \mathbf{I} \end{bmatrix} \mathbf{m}^* = \begin{bmatrix} \mathbf{d} \\ \mathbf{a} \end{bmatrix} \qquad (5.50)$$

where $\mathbf{I}$ represents the $(3M \times 3M)$ identity matrix, $\mathbf{d}$ is the $(N \times 1)$ vector of measured electrical potential data,

$\mathbf{m}^*$ denotes the computed and scaled $(3\,M \times 1)$ vector of model parameters, and $\mathbf{a}$ refers to a $(3\,M \times 1)$ vector with only zeroes.

Step 6. Because the resulting inverted model with compactness, $\mathbf{m}^*$, is depth weighted, we need to unscale the model solution to produce the inverted $(3\,M \times 1)$ model vector, $\mathbf{m}$, by

$$\mathbf{m} = \left[\mathbf{m}_x, \mathbf{m}_y, \mathbf{m}_z\right]^{\mathrm{T}} \qquad (5.51)$$

$$\mathbf{m}_{x,y,z} = \boldsymbol{\Omega}_{x,y,z}^{-1} \mathbf{m}_{x,y,z}^* \qquad (5.52)$$

This model vector solution is in terms of the current dipole moment expressed in A m. For compactness, we initially use a small support parameter of $\beta = 10^{-12}$. Then for this value, we compute the "best" value of the regularization parameter, λ, using the L-curve approach. If the solution is not compact enough, we multiply the previous value of β by 10 and repeat the process until no additional compactness is achieved.

Step 7. The solution found in step 6 represents a mathematical solution to Equation (5.46) that represents the minimum of the objective function. This solution has the smallest mismatch between the measured data and the synthetic data forward computed using the inverted model and as such contains dipoles with finite moments in every dipole point position in the model vector. However, we wish to use only the main contributing components of this model that best represents the electrical effects. Based on this, the computed model is thresholded to keep only the main dipoles that explain most of the solution. A threshold is applied to the model vector to zero all of the dipoles with moment values that were below the final value of β. The existence of a large number of low-magnitude dipoles in the model vector that are produced by the gradient inversion process represents a purely mathematical solution to the source inversion. Therefore, it is considered that these small dipoles do not represent real sources and are not likely to be physically present during the event generation process. Thresholding these widely spread small dipoles removes the nonphysical and mathematical only contributions to the solution, allowing the comparison of the principal elements of the inversion with the real data. This helps to realistically quantify the inversion process and source localization error but does not affect the main components of the solution. Indeed, we expect the solution to be rather compact, and not broadly distributed.

Thresholding does bias the solution; however, we expect the solution to be in the vicinity of Hole 9 without components spread throughout the block volume.

The grid used for the previously described gradient-based approach, combined with compactness, is actually pretty coarse (see Figure 5.21). This is to reduce the computational effort and time needed to find a preliminary position for the source. As shown in the next section (see Figures 5.22, 5.23, 5.24, 5.25, and 5.26), the inversion of the data leads to a source localized in the vicinity of Hole 9 (as expected). Once this is done, we switch our inversion to the GA on a refined grid located in the vicinity of the solution found by the gradient-based approach. The GA used for this second inversion phase is described in the next section.

5.3.5.3 Inversion phase 2: GA approach

Using the results of the gradient-based approach, we apply a single dipole GA-based search through a new and finer kernel matrix with only 360 positions (Figure 5.27). A single dipole is presumed to represent the overall effect of water flow during the leak part of experiment. This assumption is expected to be good enough to locate the centroid of the leak position within the block volume where the leak is occurring. The GA is used as follows: a population of candidate solutions is used to find the solution of the inverse problem. This population has to evolve toward solutions that minimize the data misfit function:

$$P_{\mathrm{d}}(\mathbf{m}) = \|\mathbf{d} - \mathbf{Km}\|_2 \qquad (5.53)$$

where $\|\cdot\|_2$ refers to the L2 norm. The evolution starts from a population of randomly generated individual solutions in the volume defined by the deterministic gradient-based algorithm described earlier. For each generation, the goodness of fit is evaluated through Equation (5.49), and multiple individuals are stochastically selected from the current population and modified to form a new population. This new population is then used at the next iteration. The process is continued until a predetermined number of generations have been reached or a satisfactory data misfit has been reached for the population.

The model vector, $\mathbf{m}$, contains $360 \times 3 = 1080$ elements (the number 3 represents the three components of the current dipole moment vector). The 360 positions of the kernel matrix are positioned on three concentric

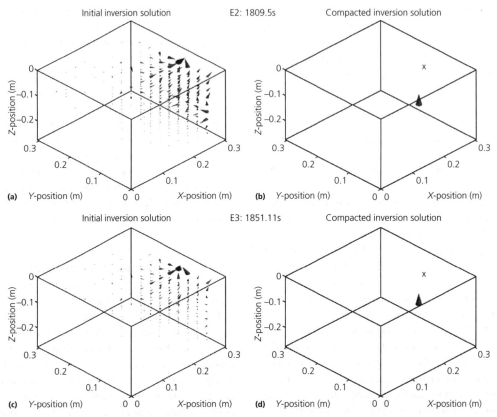

Figure 5.22 Gradient based inversion results showing the initial and compacted dipole inversion for Events E2 and E3. In these cases, the compact inversion of Events E2 and E3 yields one dipole that dominates the response through a higher magnitude. Note that Event E3 is localized at a shallower depth than Event E2, showing that the causative source is moving closer to the top surface of the block in the vicinity of Hole 9. **a)** Event E2 results of the source distribution after inversion. **b)** Same after 40 iterations in compacting the support of the source. **c)** and **d)** Same for Event E3. (*See insert for color representation of the figure.*)

cylindrical surfaces centered around Hole 9. Each cylindrical surface (radii = 0.0105, 0.0125, and 0.0145 mm) contains 10 z-axis levels with 4.4 mm spacing between levels and 12 points in the x–y plane at each z-axis level equally spaced at 30 degree increments (see details in Figure 5.27). The kernel matrix was computed in the same way as the coarse kernel matrix and resulted in a kernel matrix that was $(360 \times 3) \times 32$ or 1080×32 elements. The GA we used is the one found in the MATLAB (Global Optimization Toolbox, MathWorks (R2012a) (http://www.mathworks.com/products/global-optimization/functionga.m; see for details http://www.mathworks.com/products/global-optimization/description4.html). It should be noted that COMSOL Multiphysics renumbers the dipole points during its internal geometry point scan, and the dipole numbers listed here are aligned with this COMSOL point sequence.

5.3.6 Results of the inversion
5.3.6.1 Results of the gradient-based inversion
For phase 1 of the inversion, the model vector (2187×1 elements) represents the current dipole moment (magnitude in A m) at the 729 dipole positions in the model (at each location, the current dipole moment is characterized by three components along x-, y-, and z-axes). The initial model vector solution of the gradient inversion process generates dipoles with finite moments in every dipole position in the inverted model vector, and as such, these dipoles are distributed throughout the dipole point matrix positioned within the volume of the block (the COMSOL Multiphysics model of the block). The resulting dipole moments are mathematically consistent with the electrical potential distribution in the data. The compaction process changes the dipole moments, reducing the magnitude of most of the dipoles in the process (reduced

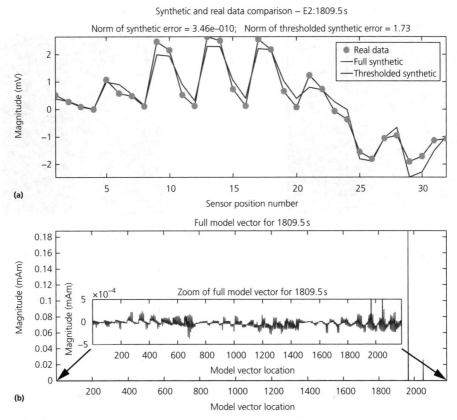

Figure 5.23 E2 compacted inversion results and forward modeled comparison with a thresholded forward model. **a)** Comparison of the real data with the forward model calculations using the full inverted model, and a thresholded model derived from the full inversion model. **b)** The full model vector that was found through inversion and compaction. The inset shows the multitude of minor dipoles spread throughout the model vector that were generated by the inversion and compaction process. (*See insert for color representation of the figure.*)

volumetric support for the source current density). This process results in a few (one or two) dominant dipoles with the rest of the points in the solution containing dipoles with very low moment magnitudes. In step 7, we threshold the small dipoles to make them equal to zero. Indeed, the large number of low-magnitude dipoles represents a nonphysical solution to the inversion as stated earlier.

To check how the thresholded solutions compare with the measured data, we examine their differences (using the L2 norm). In the following analysis, the forward modeled predictions of the data are called synthetic data.

Figure 5.22 shows the comparison between the initial inversion results after one iteration (without compaction; Figure 5.22a and c) and the compacted solutions after 40 iterations (Figure 5.22b and d) for Events E2 and

E3, respectively. Each small cone in Figure 5.22 represents a dipole at a kernel matrix point, pointing in the direction of the electrical current at that point, an expression of the dipole moment for the dipole positioned at that point. The compacted inversion (Figure 5.22b and d) results for both events show a single dipole pointing mostly in the $+z$ direction. This solution is consistent with the hypothesis of seal leakage induced upward ($+z$ direction) fluid flow based on the observations of the electrical potential distribution (see Figure 5.19). Additionally, according to the electrical current flow and the dynamic fluid coupling processes described by Equations (5.29) through (5.35), the electrical current flow is expected to occur in the direction of the fluid flow (via the drag of an excess of charge in the fluid flow direction with a current flowing in the direction of the cations).

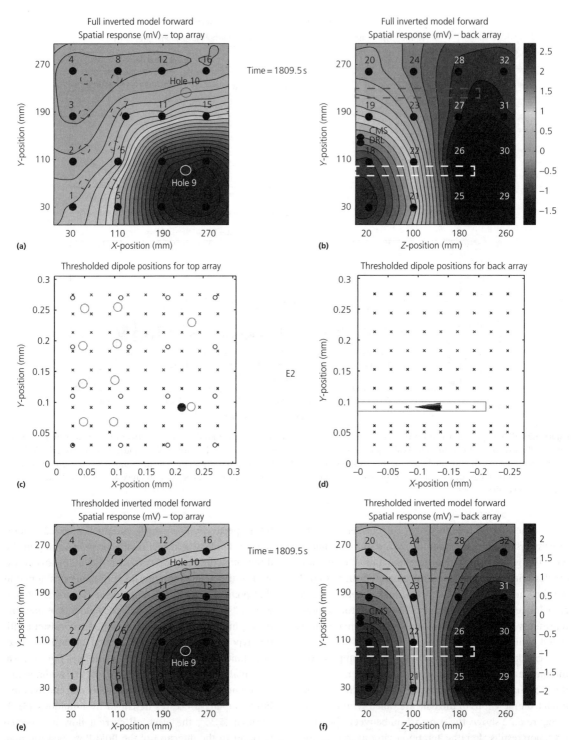

Figure 5.24 Spatial distributions and dipole location comparison for inverted and forward modeled results for Event E2. **a)** and **b)** Forward computation results using the full model vector (all the minor dipoles included). **c)** and **d)** Thresholded model vector dipole positions (there are actually two dipoles in these panels) showing the orientation and position of the main dipole (visible). **e)** and **f)** Forward computation results using the thresholded model vector. (*See insert for color representation of the figure.*)

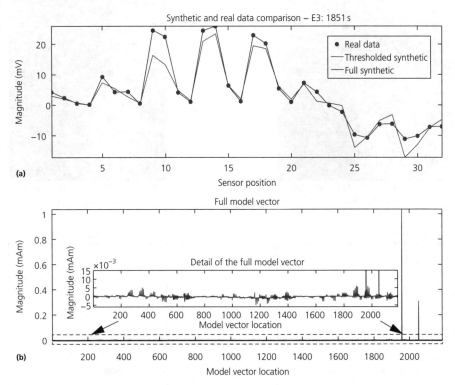

Figure 5.25 E3 compacted inversion results and forward modeled comparison with a thresholded forward model. **a)** Comparison of the real data with the forward model calculations using the full inverted model and a thresholded model derived from the full inversion model. **b)** Full model vector found through inversion and compaction. The inset shows the multitude of minor dipoles spread throughout the model vector that were generated by the inversion and compaction process. The main figure shows the thresholded model with zero values everywhere except at the model locations where the dipoles were localized. Note that the located dipoles are in close proximity to each other. (*See insert for color representation of the figure.*)

An in-depth analysis of Event E2 is shown in Figures 5.23 and 5.24. Figure 5.23a shows that the forward modeled full model vector (with compaction and without thresholding) reproduces the real data nearly perfectly as would be expected from the gradient inversion process. Figure 5.23b shows the full model vector solution, and the inset shows the magnitude range for all of the minor dipoles throughout the model vector. The thresholded model vector consists of only two dipoles (the rest of the thresholded model vector locations have zero magnitudes) and are not shown here for obvious reasons. It can be seen that the removal of all of the minor dipoles causes the L2 norm to increase substantially, making the thresholded model a poorer fit to the data, as should be expected when some of the mathematical solution support is removed. However, the main characteristics of the data are preserved.

Figure 5.24a and b shows the spatial voltage distribution generated by the full dipole model. This full forward modeled result matches extremely well with the kriged distribution of measured data as shown in Figure 5.19a and b, as should be expected with such a good inversion solution.

The use of compactness in the inversion process focuses the dipole solutions at the kernel points that are the closest to Hole 9 (Figure 5.24c and d). Figure 5.24e and f shows the voltage spatial distribution that results from the forward modeled thresholded model vector. Comparing Figure 5.24e and f with Figure 5.24a and b shows some minor differences between the distributions. However, the main features are preserved. This demonstrates that the thresholded solution has identified the correct location and general flow direction near Hole 9 for Event E2.

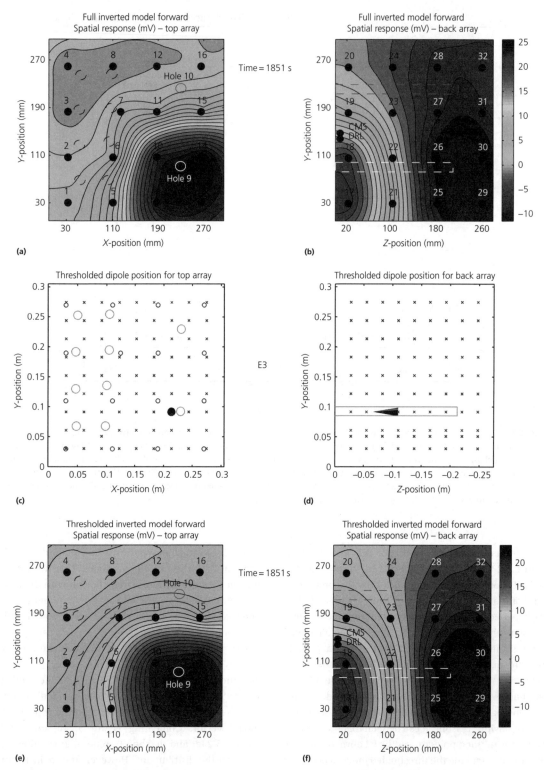

Figure 5.26 Spatial distributions and dipole location comparison for inverted and forward modeled results for Event E3. **a)** and **b)** Forward computation results using the full model vector (all the minor dipoles included). **c)** and **d)** Thresholded model vector dipole positions (there are actually two dipoles in these panels) showing the orientation and position of the main dipole (visible). **e)** and **f)** Forward computation results using the thresholded model vector. (*See insert for color representation of the figure.*)

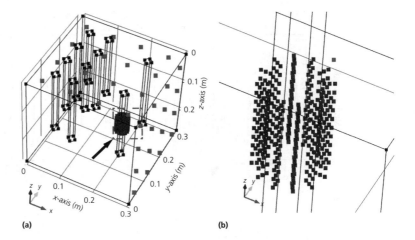

Figure 5.27 COMSOL geometry used for the fine geometry kernel matrix computations (360 positions). **a)** Finer resolution cylindrical kernel matrix point distribution along with the measurement points used with the genetic algorithm. **b)** Close-up of the kernel matrix point distribution. (*See insert for color representation of the figure.*)

The same analysis was carried out for Event E3 (Figures 5.25 and 5.26). The fundamental results for Event E2 apply to Event E3 with a notable difference. The main dipole shown in Figure 5.26c is positioned at the same x–y location as the dipole responsible for Event E2 in Figure 5.24c. However, the z-axis position of the main dipole shown in Figure 5.26d has moved up along Hole 9. This is consistent with the centroid of a hydromechanical disturbance moving upward along Hole 9 over time.

From the inverted data, it can be seen that the full model inversion vectors produce excellent matches to the real data but contain nonphysical elements because we expect a localized source. The thresholded models get rid of the nonphysical elements but yields, as expected, a solution that fits the data with a larger data RMS error. This increase in RMS error is expected because the coarse physical positions of the kernel matrix points used in this inversion do not perfectly match the true position and geometry of the source current density of the actual hydromechanical disturbance.

In conclusion, the result of the gradient-based inversion approach used in phase 1 provides reasonable estimation for the centroid of current dipole source locations at different times. We know from AE (not shown here) and pressure data that there was no fracturing of the block during the experiment reported in this paper. Therefore, all dipole solutions located far from Hole 9 are considered not to have a physical cause, given that the current source is expected to be compact. The coarse nature of the inversion process calls for refinement; therefore, this first set of solutions is used to direct the refinement of the source localization using the GA.

5.3.6.2 Results of the GA

Figure 5.28 shows the results of the GA single dipole search for Event E2. Figure 5.28a shows a plot of the data misfit error per Equation (5.45) for each dipole position (the best solution is highlighted by the small red circle, Dipole #42). The black line in the figure represents the minimum of the data misfit error for comparison with other dipole position optimizations. This figure also shows how several other dipole positions come close to the minimum value of the data misfit error. These other points are at positions 11, 12, and 41. Because of the COMSOL point sequence, these point positions represent spatially adjacent or nearly adjacent positions, indicating that the dipole point matrix may not be positioned in a manner that places a dipole in the exact position of the hydromechanical disturbance. It could also indicate that the disturbance has a volume that encompasses more than a single dipole point, requiring more than one dipole point as the solution. Other forms of inversion and analysis may be able to resolve this uncertainty. However, for this analysis, the general location of the centroid of the leak is sufficient to characterize the problem.

Figure 5.28b shows the comparison of the real data with the forward model of the dipole that represents the minimum of the objective function. The synthetic data reproduces all of the major features of the real data with an improved L2 norm relative to the coarse gradient inversion results. The model vector for Dipole #42 is shown in Figure 5.28c. Note the three positions that have nonzero values. These values represent the orthogonal components of the dipole moment found by the GA.

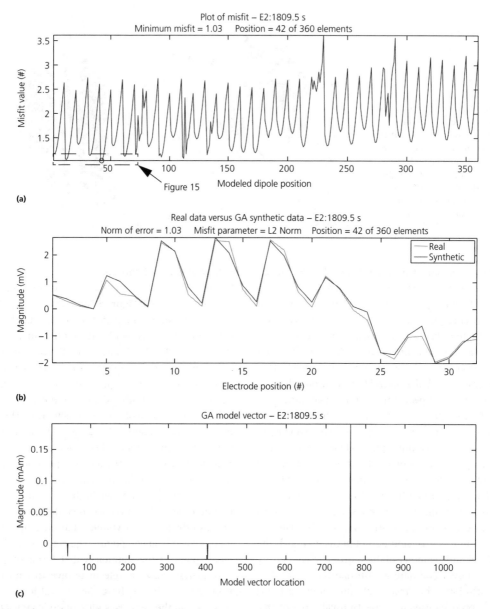

Figure 5.28 a) Shows the E2-related genetic algorithm localization misfit values for each position in the kernel matrix. The red circle represents the minimum value and therefore the best fit for the measured data, given the inversion constraints. **b)** Comparison of the real data and the genetic algorithm-based model vector forward computation. **c)** The genetic algorithm-based best fit model vector. Note that there are three points on the vector where there are values that are greater than zero. These points are exactly 360 elements apart, representing a single dipole moment orthogonal components. (*See insert for color representation of the figure.*)

It can be seen that the dipole moment orients the dipole in a direction other than exactly along the z-axis of the block, which is slightly different from the results of the gradient inversion.

As shown in Figure 5.29, there are other dipoles that have objective function values that are close to the data misfit minimum obtained with Dipole #42 (Figure 5.29). This indicates that other solutions may

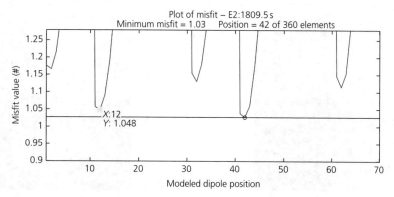

Figure 5.29 Zoom in misfit for all the observations of the genetic algorithm search algorithm of Event E2. This figure shows that there are dipole positions that are close to the minimum of the misfit function, which is shown by the small open circles. The position marked by $x = 12$ represents another possible location that is only slightly greater than the minimum found at 42. There are other points that are spatially adjacent to 42, including 11, and 41. (*See insert for color representation of the figure.*)

yield equivalent or better results if more than one dipole is considered in the inversion process (see earlier comments about the point matrix positioning for Event E2). Figure 5.30 shows the forward modeled results with Dipole #42 only. The position of Dipole #42 is located at 210° from the right side, on the second ring. The moment of the dipole (Figure 5.30e and f) generates a spatial distribution of the voltage that is tipped in a similar manner to the real data shown in Figure 5.20a and b. This orientation of the dipole implies that the axis of fluid flow is likely flowing mostly upward along the hole annulus with a slight flow direction away from the hole. This may be due to the way the epoxy is blocking the flow, causing a diversion of the flow away from Hole 9, or there could be a component of flow into the cement block (Darcy flow or microcracks).

Figure 5.31 shows the results of the GA single dipole search for Event E3. The misfit for this event shows a minimum at position 47 (Dipole #47) with a significant departure from the minimum as the dipole positions increase away from this position. There are some other dipoles that come close to the minimum; these dipoles are all spatially very close to Dipole #47 and at positions 17, 18, 48, 77, and 78 (see earlier comments about the point matrix positioning for Event E2) (Figure 5.32). The spatial voltage distribution of the inversion also shows a slight tilt on the back array that is similar to the tip shown in Figures 5.19, 5.20c and d. This GA-based inversion places the event E3 dipole higher in the block than the dipole of Event E2, with an orientation of the dipole moment that indicates the fluid flow axis is mostly

upward along the hole with a slight tipping toward a tangent-like trajectory along the hole. It should also be noted that Dipole #47 is in a position that is directly above the Dipole #42 position, indicating that the flow between Events E2 and E3 is likely to be mostly vertical. It should be clarified here that the other impulses in the data may localize to somewhat different points along the hole and may show other shifting of the flow pathway, and therefore, great care should be used to not overinterpret the results of the inversion (Figure 5.33). Because of the still relatively coarse resolution of this finer point spread and the nonadjacency of the selected time slices, the subtleties of the fluid flow along the annulus of the hole cannot be resolved from this analysis. However, the dipole moments of these events may indicate unresolved fluid flow complexity. The single dipole inversion using the GA reproduces a similar degree of fit as the thresholded gradient inversion. Analysis of the position data from the two inversions shows a high degree of consistency, with all of the GA located point coordinates falling inside the range of ±1/2 point spacing of the coarse dipole point matrix.

Figure 5.34 shows, in a notional conceptualization, how the fluid moved up along the Hole 9 annulus, breaking through numerous epoxy barriers, and in the process making direct contact with the cement block borehole wall. The electrical and pressure data supports this multiple-chamber, fluid movement concept involving plastically breached epoxy barriers as shown in Figure 5.34e. Figure 5.34a through d show the results of the GA inversion process in the same figure with Event

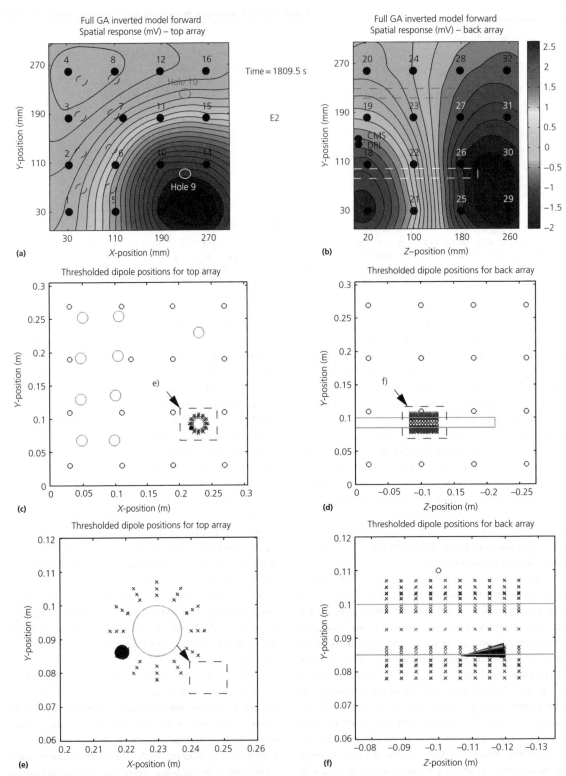

Figure 5.30 a) and **b)** Event E2 forward modeled voltage distribution of the genetic algorithm located dipole. **c)** and **d)** Spatial location of the dipole within the concrete block. **e)** and **f)** Close-up of the dipole location showing the off-vertical orientation of the dipole moment. (*See insert for color representation of the figure.*)

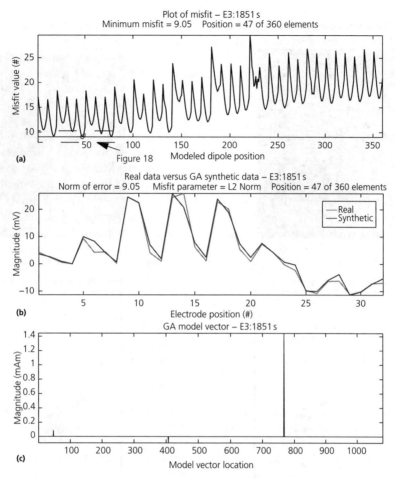

(a)

Figure 18

(b)

(c)

Figure 5.31 a) E3-related genetic algorithm localization misfit values for each position in the kernel matrix. The red circle represents the minimum value and therefore the best fit for the measured data, given the inversion constraints. **b)** Comparison of the real data and the genetic algorithm-based model vector forward computation. **c)** The genetic algorithm-based best fit model vector. Note that there are three points on the vector where there are values that are greater than zero. These points are exactly 360 elements apart, representing a single dipole moment orthogonal components. (*See insert for color representation of the figure.*)

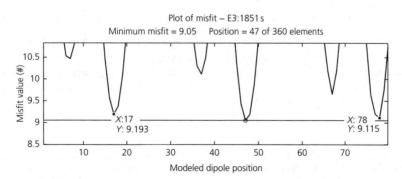

Figure 5.32 Zoom in the values of the misfit parameter obtained using the genetic algorithm-based inversion of Event E3. This figure shows that there are several dipole solutions that are close to the minimum of the data misfit function. In this figure, there are several other possible solutions, and they are all close to the point found at position 47, the minimum. These additional possible locations corresponds to dipole positions 17, 18, 48, 77, and 78. (*See insert for color representation of the figure.*)

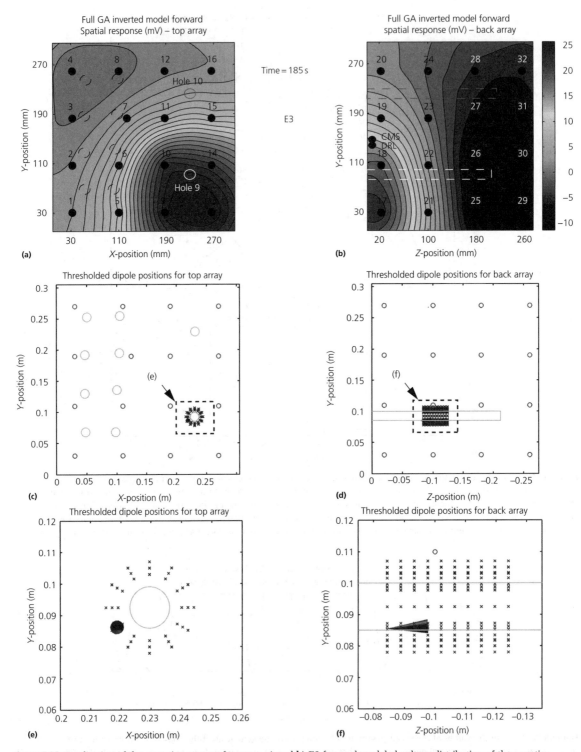

Figure 5.33 Localization of the causative source of current. **a)** and **b)** E3 forward modeled voltage distribution of the genetic algorithm located dipole. **c)** and **d)** Spatial location of the dipole within the concrete block. **e)** and **f)** Close-up of the dipole location showing the vertical orientation of the dipole moment. (*See insert for color representation of the figure.*)

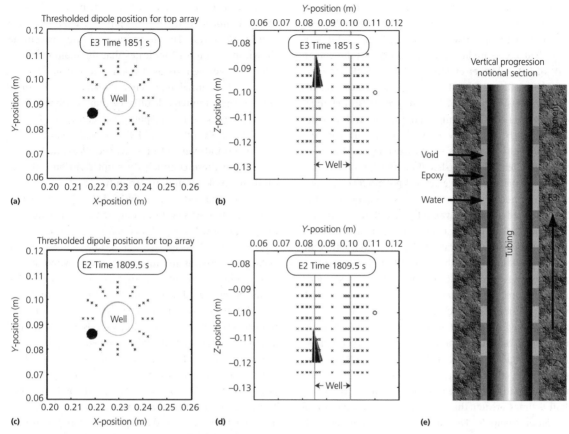

Figure 5.34 Proposed explanation of the electrical disturbances observed in the experiment. Events E2 and E3 are shown together with their spatial placement near Hole 9 in panels **a)** through **d)**. The blue *x*'s in panels **a)** through **d)** are the dipole point positions used in the genetic algorithm source inversion process. The green circle in panels **a)** and **c)** represents the position of Hole 9 as viewed from the top, and the vertical green lines in panels **b)** and **d)** represent the position of Hole 9 as seen from the side. The red circles in panels **b)** and **d)** are the position of one of the measurement electrodes. It can be clearly seen in this figure that the first event, E2, occurs lower in the model representation (and by correlation, in the test specimen) than the later event, E3. The figure in panel *e* represents a conceptualization of the movement of the fluid through multiple epoxy barriers, filling voids in between each the barriers with water. The voids represent points of open contact with the cement that allow the fracturing fluid to make direct contact, generating the observed pressure fluctuations and electrical impulses. **e)** Sketch of the tubing showing the progression of the Events E2 and E3 along the well over time.
(*See insert for color representation of the figure.*)

E2 being lower in the block than the later in time Event E3. This clearly shows the upward fluid movement that was found through the inversion of the data, and matches with the physical observation of leakage of fluid at the top of the block, later in the experiment.

5.3.6.3 Noise and position uncertainty analysis
To see how robust the inversion solutions are, a noise analysis was performed on the voltage and the pressure data sets, and the results show a very good SNR for both of these data sets. The base SNR for the pressure data was computed during the constant pressure phase of the experiment, before the pressure increases in the constant flow phase of the experiment. This portion of the signal had a mean pressure of 1169.8 kPa with an RMS noise contribution of 1.4 kPa. This noise is what is used to compute the SNR of the pressure fluctuations that caused the leakage that was detected electrically. The smallest

pressure change during an event was observed during the transition from the constant pressure period to the constant flow period (E0p) at time 1796 s (Figure 5.20). This event had a pressure increase of 13.3 kPa, resulting in an SNR of 9.8. All other events had higher SNRs, ranging from about 22 at E3p to 108 at E5p. This process results in the conclusion that all of the observed pressure changes were due to a physical change in the system and were not due to noise. Combined with the electrical data, it is clear that the pressure changes were caused by seal breakage events that lead to a burst-like fluid movement.

A noise analysis of the electrical potential signals was also performed. The noise baseline was established after DC offset and trend removal, reference channel subtraction, and after the constant flow was initiated, but before the onset of rapid electrical activity. Note that when the reference channel was subtracted, all residual correlated noises were removed, but the uncorrelated noise added in quadrature. This resulted in a net reduction of overall noise in all channels but channel 13 where the uncorrelated noise was dominant over the correlated noise component; the reference channel removal caused the noise in this channel to increase slightly. The mean RMS noise level for all channels was computed, resulting in the observation that the noise level on channel 13 was more than 9 times greater than the mean of the noise of all the other channels. The noise level on channel 13 is

computed to be 0.121 mV RMS, and the mean noise level of all channels (with the exception of channel 13) is 0.0134 mV RMS. The inspection of the waveforms shown in Figure 5.17 indicates that the main events of interest have high voltages relative to the noise level. Since each channel represents a spatial measurement point on the cement block, only the channels that contribute to the peak voltage response are relevant to the SNR calculations. Events E2 and E3 have peak channel SNRs of over 200 and over 1900, respectively. We conclude that the SNR of these channels does not contribute significantly to dipole location uncertainty.

Positional uncertainty analysis of the E2 GA current dipole solution was performed by adding Gaussian noise to the computed forward solution using the E2 dipole moment model vector. This new noisy measurement vector was used as the input to the GA, where a new dipole solution would be computed. The new solution would be compared with the initial solution. Three levels of noise were used in this analysis: 1, 5, and 10% noise levels were used. These noise levels were computed at as a random Gaussian additive voltage computed from the mean voltage of each channel. Solutions for the three noise cases were found at the following dipole point positions: 2 for 1%, 62, for 5%, and 61 for 10% noise. These solution points have a displacement from the initial solution of one radial dipole point in the position

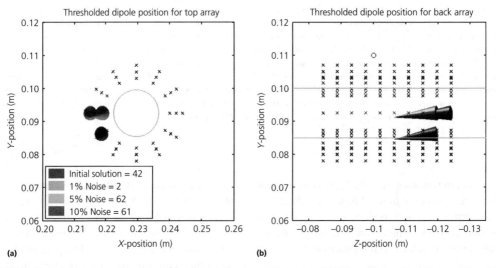

Figure 5.35 Localization of the causative source of current and noise analysis. **a)** and **b)** These figures show the spatial positions of the dipoles found during the localization uncertainty test. Note that the solutions cluster near the initial solution found during the inversion process. The bias in the +y and −z directions may indicate that the true solution may be between the solutions with noisy data and the solution found initially. (*See insert for color representation of the figure.*)

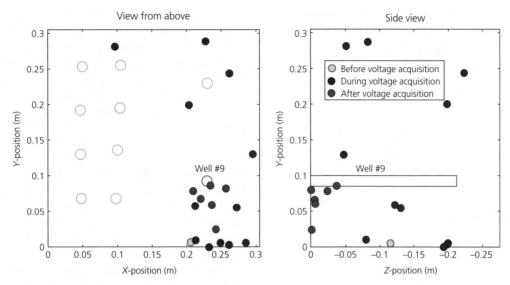

Figure 5.36 Localization of the acoustic emissions with respect to the time window shown in Figure 5.20. The events localized far from Well #9 are probably associated with the reactivation of small cracks. Note that only a tiny fraction of the AE hits shown in Figure 5.20 are localizable. (*See insert for color representation of the figure.*)

from the initial solution point in the matrix. The new point solutions are generally biased in the positive y direction (see Figure 5.35). The result of this noise analysis demonstrates the robustness of the solution method by showing the volumetric clustering nature of all of the solutions in the same general area. The bias in the computed noise solutions may indicate that the true solution for the problem resides somewhere in between the initial solution and the noise-based solutions or that more than one closely spaced dipoles may also be a satisfactory solution to the problem. The coarse nature of the point matrix forces solutions that are on the point matrix grid, and therefore, this is a significant contribution to the degree of error in the solution misfit, as well as the simple dipole approximation assumption.

5.3.7 Discussion

The AE hit count data (see Figure 5.20a) indicates nominally progressive increases in breakage intensity around the times of the pressure changes. The localization of the AE is shown in Figure 5.36. Some of the AE hits are due to activity far from Well #9. These AE hits could be associated with stress changes and the reactivation of existing fractures in the block. Indeed, because the block is unconfined, there may be numerous surface events that would be generated by the cracking of the block surface

as it expands under increased internal fluid pressure. This type of event is highly localizable because it is associated with clear arrivals in the AE.

The AE events associated with the seal rupture along the Hole 9 are difficult to localize for two reasons: (i) there may be too many overlapping events to localize them, and (2) some events likely have a tremor-type signature which makes localization difficult. It seems, however, that a number of events localized close to Well #9 progress further up over time along the well in agreement with the localization of the source current density associated with the burst in the electrical field. Note that the latest AE localizations occur near the top of the injection hole at the time where the seal failure was confirmed through the appearance of water on the surface of the block and indicate surface breakthrough. This shows the possibility, in the future, to perform a joint localization of the electrical and AE data to improve the localization of the hydromechanical disturbances.

The second point to address concerns the nature of the coupling mechanism. We define the sensitivity coefficient of the voltage with respect to the fluid pressure changes as

$$C_0 = \left(\frac{\partial \psi}{\partial p}\right)_{\mathbf{J}=0} \tag{5.54}$$

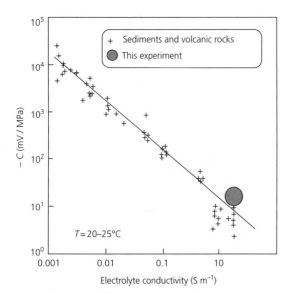

Figure 5.37 Comparison between the coupling coefficient inferred from the block experiment (this study) and the experimentally measured streaming potential coupling coefficients reported by Revil et al. (2003). The consistency between the data indicates that the observed coupling mechanism is likely to be electrokinetic in nature.

When the coupling process is electrokinetic in nature (i.e., related to a relative displacement between the solid skeleton and the pore water), the coupling coefficient scales with the pore water conductivity (e.g., Revil et al., 2003). In the present case, the coupling coefficient is roughly estimated to be on the order of $-20\,\text{mV}\,\text{MPa}^{-1}$ from the data shown in Figure 5.20b and c (typically 3–6 mV variations for 0.2–0.3 MPa of pore fluid pressure changes) at a pore water conductivity of $1.7\,\text{S}\,\text{m}^{-1}$. In Figure 5.37, we plotted the data of Revil et al. (2003) with the result of the present estimate (roughly $-20\,\text{mV}$ MPa^{-1}). The present estimate of the coupling coefficient matches the trend for electrokinetic data, implying that the mechanism we observed is likely to be electrokinetic in nature.

The results presented here have several applications. In the oil and gas industry, the extension of these laboratory observations to field applications can help close the knowledge gap associated with the risks of drilling, completion, and hydraulic fracturing operations. This can be accomplished by electrically monitoring drinking water aquifers with an aquifer safety system designed to detect undesirable leaks of fluids or gases in their very early

stages, before damage can occur. This type of system would also be useful for overall well integrity monitoring, and in the process would be able to, at a minimum, detect and possibly localize annulus flow of fluids upward in the well system, or the displacement of fluids by gases. In older oil and gas wells, it may be possible to assess the presence of microannulus and its connectivity to protected aquifer formations. For old, reentrant oil and gas fields, old plugged and abandoned wells may be assessed for integrity and microannulus issues associated with protected aquifer formations. Additionally, it may be possible to monitor aquifer systems for potential long-term problems by employing a continuous aquifer monitoring based on the physical principles and some of the measurement concepts used for these experiments.

These results also indicate that fluid flowing in the annulus of the well (in contact with porous media such as concrete or natural formations) can be connected to pore water flowing in a network of cracks. As such, this pore water flow would produce electrical signals similar to those observed in these experiments. This could lead to the ability to help characterize fractured rock systems through the movement of the fluids within them. This opens up a new area of fractured rock characterization through passive electrical potential measurement and may lead to the characterization of the fracture networks that opens up in hydraulically fractured formations.

A number of very near-surface civil and environmental applications could also benefit from the findings presented here. For example, injection grouting is commonly used to fill voids in soil and rock, strengthen weak soil, and slow water seepage, for example, in dams and levees. One significant limitation in current practice is the inability to track the movement of grout during injection. Passive electrical potential measurements could be employed to address this challenge. Further, environmental applications involve tracking fluid flow associated with injection and pumping gallery remediation techniques. Passive electrical potential measurements could be used to check for biofouling of the screens or ports on the relevant wells or other relevant changes to the flow field. This can improve system efficiency by detecting reduction in flow rates or changes in the flow pattern through ports in gallery wells or within the formation. It is also possible to apply these techniques to water wells of a diverse variety to check for a variety of situations such as determining the source or direction of the water, fouling or failure of water uptake or injection systems, and other issues related to well integrity.

5.4 Haines jump laboratory experiment

5.4.1 Position of the problem

The dual and multiphase fluid and gas flow through porous media is a very common phenomenon that occurs in multiple types of settings such as industrial processes, aquifer drainage and recharge, oil extraction, and surf action on beaches as well as in many other types of settings. This phenomenon is so common that it is often ignored with respect to the information that can be extracted from the physical processes that occur with this type of fluid flow. In the following sections, a very simple experiment will be described and discussed. This experiment is simply the draining and imbibing of an aquarium filled with sand that starts with 100% tap water saturation. This simple experiment has substantial similarity to several common yet important processes, as described earlier, as well as everyday experiences at the beach. This experiment focused only on the two-phase flow phenomena associated with general fluid flow in an unconsolidated porous medium. This was the process under investigation, and therefore, we describe here the discovery of an electrical phenomenon that is directly and uniquely associated with this type of fluid flow.

Drainage of a porous medium with an invading immiscible nonwetting phase like air, which is initially saturated by a wetting phase like water, takes place under a process called invasion percolation (Aker et al., 2000; Crandall et al., 2009). In a granular material like a sand, large pores are connected by smaller throats. The invading nonwetting fluid (air) will get held up at the throats between the pores where the capillary forces are greatest. As the wetting phase (water in the experiments reported in the following text) is continuously removed from below, the air/water interface at the throat with the largest radius will become unstable. Finally, when the capillary entry pressure of the throat is exceeded, the throat spontaneously fills with air. All the larger pores connected to it fill with the air along with it in an avalanche process. The wetting/nonwetting interface then restabilizes at pore throats that have capillary meniscus forces greater than the drainage pore pressures. Consequently, if one or many of the other pore throat capillary meniscus forces connected to the newly filled pore are smaller than the initial throat holding the interface, these throats and subsequent pores spontaneously fill with the nonwetting phase also. Therefore, once the interface at a particular throat becomes unstable, it causes an avalanche of pore and throat fillings with the nonwetting phase that only stops at throats with capillary meniscus forces greater than the previous drainage pore pressures. This jump of the position of the meniscus is called a Haines jump (Haines, 1930; Aker et al., 2000; Crandall et al., 2009).

Imbibition is also characterized by Haines jumps. However, because imbibition is a collective process and because water is a mineral wetting fluid, capillary forces will hold up the invading water at the larger pores. Once a certain pore fills, all of the throats connected to that pore fill as they are smaller in size. However, during imbibition, even if a small pore lies behind a large pore, the small pore may not spontaneously fill after the filling of the large pore. This limits the size of the potential avalanches on imbibition when compared to drainage. Another process called "snapoff" is also acting during imbibition to decrease the size of the avalanches. Water layers move along the edges of the pore space, allowing small throats ahead of the wetting front to fill spontaneously (Tuller & Or, 2001). These filling events would consist of a single throat and may therefore not emit enough energy to be detectable.

Because fluid is moving with respect to mineral grain surfaces, it is reasonable to expect that Haines jumps would generate bursts of electrical currents. Indeed, in a porous material, all minerals in contact with water develop a surface charge through chemical reactions. Part of this charge is neutralized by the sorption of counterions in the so-called Stern layer. Neutrality is achieved by the presence of an excess of charge in the vicinity of the mineral surface through coulombic interaction with the charge attached to the mineral framework, known as the electrical double layer (see description in Chapter 1). This coulombic interaction implies that the pore water carries a net amount of charge, which is generally positive (Leroy et al., 2007). The displacement of this positive charge by the flow of pore water is equivalent to a source of electrical current density (charge moving per unit surface area per unit time) called the streaming current. The associated electrical field is called the streaming potential and is generated from

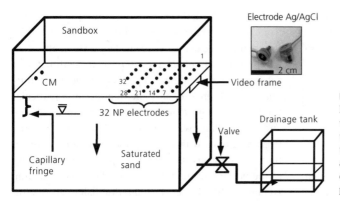

Figure 5.38 Sketch of the experiment for the localization of the Haines jumps. The sandbox is filled with sand saturated by tap water. CM corresponds to the common-mode range control node (reference) of the nonpolarizing (NP) electrodes. The NP electrodes are Ag–AgCl electrodes with amplifiers. All the electrical potentials are recorded relative to the potential of the reference (CM) electrode.

the flow of the streaming current through a porous medium that has a finite resistivity.

Here, we observe, for the first time, the streaming potentials associated with Haines jumps and briefly compare the electrical burst trends we observed with acoustic burst trends for the same basic process, previously studied by DiCarlo et al. (2003). We used a test apparatus that was initially developed for EEG and was used by Crespy et al. (2008) to observe, in real time, the spatial distribution of very small voltages over the surface of a saturated sandbox during very small fluid pumping and infiltration tests. In this section, we apply this measurement approach to the detection of the temporally dynamic electrical bursts potentially associated with Haines jumps during drainage and imbibition experiments.

5.4.2 Material and methods

The test system consists of a 20 gallon (0.075 m^3) sandbox filled with an unconsolidated sand (mean diameter in the range 425–600 µm) filled to saturation with tap water (electrical resistivity of 650 ohm m at 25°C, measurements made at 22°C). The data acquisition system is composed of 32 nonpolarizing silver–silver chloride electrodes placed on a 5.1 cm by 5.1 cm grid forming a 5 × 7 network of electrodes at the top surface of the sand-filled aquarium (Figure 5.38). The data acquisition system used in this experiment is the same system used in Section 5.3 (Hydraulic Fracturing Laboratory Experiment), subsection 5.3.2 (Material and Method), and will not be described again here.

The voltage reference for the measurements was contained within the measurement area and was designed into the measurement system as a DLR and voltage reference (CM in Figure 5.38). All voltages measured by this system are relative to the CM electrode. The entire system, including the computer, was operated on batteries to minimize conductive coupling with the electrical power system (see Crespy et al., 2008, for further explanations). It should be noted that this instrument has also been used in numerous other investigations, including Crespy et al. (2008), Ikard et al. (2012), and recently Haas et al. (2013) for laboratory experiments.

After turn on, the amplified electrodes and data acquisition system were allowed to stabilize thermally for about 15 min. This minimized the thermal drift in the signal due to electronics temperature stabilization. Overall, the experiment was accomplished relatively quickly, typically less than 5 minutes, and the thermal mass of the sand and water was very large and responds very slowly relative to any room temperature fluctuations. Therefore, thermal effects on the data were minimal.

The drainage was very simply driven by gravity through a head difference and was started by turning on the control valve, allowing water to flow out of the main tank into the drainage tank (the opposite for imbibition). The drainage tube (0.6 cm vinyl tube) was connected near the bottom of the main tank and near the bottom of the drainage tank (Figure 5.38), generating a head difference between the saturated surface of the sand and the water level in the drainage tank of 30.5 cm. Water was imbibed into or drained from the sand through a porous body at the bottom of the sandbox.

MATLAB code was written to read the data file generated by the data acquisition system (a BDF file format), perform digital signal processing (DSP, using various functions from the MATLAB Signal Processing Toolbox

(MSPT)), and display it in a variety of graphical formats for visual analysis. The DSP portion of the code was accomplished on a channel-by-channel basis and included DC normalization (not with MSPT), window function application for Fourier transform analysis, Butterworth-based infinite impulse response (IIR) narrowband reject filtering (to suppress 60 Hz and related harmonics), and a high-order linear-phase finite impulse response (FIR) 1Hz high-pass filter for low-frequency trend removal process. MATLAB nonlinear curve fitting tools were used to generate the fits to the thresholded and counted data shown in the following text.

The post-data acquisition signal processing and analysis sequence is as follows: (1) the data in the BDF file is read into memory; (2) a single channel was selected and used to perform spectral analysis of the data to determine what filters were needed to filter the power system related and other undesirable sinusoidal spectral content and low-frequency trends from the data; (3) the needed filters were designed; (4) the data was filtered to remove the undesirable sinusoidal spectral content; (5) the data was then DC normalized; (6) the resulting filtered and DC normalized data was spectroscopically assessed to confirm filtering results, and the time series data was plotted versus voltage; (7) the high-pass filter was applied to the data to remove all low-frequency trends leaving only the impulses in the data, and this filtered time series data was plotted as a function of voltage; and (8) the data was then passed to a thresholding process where the signal voltage level was compared to a threshold voltage level (positive and negative thresholds). Every time the signal voltage passed a threshold value, a counter incremented a threshold value bin. This process generated a list of counts versus threshold that was used to determine the histogram trend through nonlinear regression and then the result was plotted. This process was applied to both drainage and imbibition data.

We sought to visually confirm Haines jumps during the drainage process. To accomplish that, video recordings of drainage-based processes show that individual pore dewatering takes place in a discontinuous process called Haines jumps (Lu et al., 1994). Figure 5.39 shows, for example, Haines jumps during the drainage process for the sand used in these experimental investigations. The pictures are frame grabs from the 15 frame per second digital video stream and show that there are places where there are sudden jumps in the displacement of the meniscus during drainage, while there were other areas where

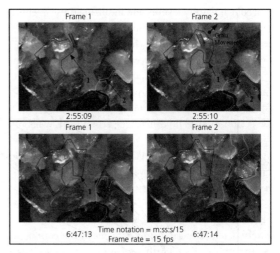

Figure 5.39 Haines jumps. Pictures showing instabilities in the position of the meniscus (outlined by the plain lines) during drainage. The two pictures show that the meniscus labeled 1 is jumping, while the meniscus labeled 2 is stable. The opposite situation occurs for the two bottom pictures (fps stands for frame per second).

the position of the meniscus is stable over the same period of time. The top two pictures in Figure 5.39 and then the bottom two pictures in the same figure represent a single frame step for each pair, or 66.7 ms between frames. This means that the visualized Haines jumps occur faster than the frame rate of the digital video camera.

5.4.3 Discussion

Drainage was begun by opening the control valve and terminated by closing the control valve (Figure 5.38). Figure 5.40 is broken down into three different phases, where the phase preceding the drainage is labeled phase I, while phase II corresponds to the drainage time of the experiment, and phase III corresponds to the postdrainage relaxation time. It can be seen in Figure 5.40 that each phase has different electrical characteristics. The channel-by-channel DC properties of the records were normalized to the DC levels in phase I, making most channel responses near zero in this region. The phase I region is characterized by relatively stable, near-zero trends with a low RMS noise.

The drainage region (phase II) shows an expected positive self-potential trend during water table lowering (drainage). This low-frequency dynamic behavior is due

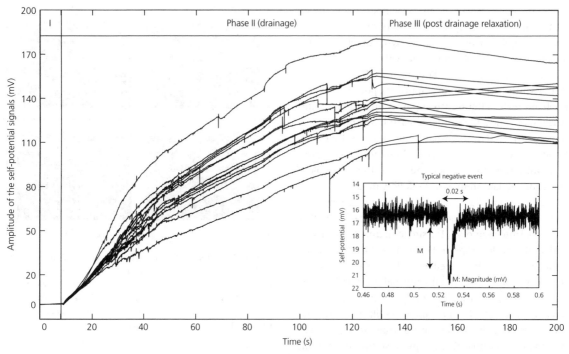

Figure 5.40 Electrograms for a selected set of channels. Phase I corresponds to the data recorded before starting the drainage. Phase II corresponds to the drainage experiment. Phase III corresponds to the postdrainage data and the relaxation of the water table. A typical negative event is shown in the insert (voltages are in mV). These anomalies are characterized by a drop in the electrical potential (characterizing the magnitude of the event) and a relaxation tail.

to the large convergence of flow in the vicinity of the drain and was modeled recently by Revil et al. (2008) for a similar situation. However, bursts in the electrical potential were also observed on top of these expected low-frequency behaviors. It is these transient signals that are of most interest. It should be noted that the drainage was started from a completely saturated sand volume (saturated to the surface) and that the electrical characteristics at the beginning of the drainage phase are different from the electrical characteristics displayed later in the drainage phase. The drainage phase times of 10–25 s are electrically and acoustically quieter than later time sections. Since the water was drained through one port, it is expected that the water table will achieve a nonflat surface that is higher at the end farthest from the drain and lowest directly above the drain. Enough water drains away from the surface of the sand by 30 s into the data set for the onset of the electrical and acoustic noise behavior of interest. This noisy behavior continues until drainage was terminated.

The relaxation region (phase III) shows a significant DC offset with a negative trend on most channels, indicating that the water table surface is relaxing to a constant hydraulic potential surface. Full relaxation of the surface takes a long time, and it can be seen in Figure 5.40 that data acquisition was terminated before relaxation was completed. It should be noted that the AC electrical and acoustic properties of this region are quiet except for a few straggling transient events (Figure 5.40). These straggling events suggest that the sand volume may be undergoing a combined process of imbibition with some drainage, while the phreatic surface equalizes to a constant level possibly including some localized consolidation of the sand during this phase of the experiment.

Figure 5.40 shows how the signal characteristics change as drainage proceeds from being dominated by purely fluid flow to the invasion of air as it begins to enter pores initially containing water. Figure 5.41 shows the electrograms with the low-frequency components

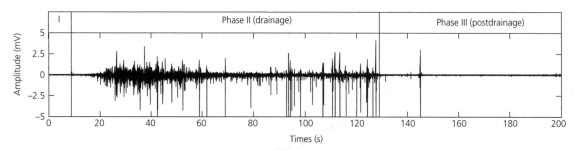

Figure 5.41 High-pass filtered self-potential data showing the level of electrical noise produced during drainage. The full 195 s electrogram is shown. In phase II, the electrogram exhibits large negative voltage spikes with decaying tails. The negative voltage spikes typically involve only few channels, implying that they are located not too far from these electrodes.

removed with a high-pass filter. It can be seen clearly in Figure 5.41 that in phase II, the "noise" clearly begins at a much lower level and increases dramatically with mostly negative excursions (Figure 5.41), which dominate as time increases (Figure 5.41). The high-pass filter causes a differentiation-like effect converting negative-only pulses into pulses with both positive and negative excursions. Phase II is dominated by a series of different types of electrical events, but most prominent are the negative-going voltage leading edge and a long (up to 1 s) tail. Negative-going voltage spikes are consistent with water impulsively descending during drainage. Most of the negative voltage spikes do not have a coincident signal offset; however, some of them show various types and magnitudes of offsets (Figure 5.40). The cause of the variety of these offsets has not been determined, but could be suggestive of a localized process of consolidation of the sand near the lesser compacted surface or within the volume, altering the resistivity of the local sand volume. Video recordings show movement of sand grains near the surface of the sand during drainage, lending support to the consolidation suggestion as an explanation of the observed signal offsets.

Similar data have been obtained for an imbibition experiment (not shown here) using the same test apparatus. However, the amplitude and the number of electrical bursts were observed to be much smaller than for drainage. During imbibition, the most important events are positive excursions in the self-potential signals. These observations are consistent with Haines jumps during the imbibition process described earlier and the upward movement of water within the sand volume.

As explained earlier, the sudden movement of the meniscus during drainage experiments has been documented in a number of prior studies (Aker et al., 2000). In these prior works, the invading fluid (air in the present case) was found to suddenly invade a larger region of the pore space. Those studies demonstrated that the observed bursts of fluid flow (Haines jumps) were characterized by large fluctuations in the pore water pressure and AE (Sethna et al., 2001; DiCarlo et al., 2003). In our drainage experiment, we did not measure AE, but a crackling AE was easily discerned by human ears during drainage and was heard at the same time that the electrical impulses occurred in the data. However, the acoustic and electrical noise phenomena have not yet been correlated with a simultaneous data acquisition. We attribute Haines jumps as the source of the signals we observed.

We studied the nature of the imbibition and drainage processes using a statistical analysis of the occurrences of impulsive events in different data sets involving the two different processes. As described earlier, the impulses from each data set were tabulated for a set of voltage thresholds and displayed in histogram form (Figure 5.42). The difference in voltage magnitude distributions for drainage and imbibition is a result of the difference in the types of pore-filling processes. Note that the magnitude of the self-potential events is much larger on drainage than on imbibition (x-axis on both panels in Figure 5.42). During drainage, the magnitude of the electric events shows a power law distribution with an exponent of -1.7 ± 0.1 fitting the data well over the complete range of event magnitudes. These results are in agreement with the seismic data analysis of DiCarlo et al. (2003) who also found a power law exponent of -1.7 ± 0.2.

During imbibition, no large events observed, and the histogram of event sizes drops off much steeper with a

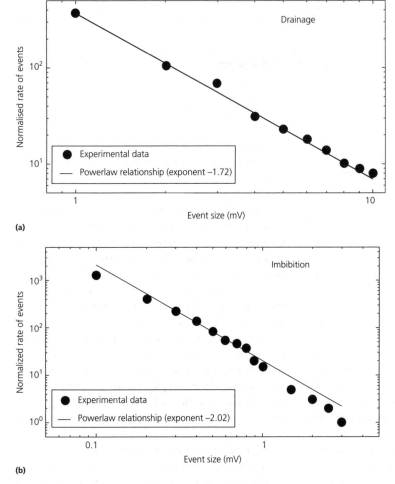

(a)

(b)

Figure 5.42 Frequency of electrical burst for both drainage and imbibition versus the magnitude of the events. **a)** In the case of drainage, the data follow a power law distribution with a fitting exponent of −1.72. **b)** The imbibition data show a drop-off that is much faster with increasing event magnitude because fewer numbers of pores are involved in the formation of Haines jumps.

fit exponent of -2.0 ± 0.4 (the seismic data of DiCarlo et al. (2003) show an exponent of -2.6 ± 0.4), which overlaps with the data of DiCarlo. Additionally, our results are comparable with the pressure fluctuations modeled by Aker et al. (2000a, b). In the work of Aker (2000), the pressure fluctuations have a power law distribution during slow drainage with an exponent of −1.9. It is interesting to note that the Haines jump phenomenon works down to liquid helium temperatures, as shown by Lilly et al. (1996). They observed avalanches in He4 at temperatures of about 1.5 K.

Another possibility to explain the electrical fluctuations we observed would be that the observed electrical disturbances may result from a seismoelectric conversion of the acoustic waves caused by the AE during the fluid movement processes. This possibility can be dismissed,

because acoustic signals are typically on the order of 0.1 Pa (DiCarlo et al., 2003). With a typical streaming coupling coefficient of -10^6 V/Pa for tap water, the order of magnitude of the coseismic electrical signals would be 10^{-4} mV, much below the amplitude of the signals we detected (above 0.1 mV). Additionally, the observed impulses did not exhibit wavelet-like characteristics that are typical of AE. Instead, the impulses exhibited characteristics like the inset shown in Figure 5.40.

5.5 Small-scale experiment in the field

A small-scale field experiment was devised to localize the source of a near-surface water injection pulse into an unconsolidated soil.

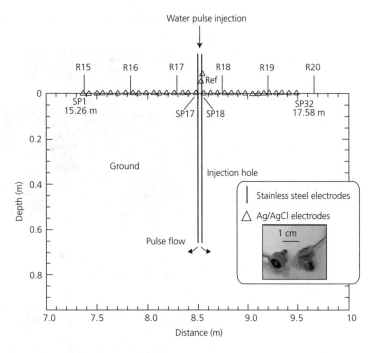

Figure 5.43 Experimental setup. The nonpolarizing electrodes were emplaced around the injection well. The well consists of a plastic tube with an open end localized at a depth of 65 cm from the ground surface. The self-potential data were collected with a 32-channel ultrasensitive voltmeter (BioSemi). The distance between the Ag and AgCl electrodes was 7.5 cm, and the well is localized between electrodes SP17 and SP18. The two reference electrodes are located roughly 10 cm close to the well on the side with respect to the profile shown in this figure.

5.5.1 Material and methods

The experimental setup is shown in Figure 5.43 and consists of a plastic tube in which water was injected for 4 s with a flow rate of $0.6 \, \mathrm{L \, s^{-1}}$. The self-potential response during the experiment was recorded using the very sensitive multichannel voltmeter manufactured by BioSemi, Inc., and was originally designed for EEG (http://www.biosemi.com/). This instrument is the same one used in Section 5.3 (Hydraulic Fracturing Laboratory Experiment), subsection 5.3.2 (Material and Method), and in Section 5.4 (Haines Jump Laboratory Experiment), subsection 5.4.2 (Material and Methods), and will not be described again here. Also, this instrument has been used in numerous other investigations, including Crespy et al. (2008), Ikard et al. (2012), and recently Haas et al. (2013) for laboratory experiments.

Two DC resistivity profiles were also performed prior to and after the water injection test with the ABEM SAS-4000 Terrameter impedance meter. The DC resistivity profile comprised a total of 32 stainless steel electrodes (50 cm spacing) and 118 measurements with the Wenner array data acquisition profile. The apparent resistivity data were inverted with RES2DINV 3.4 (Loke and Barker, 1996).

5.5.2 Results

The resistivity tomograms are shown in Figure 5.44. The data shows small changes in the resistivity close to the well. These resistivity data are essential for the inverse calculation of the self-potential to localize the causative source in the subsurface.

The full time series of the self-potential measurements (called electrograms; see Crespy et al., 2008) are shown in Figure 5.45. They are taken for a time window comprised between 140 and 148 s (Figure 5.45). We applied a linear trend and DC offset removal process and brought the time series of all 32 channels of the data acquisition system to the same baseline during the preinjection time window shown in Figure 5.45. The result of this process is shown in Figure 5.45b and c. It can be seen that this offset and trend removal process allows the characteristics of the electrical pulses to be seen much more clearly across all channels together, along with their correlations. A snapshot of the resulting self-potential distribution along with the least-squares inversion result is shown in Figure 5.46b. It shows a negative baseline with a positive trending anomaly on the top of this baseline. The negative baseline corresponds to the fact that the CMS and DRL reference electrodes are located close to the injection well. What is really important is the positive

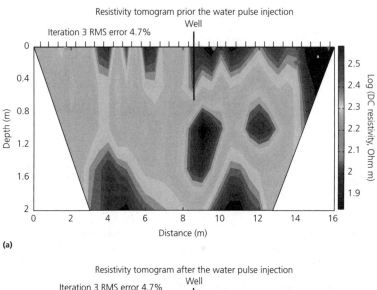

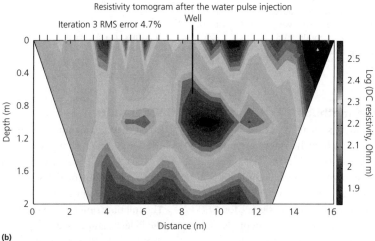

Figure 5.44 Inverted DC resistivity section using the Gauss–Newton method. The data are collected using 32 electrodes with 50 cm spacing. **a)** Tomogram prior the pulse injection. **b)** Tomogram after the pulse injection. (*See insert for color representation of the figure.*)

anomaly, which amount 6–8 mV, and that it is centered on the injection well. This anomaly is inverted in the next section to localize the source current density responsible for this anomaly.

5.5.3 Localization of the causative source of the self-potential anomaly

The injection of water into a porous material produces a source current density, $\mathbf{J}_S$. The pulse injection experiment is therefore responsible for such a source current

density, especially at the tip of the injection well, at a depth of 65 cm. We used Tikhonov regularization to perform the inverse computation. The value of the regularization parameter, α, was determined using the L-curve approach with a regularization parameter in the range of $[10^{-3}, 10^3]$. We obtain $\alpha = 0.121$ as the optimized regularization parameter. In this case, we have $M = 8241$ and $N = 28$ (4 noisy channels were removed) from the data set. The kernel is computed in 3D by extending the resistivity distribution in the strike

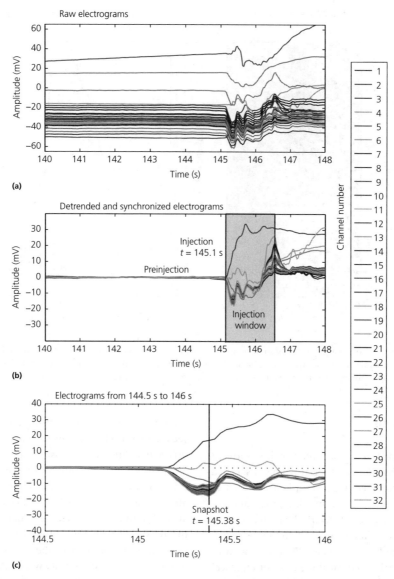

Figure 5.45 Self-potential time series from the water injection experiment. **a)** Raw electrograms. **b)** Detrended and background corrected electrograms. **c)** Magnification of the electrograms showing the preinjection and injection time windows. (*See insert for color representation of the figure.*)

direction. We applied the classical inversion algorithm described earlier to a temporal snapshot of self-potential data. The kernel was computed using the resistivity distribution prior to the injection of the water pulse. The result of the inversion is shown in Figure 5.46a at the third iteration. The tomogram shows a positive distribution of the volumetric current density, $\Im$ (as

defined in Section 5.3), located close to the end of the open well where the pulse water injection takes place (Figure 5.46a). This source current density reproduces the data reasonably well, as shown in Figure 5.46b. Therefore, we have shown that we successfully determined the location of the pulse injection of water into the ground.

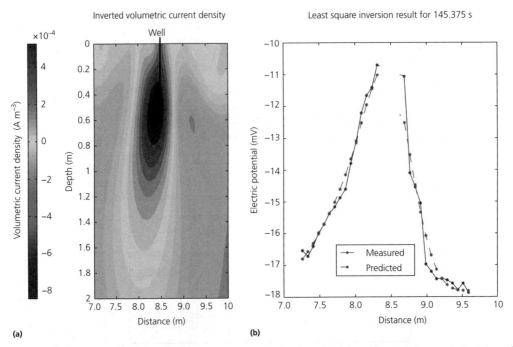

Figure 5.46 Source localization of a single snapshot at time 145.38 s the localization of the self-potential anomaly (positive pole associated with the pulse injection) is located very close to the end of the open well. The electrical resistivity has been accounted for this inversion. Results shown are at the third iteration. (*See insert for color representation of the figure.*)

5.6 Conclusions

We have developed a framework for the forward modeling of seismograms and electrograms associated with a localized hydromechanical disturbance characterized by a moment tensor. The numerical propagation of seismic waves and associated electrical disturbances are done with the finite-element approach in the frequency domain. The forward model couples Biot theory with an electrokinetic model based on the excess of electrical charges per unit pore water. We looked at the value of the information content of the electrical data with respect to the seismic data to see the value of performing a full waveform joint inversion with both types of data. The inverse is performed using the AMA, an efficient MCMC sampler. For noise-free data, the electrical data contains, in principle, more information than the seismic data. For noisy data, the joint inversion provides better constraints for obtaining model parameters than the inversion of the seismic data alone. One of the reasons for this improvement is that the electrical information can be used to estimate the time of occurrence of the seismic source (type I disturbance). Therefore, the record of the electrical disturbances of electrokinetic origin in addition to the seismic information can generate a clear advantage in localizing hydraulic fracturing events in time and space and characterizing the associated moment tensor. We have also used a deterministic algorithm to invert the electrical data alone. Assuming the resistivity model is known, this algorithm is used to retrieve the position of the seismic source and the position of any heterogeneity within the model domain. Then, the moment tensor is successfully inverted using the AMA.

Two important observations have come from the experimental data presented in Section 5.3. First, borehole seal failure during hydraulic fracturing generated electrical current densities and measurable electrical potentials when fracturing fluid flowed into the annulus between the wall of the well and the casing and came in contact with the borehole wall. Second, we have shown that the source current density associated with the seal leak can be located using a self-potential tomography technique.

References

Aker, E., Maløy, K.J., Hansen, A. & Basak, S. (2000a) Burst dynamics during drainage displacements in porous media: Simulations and experiments. *Europhysics Letters*, **51**, 55–61, doi:10.1209/epl/i2000-00331-2

Aker, E., Maløy K. J., Hansen A. & Basak S. (2000b) Burst dynamics during drainage displacements in porous media: Simulations and experiments. *Europhysics Letters*, **51**, 55–61.

Aki, K. & Richard, P.G. (2002) *Quantitative Seismology*. 2nd edition, University Science Book, Sausalito, CA, 700 pp.

Araji A.H., Revil, A., Jardani, A., Minsley, B.J. & Karaoulis, M. (2012) Imaging with cross-hole seismoelectric tomography. *Geophysical Journal International*, **188**, 1285–1302, doi: 10.1111/j.1365-246X.2011.05325.x.

Agarwal, R.G., Carter, R.D. & Pollock, C.B. (1979) Amoco evaluation and performance prediction of low-permeability gas wells stimulated by massive hydraulic fracturing. *Journal of Petroleum Technology*, **31**, 362–372.

Bolève, A., Crespy, A., Revil, A., Janod, F., & Mattiuzzo, J.L. (2007a) Streaming potentials of granular media: Influence of the Dukhin and Reynolds numbers. *Journal of Geophysical Research*, **112**, B08204, doi:10.1029/2006JB004673.

Bolève, A., Revil, A., Janod, F., Mattiuzzo, J.L. & Jardani, A. (2007b) Forward modeling and validation of a new formulation to compute selfpotential signals associated with ground water flow. *Hydrology and Earth System Sciences*, **11**, 1661–1671, doi:10.5194/hess-11-1661-2007

Bolève, A., Janod, F., Revil, A., Lafon, A. & Fry, J.-J. (2011) Localization and quantification of leakages in dams using time-lapse self-potential measurements associated with salt tracer injection. *Journal of Hydrology*, **403**(3–4), 242–252.

Byrdina, S., Friedel, S., Wassermann, J. & Zlotnicki, J., (2003) Self-potential variations associated with ultra-long period seismic signals at Merapi Volcano. *Geophysical Research Letters*, **30**(22), 2156, doi:10.1029/2003GL018272.

Carcione, J.M. (2007) *Wave Fields in Real Media: Wave Propagation in Anisotropic, Anelastic, Porous and Electromagnetic Media*, 2nd ed.. Elsevier Science, Amsterdam

Castermant, J., Mendonça, C.A., Revil, A., Trolard, F., Bourrié, G. & Linde, N. (2008) Redox potential distribution inferred from self-potential measurements during the corrosion of a burden metallic body. *Geophysical Prospecting*, **56**, 269–282.

Chave, A.D., Constable, S.C. & Edwards, R.N. (1991) Electrical exploration methods for the seafloor, Chapter 12, In *Electromagnetic Methods in Applied Geophysics*, Vol. 2, Applications, Parts A and B, Investigations in Geophysics 3, edited by M. N. Nabighian, Society of Exploration Geophysicists, Houston, TX, pp. 931–966.

Chen, M.Y., Raghuraman, B., Bryant, I., Supp, M.G. & Navarro, J. (2005) Wireline apparatus for measuring steaming potentials and determining earth formation characteristics: December. U.S. Patent 6,978,672.

Chen, M.Y., Raghuraman, B., Bryant, I., Supp, M.G. & Navarro, J. (2011) Completion apparatus for measuring streaming potentials and determining earth formation characteristics: February. U.S. Patent 7,891,417 B2.

Chen, S. & Wang, E. (2011) Electromagnetic radiation signals of coal or rock denoising based on morphological filter. *Procedia Engineering*, **26**, 588–594.

Cihan, A., Zhou, Q. & Birkholzer, J.T. (2011) Analytical solutions for pressure perturbation and fluid leakage through aquitards and wells in multilayered-aquifer systems. *Water Resources Research*, **47**(10), W10504.

Crandall, D., Ahmadi, G., Ferer, M. & Smith, D.H. (2009), Distribution and occurrence of localized-bursts in two-phase flow through porous media. *Physica A*, **388**, 574–584.

Crespy, A., Revil, A., Linde, N., et al. (2008) Detection and localization of hydromechanical disturbances in a sandbox using the self-potential method. *Journal of Geophysical Research*, **113**, B01205, doi: 10.1029/2007JB005042.

DiCarlo, D.A., Cidoncha, J.I.G. & Hickey, C. (2003) Acoustic measurements of pore-scale displacements. *Geophysical Research Letters*, **30**(17), 1901, doi:10.1029/2003GL017811.

Entov, V.M., Gordeev, Y.N., Chekhonin, E.M. & Thiercelin, M.J. (2010) Method and an apparatus for evaluating a geometry of a hydraulic fracture in a rock formation; October. U.S. Patent 7,819,181 B2.

Frash, L.P. & Gutierrez, M. (2012) Development of a new temperature controlled true-triaxial apparatus for simulating enhanced geothermal systems (EGS) at the laboratory scale. Proceedings of the 37th Workshop on Geothermal Reservoir Engineering, Stanford University, Stanford, CA, January 30–February 1, 2012, SGP-TR-194.

Gaffet, S., Guglielmi, Y., Virieux, J., et al. (2003) Simultaneous seismic and magnetic measurements in the Low-Noise Underground Laboratory (LSBB) of Rustrel, France, during the 2001 January 26 Indian earthquake. *Geophysical Journal International*, **155**, 981–990.

Gao, Y. & Hu, H. (2010) Seismoelectromagnetic waves radiated by a double couple source in a saturated porous medium. *Geophysical Journal International*, **181**, 873–896.

Garambois, S. & Dietrich, M. (2001) Seismoelectric wave conversions in porous media: Field measurements and transfer function analysis. *Geophysics*, **66**(5), 1417–1430.

Gelinsky, S., Shapiro, S.A., Müller, T.M. & Gurevich, B. (1998) Dynamic poroelasticity of thinly layered structures. *International Journal of Solids and Structures*, **35**, 4739–4751, doi:10.1016/S0020-7683(98)00092-4

Gelman, A.G., Roberts, G.O. & Gilks, W.R. (1996) Efficient Metropolis jumping rules, in Bernardo, J.M., Berger, J.O., Dawid, A.P. and Smith, A.F.M., eds. Bayesian statistics, vol. **5**. Oxford University Press Inc., Oxford, 599–608.

Grech, R., Cassar, T., Muscat, J., et al. (2008) Review on solving the inverse problem in EEG source analysis. *Journal of NeuroEngineering and Rehabilitation*, **5** (**25**), 1–33.

Haario, H., Saksman, E. & Tamminen, J. (2001) An adaptive Metropolis algorithm. *Bernoulli*, **7**, 223–242.

Haario, H., Laine, M., Lehtinen, M., Saksman, E., & Tamminen, J. (2004) MCMC methods for high dimensional inversion in remote sensing. *Journal of the Royal Statistical Society, Series B*, **66**, 591–607, doi:10.1111/rssb.2004.66.issue-3

Haas, A. & Revil, A. (2009) Electrical signature of pore scale displacement. *Water Resources Research*, **45**, W10202, doi: 10.1029/2009WR008160.

Haas, A.K., Revil, A., Karaoulis, M., et al. (2013) Electrical potential source localization reveals a borehole leak during hydraulic fracturing. *Geophysics*, **78**(2), D93–D113, doi:10.1190/GEO2012-0388.1.

Haines, W.B. (1930) Studies in the physical properties of soils. The hysteresis effect in capillary properties, and modes of moisture distribution associated therewith. *The Journal of Agricultural Science*, **20**, 97–116.

Hampton, J. (2012) Laboratory hydraulic fracture characterization using acoustic emission. M.S. Thesis. Colorado School of Mines, Golden, CO.

He, X., Chen, W., Nie, B. & Mitri, H. (2011) Electromagnetic emission theory and its application to dynamic phenomena in coal-rock. *International Journal of Rock Mechanics and Mining Sciences*, **48**(8), 1352–1358.

He, X., Nie, B., Chen, W., et al. (2012) Research progress on electromagnetic radiation in gas-containing coal and rock fracture and its applications. *Safety Science*, **50**(4), 728–735.

Honkura, Y., Ogawa, Y., Matsushima, M., Nagaoka, S., Ujihara, N. & Yamawaki, T. (2009) A model for observed circular polarized electric fields coincident with the passage of large seismic waves. *Journal of Geophysical Research*, **114**, B10103, doi:10.1029/2008JB006117.

Huang, Q. (2002) One possible generation mechanism of co-seismic electric signals. *Proceedings of the Japan Academy*, **78**(7B), 173–178.

Huang, Q. & Liu, T. (2006) Earthquakes and tide response of geoelectric potential field at the Niijima station. *Chinese Journal of Geophysics*, **49**(6), 1745–1754.

Ikard, S.J., Revil, A., Jardani, A., Woodruff, W.F., Parekh, M., & Mooney, M. (2012) Saline pulse test monitoring with the self-potential method to non-intrusively determine the velocity of the pore water in leaking areas of earth dams and embankments. *Water Resources Research*, **48**, W04201, doi:10.1029/2010WR010247.

Iuliano, T., Mauriello, P. & Patella, D. (2002) Looking inside Mount Vesuvius by potential fields integrated probability tomographies. *Journal of Volcanology and Geothermal Research*, **113**, 363–378, doi:10.1016/S0377-0273(01)00271-2

Jardani, A., Revil, A., Bolève, A., Dupont, J.P., Barrash, W. & Malama, B. (2007) Tomography of groundwater flow from self-potential (SP) data. *Geophysical Research Letters*, **34**, L24403.

Jardani, A., Revil, A., Bolève, A. & Dupont, J.P. (2008) 3Dinversion of self-potential data used to constrain the pattern of ground water flow in geothermal fields. *Journal of Geophysical Research*, **113**, B09204, doi: 10.1029/2007JB005302.

Jardani, A., Revil, A., Slob, E. & Sollner, W. (2010) Stochastic joint inversion of 2D seismic and seismoelectric signals in linear poroelastic materials. *Geophysics*, **75**(1), N19–N31, doi: 10.1190/1.3279833.

Jia, H., Wang, E., Song, X., Zhang, H. & Li, Z. (2009) Correlation of electromagnetic radiation emitted from coal or rock to supporting resistance. *Mining Science and Technology (China)*, **19**(3), 317–320.

Jougnot, D., Linde, N., Revil, A. & Doussan, C. (2012) Derivation of soil-specific streaming potential electrical parameters from hydrodynamic characteristics of partially saturated soils. *Vadose Zone Journal*, **11**(1), doi:10.2136/vzj2011.0086.

Keck, R.G. & Withers, R.J. (1979) A field demonstration of hydraulic fracturing for solids waste injection with real-time passive seismic monitoring. Conference paper, SPE Annual Technical Conference and Exhibition, September 25–28, 1994, New Orleans, Louisiana. doi: 10.2118/28495-MS.

Kohl, T., Evansi, K.F., Hopkirk, R.J. & Rybach, L. (1995) Coupled hydraulic, thermal and mechanical considerations for the simulation of hot dry rock reservoirs. *Geothermics*, **24**(3), 345–359.

Kuriyagawa, M., Kobayashi, H., Matsunaga, I., Yamaguchi, T. & Hibiya, K. (1989) Application of hydraulic fracturing to three-dimensional *in situ* stress measurement. *International Journal of Rock Mechanics and Mining Sciences and Geomechanics*, **26**(6), 587–593.

Kuznetsov, V.V., Plotkin, V.V., Khomutov, S.Y., Grekhov, O.M., Pavlov, A.F. & Fedorov, A.N. (2001) Powerful seismovibrators as a possible source of acoustic and electromagnetic disturbances. *Physics and Chemistry of the Earth, Part A: Solid Earth and Geodesy*, **25**(3), 325–328.

Last, B.J. & Kubik, K. (1983) Compact gravity inversion. *Geophysics*, **48**(6), 713–721.

Lee, J.S., Bang, C.S., & Choi, I.-Y. (1999) Analysis of borehole instability due to grouting pressure: Conference Paper. The 37th U.S. Symposium on Rock Mechanics (USRMS), June 7–9, 1999, Vail, CO.

Leroy, P. & Revil, A. (2004) A triple layer model of the surface electrochemical properties of clay minerals. *Journal of Colloid and Interface Science*, **270**(2), 371–380.

Leroy, P., Revil, A., Altmann, S. & Tournassat, C. (2007) Modeling the composition of the pore water in a clay-rock geological formation (Callovo-Oxfordian, France). *Geochimica et Cosmochimica Acta*, **71**(5), 1087–1097, doi:10.1016/j.gca.2006.11.009.

Lilly, M.P., Wootters, A.H. & Hallock, R.B. (1996) Spatially extended avalanches in a hysteretic capillary condensation system: Superfluid 4He in nuclepore, *Physical Review Letters*, **77**(20), 4222–4225, doi:10.1103/PhysRevLett.77.4222

Loke, M.H. & Barker, R.D. (1996) Practical techniques for 3D resistivity surveys and data inversion. *Geophysical Prospecting*, **44**, 499–523.

Lu, T.X., Biggar, J.W. & Nielsen, D.R. (1994) Water movement of glass bead porous media: 1. Experiments of capillary rise and hysteresis. *Water Resources Research*, **30**, 3275–3281, doi:10.1029/94WR00997.

Mahardika, H., Revil, A. & Jardani, A. (2012) Waveform joint inversion of seismograms and electrograms for moment tensor characterization of fracking events. *Geophysics*, **77(5)**, ID23–ID39, doi:10.1190/GEO2012-0019.1

Martínez-Pagán, P., Jardani, A., Revil, A. & Haas, A. (2010) Self-potential monitoring of a salt plume. *Geophysics*, **75(4)**, WA17–WA25, doi:10.1190/1.3475533.

McGarr, A., Simpson, D. & Seeber, L. (2002) Case Histories of Induced and Triggered Seismicity, in Lee, W.H.K., Jennings, P., Kisslinger, C., and Kanamori, H., eds., *International Handbook of Earthquake and Engineering Seismology*, Part A, Chapter 40: Academic Press, 637–661.

Mendonça, C.A. (2008) Forward and Inverse self-potential modeling in mineral exploration. *Geophysics*, **73(1)**, F33–F43.

Moore, J.R. & Glaser, S.D (2007) Self-potential observations during hydraulic fracturing. *Journal of Geophysical Research*, **112**, B02204, doi:10.1029/2006JB004373. 23.

Mosegaard, K. (2011) Quest for consistency, symmetry, and simplicity — The legacy of Albert Tarantola. *Geophysics*, **76(5)**, W51–W61, doi:10.1190/geo2010d-0328.1

Mosegaard, K. & Tarantola, A. (2002) Probabilistic Approach to Inverse Problems, in Lee, W.H.K., Jennings, P., Kisslinger, C. & Kanamori, H., eds., *International Handbook of Earthquake and Engineering Seismology*, Part A, Chapter 16: Academic Press, 237–265.

Müller, T.M. & Rothert, E. (2006) Seismic attenuation due to waveinduced flow: Why Q in random structures scales differently. *Geophysical Research Letters*, **33**, L16305, doi:10.1029/2006GL026789.

Nie, B., He, X., & Zhu, C. (2009) Study on mechanical property and electromagnetic emission during the fracture process of combined coal-rock. *Procedia Earth and Planetary Science*, **1(1)**, 281–287.

Onuma, K., Muto, J., Nagahama, H. & Otsuki, K. (2011) Electric potential changes associated with nucleation of stick-slip of simulated gouges. *Tectonophysics*, **502(3–4)**, 308–314.

Pride, S.R. & Haartsen, M.W. (1996) Electroseismic wave properties. *Journal of the Acoustical Society of America*, **100(3)**, 1301–1315.

Ravaut, C., Operto, S., Improta, L., Virieux, J., Herrero, A. & Dell'Aversana, P. (2004) Multiscale imaging of complex structures from multifold wide-aperture seismic data by frequency-domain full-waveform tomography: application to a thrust belt. *Geophysical Journal International*, **159(3)**, 1032–1056, doi:10.1111/gji.2004.159.issue-3.

Revil, A. & Linde, N. (2006) Chemico-electromechanical coupling in microporous media. *Journal of Colloid and Interface Science*, **302**, 682–694.

Revil, A., Pezard, P.A. & Glover, P.W.J. (1999a) Streaming potential in porous media. 1. Theory of the zeta-potential. *Journal of Geophysical Research*, **104(B9)**, 20,021–20,031.

Revil, A., Schwaeger, H., Cathles, L.M. & Manhardt, P. (1999b) Streaming potential in porous media. 2. Theory and application to geothermal systems. *Journal of Geophysical Research*, **104(B9)**, 20,033–20,048.

Revil, A., Ehouarne, L. & Thyreault, E. (2001) Tomography of self-potential anomalies of electrochemical nature: *Geophysical Research Letters*, **28(23)**, 4363–4366.

Revil, A., Naudet, V., Nouzaret, J. & Pessel, M. (2003) Principles of electrography applied to self-potential electrokinetic sources and hydrogeological applications. *Water Resources Research*, **39(5)**, 1114, doi: 10.1029/2001WR000916.

Revil, A., Gevaudan, C., Lu, N. & Maineult, A. (2008) Hysteresis of the self-potential response associated with harmonic pumping tests. *Geophysical Research Letters*, **35**, L16402, doi:10.1029/2008GL035025.

Revil, A., Mendonça, C.A., Atekwana, E., Kulessa, B., Hubbard, S.S. & Bolhen, K. (2010) Understanding biogeobatteries: where geophysics meets microbiology. *Journal of Geophysical Research*, **115**, G00G02, doi:10.1029/2009JG001065

Revil, A. (2007) Thermodynamics of transport of ions and water in charged and deformable porous media. *Journal of Colloid and Interface Science*, **307(1)**, 254–264.

Revil, A. & Jardani, A. (2010) Seismoelectric response of heavy oil reservoirs. Theory and numerical modelling. *Geophysical Journal International*, **180**, 781–797, doi: 10.1111/j.1365-246X.2009.04439.x.

Revil, A., Woodruff, W.F. & Lu, N. (2011) Constitutive equations for coupled flows in clay materials. *Water Resources Research*, **47**, W05548, doi:10.1029/2010WR010002.

Ritsema, J. & Lay, T. (1995) Long-period regional wave moment tensor inversion for earthquakes in the western United States. *Journal of Geophysical Research*, **100**, 9853–9864.

Sato, M. & Mooney, H.M. (1960) The electrochemical mechanism of sulfide self-potentials. *Geophysics*, **25**, 226–249.

Schenk, O. & Gärtner, K. (2004) Solving unsymmetric sparse systems of linear equations with PARDISO. *Journal of Future Generation Computer Systems*, **20(3)**, 475–487.

Schenk, O. & Gärtner, K. (2006) On fast factorization pivoting methods for symmetric indefinite systems, *Electronic Transactions on Numerical Analysis*, **23**, 158–179.

Schenk, O., Waechter, A. & Hagemann, M. (2007) Matching-based preprocessing algorithms to the solution of saddle-point problems in large-scale non-convex interior-point optimization, *Journal of Computational Optimization and Applications*, **36(2–3)**, 321–341.

Schenk, O., Bollhoefer, M. & Roemer, R. (2008) On large-scale diagonalization techniques for the Anderson model of localization, *SIAM Review*, **50**, 91–112.

Sekiguchi, H., Irikura, K. & Iwata, T. (2002) Source inversion for estimating the continuous slip distribution on a fault—introduction of Green's functions convolved with a correction function to give moving dislocation effects in subfaults, *Geophysical Journal International*, **150**, 377–391.

Sethna, J.P., Dahmen, K.A. & Myers, C.R. (2001) Crackling noise. *Nature*, **410**, 242–250.

Soloviev, S.P. & Sweeney, J.J. (2005) Generation of electric and magnetic field during detonation of high explosive charges in boreholes. *Journal of Geophysical Research*, **110**, B01312.

Stein, S. & Wysession, M. (2003) *An introduction to seismology, earthquakes and earth structure: Blackwell Publishing*, Oxford.

Tikhonov, A.N. & Arsenin, V.Y. (1977) *Solution of Ill-posed Problems*. Washington, DC: Winston & Sons.

Tuller, M. & Or, D. (2001) Hydraulic conductivity of variably saturated porous media: Film and corner flow in angular pore space. *Water Resources Research*, **37**(5), 1257–1276, doi:10.1029/2000WR900328

Ushijima, K., Mizunaga, H. & Tanaka, T. (1999) Reservoir monitoring by a 4-D electrical technique. *The Leading Edge*, **12**, 1422–1424.

Wang, E., He, X., Wei, J., Nie, B. & Song, D. (2011) Electromagnetic emission graded warning model and its applications against coal rock dynamic collapses: *International Journal of Rock Mechanics and Mining Sciences*, **48**(4), 556–564.

Warpinski, N.R. (1991) Hydraulic fracturing in tight, fissured media. *Journal of Petroleum Technology*, **43**(2), 146–151, 208–209.

Webbs, S.C., Constable, S.C., Cox, C.S. & Deaton, T.K. (1985) A seafloor electric field instrument. *Journal of Geomagnetism and Geoelectricity*, **37**, 1115–1129, doi:10.5636/jgg.37.1115

Wishart, D.N., Slater, L., Schnell, D.L. & Herman, G.C. (2008) Hydraulic anisotropy characterization of pneumatic-fractured sediments using self-potential gradient. *Journal of Contaminant Hydrology*, **103**, 133–144.

Wurmstich, B. & Morgan, F.D. (1994) Modeling of streaming potential responses caused by oil well pumping. *Geophysics*, **59**(1), 46–56.

Yoshida, S. (2001) Convection current generated prior to rupture in saturated rocks. *Journal of Geophysical Research*, **106**, 2103–2120.

Yoshida, S. & Ogawa, T. (2004) Electromagnetic emissions from dry and wet granite associated with acoustic emissions. *Journal of Geophysical Research*, **109**, B09204.

Zhao, L.S. & Helmberger, D.V. (1994) Source estimation from broadband regional seismograms. *Bulletin of the Seismological Society of America*, **84**, 91–104.

CHAPTER 6

The seismoelectric beamforming approach

As seen in Chapter 4, the seismoelectric conversion can be rather weak with respect to the coseismic electrical fields. "Seismoelectric beamforming/focusing" allows for the enhancement of the electrical field associated with seismoelectric conversions over the spurious coseismic signals. We present in this chapter the basic ideas of this method, followed by numerical tests in piecewise constant and heterogeneous materials. We demonstrate how this method can be used to improve cross-well electrical resistivity tomography (ERT) through a technique called "image-guided inversion." The basic idea of image-guided inversion is to use information from an image to impose the structure in a way that shapes the model covariance smoothness matrix that imposes this structural information in the inversion of the geophysical data. Because the computations performed for the examples shown in this chapter are rather intense, we demonstrate this method using the acoustic approximation only (see Chapter 1 and its extension in partially saturated conditions in Chapter 3). Nevertheless, the same general idea can be applied to more general poroelastic wave fields.

6.1 Seismoelectric beamforming in the poroacoustic approximation

6.1.1 Motivation

Our first goal is to explain how the principle of seismoelectric beamforming works with a simple example. For this purpose, we consider a 2D model with a porous background material using constant mechanical, hydraulic, and electrical properties. Two rectangular heterogeneities (labeled Anomalies 1 and 2) are embedded in a homogeneous background (Figure 6.1). The background and the two anomalies are characterized by the same mechanical properties. They differ only by their hydraulic and electrical properties, for instance, a change in water saturation. Indeed, an increase of the nonwetting and insulating fluid saturation (e.g., gas or oil) would decrease the permeability of the water phase, thereby reducing the electrical conductivity of the porous material. The material properties used in this test are reported in Table 6.1. The background material is supposed to be fully water saturated (electrical conductivity of $1\,\mathrm{S\,m^{-1}}$). The anomalies are more resistive ($10^{-2}\,\mathrm{S\,m^{-1}}$ for Anomaly 1 and $10^{-3}\,\mathrm{S}$ $\mathrm{m^{-1}}$ for Anomaly 2).

Seismic sources are located in two wells. The seismoelectric equipment (virtual geophones, seismic sources, and electrodes) are set every 5 m along the ground surface and in the two wells (see Figure 6.1; 19 in each well and 8 along the ground surface). The objective of this experiment is to focus seismic waves at two specific points, A and B (Figure 6.1), and evaluate whether seismoelectric conversions due to the presence of a heterogeneity can be recorded remotely at the electrodes located in the wells. When the seismic energy is focused on a heterogeneity such as an interface characterized by a drop in permeability and resistivity, we expect to record a seismoelectric conversion away from this interface (interface response).

The Seismoelectric Method: Theory and Applications, First Edition. André Revil, Abderrahim Jardani, Paul Sava and Allan Haas.
© 2015 John Wiley & Sons, Ltd. Published 2015 by John Wiley & Sons, Ltd.

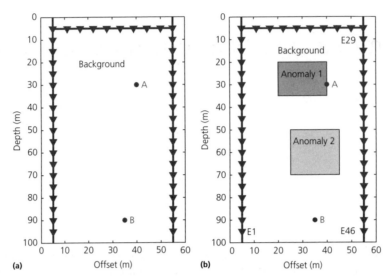

Figure 6.1 Geometry used for the beamforming problem. The medium consists of a homogeneous background model (reference model, fully saturated) plus two anomalies termed Anomaly 1 and Anomaly 2. These anomalies correspond to areas that are unsaturated (see Table 6.3). The survey area is surrounded by two vertical wells located on each side. The triangles correspond to the location of the seismic sources/geophones/electrodes. The spacing between two consecutive sensors is 5 m. The two boreholes have 19 dipoles of electrodes each, and sets of sensors are located close to the ground surface (5 m deep). The two red-filled circles correspond to the focusing points used for our numerical experiments. Ei corresponds to the position of electrode i. There are 46 set of sensors in total with E1 and E46 at the bottom of the two wells. **a)** Reference model (without the two heterogeneities). **b)** Model used for the numerical simulation with the two heterogeneities. (*See insert for color representation of the figure.*)

Table 6.1 Petrophysical properties for the background and Anomalies 1 and 2.

Property	Background	Anomaly 1	Anomaly 2
Undrained bulk modulus K_u (Pa)	22×10^9	22×10^9	22×10^9
P-wave velocity V_p (m s^{-1})	3093	3093	3093
Excess charge density $\hat{Q}_V$ (C m^{-3})	0.20	2.0	6.7
Log (permeability, k m^2)	−12	−14	−16
Skempton coefficient B	0.65	0.65	0.65
Average density ρ (kg m^{-3})	2300	2300	2300
Hydraulic viscosity of pore fluid η_f (Pa s)	10^{-3}	10^{-3}	10^{-3}
Conductivity σ (S m^{-1})	1	0.01	0.001
Saturation s_w (—)	1	0.10	0.03

If we scan various points in the medium, this technique would enable us to identify whether the point of focus is located in the vicinity of such heterogeneity or not.

Going further, we could scan the subsurface and "map" in 3D these heterogeneities.

We solve the partial differential equations for the mechanical and electrical problems in the frequency domain as explained in Chapter 4.

6.1.2 Beamforming technique

The beamforming technique enables us to focus seismic energy at a desired location and at a known time. As discussed in Sava and Revil (2012), the velocity model does not need to be perfectly known. In the present case, however, we will assume that it is perfectly known. Seismic beamforming is based on time reversal process and is accomplished in two steps:

Step 1: On a finite-element grid, we choose the point of focus and we construct a fictitious seismic point source at that location. In this case, the seismic source is a Ricker wavelet with a dominant frequency of 500 Hz (Figure 6.2). It contains energy up to about 1500 Hz. Using a constant seismic P-wave velocity of 3100 m s^{-1}, we obtain a dominant wavelength of 6.2 m and a

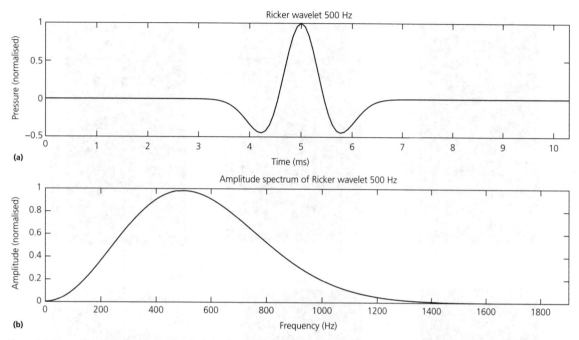

(a)

(b)

Figure 6.2 Pressure source $f(\mathbf{x}, t)$ used for the beamforming experiment. The pressure source is a Ricker wavelet with a 500 Hz dominant frequency and a time shift of 5 ms. **a)** Pressure time series of the source. **b)** Amplitude spectrum of the source.

minimum wavelength of 2 m. That gives us a seismic resolution of 1.55 and 0.51 m, respectively. All numerical modeling shown in this section is performed using the finite-element package COMSOL Multiphysics with triangular meshing of nonconstant element size (minimum element size of 0.0024 m and maximum element size of 1.2 m). This choice was driven by the need to have 5 mesh elements per wavelength. In order to model the study areas without any seismic reflection at the boundaries (i.e., simulate an infinite medium), we use the convolution perfectly matched layer (C-PML) developed by Jardani et al. (2010). The thickness of the C-PML is 10 m around the area of interest. The seismic P-waves are propagated from the source to the fictitious electrodes located in place of the true seismic sources in the wells and along the ground surface. For every shot, we record the macroscopic pressure field, $P_i(t)$, at each geophone identified by index i located in the wells.

Step 2: Once the pressure fields have been recorded at each geophone, we back propagate the pressure fields as shown in Figure 6.3. Next, we time-reverse the signal recorded at each geophone $((P_{\text{shifted}}(t) = P_{\text{recorded}}(T - t)$, where T is the total recording time or listening time). Then, we create seismic point sources located at all of the positions of the virtual geophones and reinject the reversed seismic signals into the medium. The outgoing pressure field propagates and interferes constructively at the original source location (see Figure 6.3). During this backpropagation, we record the electrical potential at the electrodes placed in the wells and along the ground surface. As indicated earlier, the 46 electrodes colocated with the seismic sources; however, this is not at all a requirement for this method.

The strength of this technique comes from the fact that we know exactly at what time and what location the seismic wave fields focus and interfere constructively. If the point of focus is located on an interface, we record an interface response with a greater amplitude than we would have recorded for a passing wave field crossing the interface. In fact, the electric potential, caused by a seismoelectric conversion from a discontinuity in medium properties, is usually orders of magnitude smaller than the coseismic field (electrical field giving rise to an electrical potential only detectable inside the support of a seismic wave). An example is shown in Figure 6.2 depicting the dipolar field created by the beamforming of the seismic waves at point A. This technique forces the interface response conversion to

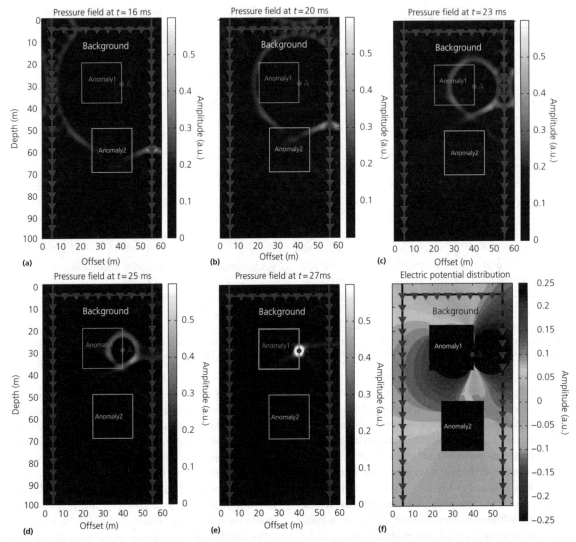

Figure 6.3 Seismic beamforming and resulting electrical potential distribution. **(a–e)** Snapshots show the pressure field (coming from the 46 seismic sources shown in Figure 6.1) focusing at point A. At $t = 27$ ms, the wave field interferes constructively at point A producing a strong seismoelectric response that is recorded at the receiver electrodes. **f)** The distribution of the electrical potential corresponds to the case where the seismic energy is focused at point A. The seismoelectric conversion at the time of focusing is characterized by a strong dipolar behavior. (*See insert for color representation of the figure.*)

increase in amplitude due to the focusing of the pressure field originating at the multiple seismic sources.

By applying this technique to a grid of points within the survey area (i.e., by scanning the area point by point), we can then use the electrical response to map the heterogeneities in terms of electrical and hydraulic properties of the material without being able to distinguish what factors are responsible for such heterogeneities. We can also adjust and increase the resolution of our mapping

by scanning over a denser grid of points around certain areas. The limitation (resolution) of the final map is dictated by the wavelength of the seismic wave field with a maximum resolution close to the dominant wavelength divided by 8.

6.1.3 Results and interpretation

We focus the seismic energy on few points of interests, and we record the electrical potential at the 46 electrodes

(19 in each borehole and 8 along the ground surface). The first point of focus (point A; see Figure 6.1) is located at an interface with a sharp discontinuity in both electrical conductivity and permeability. Therefore, we expect to see a seismoelectric conversion; our results confirm this assumption (Figure 6.4). However, we need to remove the distribution of potential recorded with only the background to see the seismoelectric conversion associated with the interface response. We can clearly see that the dominant spike in the electric potential time series (background removed) is perfectly synchronized with the focusing time of the seismic wave field at point A. This enables us to detect the presence of a heterogeneity at point A.

Figure 6.5 shows the electric potential at an electrode located at electrode E45 when the seismic waves are focused at point B which is not located close to an interface. As expected, we see no spike at the focus time, which indicates a lack of heterogeneity at this focus point. We conclude that this method can be used in time lapse to follow the evolution of a saturation front over time, for example, for CO_2 sequestration or for monitoring water flood experiments.

6.2 Application to an enhanced oil recovery problem

The example discussed in the preceding section uses several strong assumptions: constant elastic background and blocky anomalies with constant physical properties. In this section, we relax these assumptions, and we demonstrate that the method is applicable to heterogeneous geological structures with a contrast in the water saturation. A lot of work has been done recently using low-frequency electrical signals to detect oil water encroachment fronts (see, for instance, Saunders et al., 2008). We want to see if seismoelectric beamforming can be used to localize, in a very simple way, the position of the front. We consider two wells crossing a heterogeneous reservoir (see Figure 6.6). Well B is located 250 m away from Well A, and the total geometry of the model covers an area of 410×250 m. The reference position, $O(0,0)$, is located at the upper left corner of this domain. The reservoir is initially saturated with oil (oil saturation of 80%). During water flooding operations, water is injected into Well A and oil is produced from Well B.

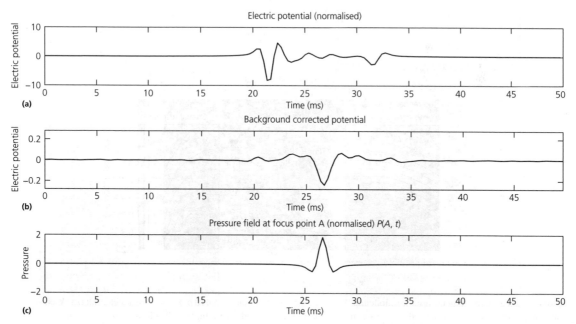

Figure 6.4 Beamforming at point A. **a)** Time series of the electrical potential at electrode E35 for beamforming the seismic wave at point A. This time series has both the seismoelectric conversions and the coseismic field. The interface response is not detectable. **b)** Electrical potential recorded when the medium contains the two anomalies minus the potential recorded if we only had the background. The interface response is now clearly visible. **c)** Time series of the pressure field at point B, $P(A, t)$.

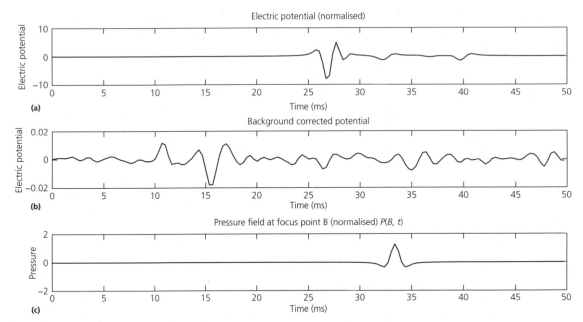

Figure 6.5 Beamforming at point B. **a)** Time series of the electrical potential at electrode E45. The time series shows both the seismoelectric conversions and the coseismic field. The interface response is not detectable. **b)** Difference of two time series: the electric potential in the top graph minus the electric potential recorded at the same location but for a homogeneous background (reference model). There is no conversion which is consistent with the fact that point B is not associated with a heterogeneity. **c)** Time series of the pressure field at point B, $P(B, t)$.

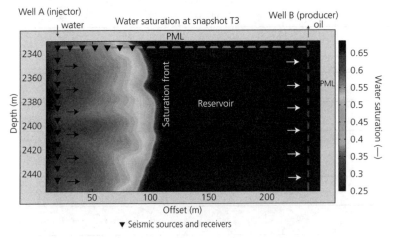

Figure 6.6 Sketch of the domain used for modeling. The total modeling domain is a 410 m × 250 m rectangle. Injector Well A, located at position $x = 0$ m, is also used for the seismic source. Production and recording Well B is located at $x = 250$ m. The discretization of the domain comprises a finite-element mesh of 205 × 125 rectangular cells. Twenty-eight receivers are located in Well B, approximately 30 m away from the nearest PML boundary (the PML boundary layers are shown in gray). (*See insert for color representation of the figure.*)

To generate the porous medium, we use a random simulator to create a stochastic realization for the clay content and the model of Revil and Cathles (1999) to determine the porosity and the permeability at saturation. A multiphase flow simulator computes the saturation profiles, which are used to compute the P-wave velocity and resistivity distributions. The P-wave velocity and the permeability of the water phase are shown in Figure 6.7 for snapshot 3, while the electrical conductivity and the porosity are shown in Figure 6.8. P-wave velocity does not depend much on the saturation. In this example, the velocity distribution is set to be between 4050 and 4300 m s^{-1}, and the electrical conductivity distribution varies over an order of magnitude. The influence of water saturation on resistivity is much greater than its influence on P-wave velocity.

We use a Gaussian seismic source (magnitude 1.0×10^4 N m, delay time $t_s = 30$ ms, dominant frequency $f_c = 160$ Hz). We then solve the poroacoustic equations described in Chapter 1 (see extension to the partially saturated case in Chapter 3) and the Poisson equation for the electrical potential in the frequency domain, based on the material properties provided in Table 6.2. Solve for the confining pressure, $P(\mathbf{r}, t)$, of the solid phase and pore fluid pressure $p(\mathbf{r}, t)$. Finally, we compute the electrical potential by solving the Poisson equation coupled to the poroacoustic problem for the frequency range 8 to 800 Hz. This choice of frequencies is valid

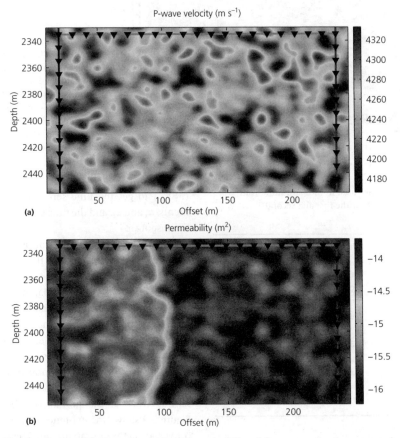

Figure 6.7 Sketch of the distribution of the P-wave velocity and permeability of the pore water phase at snapshot 3. Note that the saturation front is characterized by a sharp contrast in permeability. **a)** Velocity distribution for the compressional wave. **b)** Distribution of the permeability. (*See insert for color representation of the figure.*)

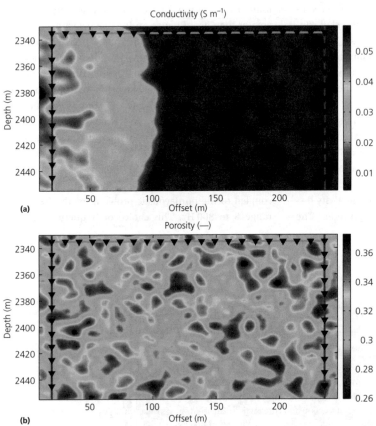

(a)

(b)

Figure 6.8 Sketch of the distribution of the electrical conductivity and porosity at snapshot 3. Note that the saturation front is characterized by a sharp contrast in electrical conductivity. **a)** Electrical conductivity distribution. **b)** Porosity distribution. (*See insert for color representation of the figure.*)

Table 6.2 Material properties used in the saturation front detection. We use $m = n = 2$ for the first and second Archie's exponents.

Parameter	Value	Units
ρ_s	2650	kg m^{-3}
ρ_w	1000	kg m^{-3}
ρ_o	900	kg m^{-3}
K_s	36.5	GPa
K_{fr}	18.2	GPa
G	13.8	GPa
K_w	2.25	GPa
K_o	1.50	GPa
η_w	1×10^{-3}	Pa s
η_o	50×10^{-3}	Pa.s

for this problem because the seismic wave and the associated electrical field operate in the same frequency range. Then we compute the inverse Fourier transform

(FFT^{-1}) to get the time series of the seismic displacements, u_x and u_z, and the time series of the electric potential response.

The mesh for this example consists of 2×2 m cells. The size of the mesh is smaller than the smallest wavelength of the seismic wave. For this mesh size, we checked that the solution of the partial differential equations governing the seismoelectric problem is mesh independent. At the four external boundaries of the domain, we apply a 50 m-thick C-PML (see Chapter 4). Figure 6.9 shows a pure transmission experiment between a seismic source located in Well A and an electrical receiver located in Well B. We also record the fluid pressure at the first Fresnel zone of the seismic wave at the position of the saturation front. Figure 6.9 shows the existence of a strong seismoelectric conversion at the saturation front. This confirms the results of Revil and Mahardika (2013) who used poroelastic theory in their work. Figure 6.10a–e shows the seismic beamforming using the seismic sources located in three

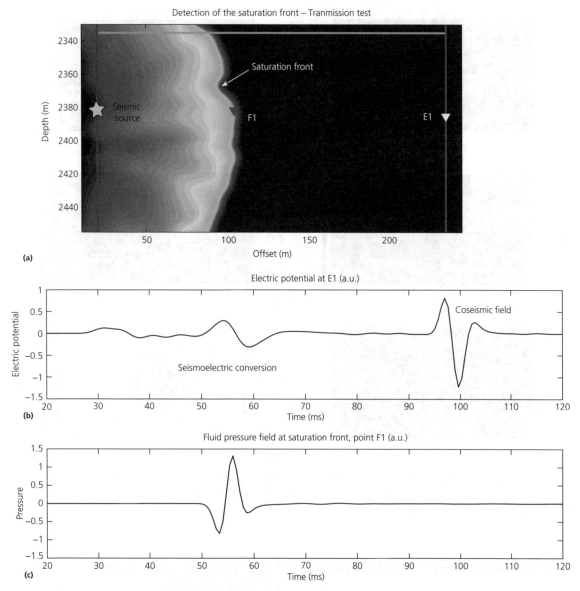

Figure 6.9 Transmission experiment with the seismic source in Well A and the electrical receiver in Well B. **a)** Geometry of the test. **b)** Time series for the electrical potential showing the seismoelectric conversion occurring at the interface and the coseismic field. **c)** Fluid pressure field at a point located at the saturation front at the center of the Fresnel zone. (*See insert for color representation of the figure.*)

wells. Figure 6.10f shows the resulting electrical potential distribution at the time for which the seismic energy is beamformed. This shows a very strong dipolar field associated with the beamforming.

In Figure 6.11, we repeat this operation for a set of scanning points located at the same depth that cross the position of the saturation front. Our analysis shows that the strongest seismoelectric conversion is located at the position of the saturation front. Therefore, we can conclude that the scanning of the reservoir could be used to determine the position of the saturation front and to monitor its progression over time.

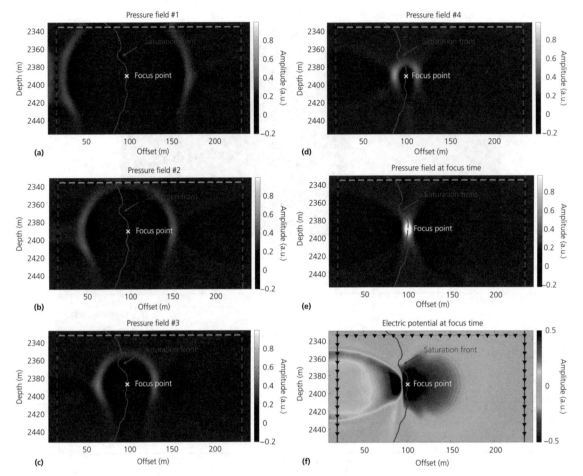

Figure 6.10 Seismic beamforming at the saturation front and resulting electrical potential distribution. **(a–e)** Evolution of the confining pressure $P(r, t)$ for 5 snapshots. The propagating wave fields interfere constructively at the focus point. **f)** Electrical potential distribution corresponding to the seismic energy focused at the saturation front (snapshot e). (*See insert for color representation of the figure.*)

6.3 High-definition resistivity imaging

The goal of this section is to show that we can use the beamforming approach described in Section 6.2 to produce a high-definition map of the heterogeneities of a formation. Then this map can be used in image-guided inversion to help the inversion of apparent resistivity data to converge to a more realistic model, producing resistivity values close to the real ones. This is especially useful if we want to monitor resistivity changes associated with changes in the water saturation of a reservoir.

6.3.1 Step 1: the seismoelectric focusing approach

We first illustrate our method using a heterogeneous material that is bimodal in terms of the distribution of its material properties (Figure 6.12). For instance, one phase represents the porous material saturated with water, and the other the porous material partially saturated with oil. The properties of these "reservoir" and "nonreservoir" rocks are provided in Table 6.3. Figure 6.12c shows the position of two wells in which seismic receivers are placed. The goal is to scan all the space indicated by the yellow rectangle in Figure 6.12c ($\mathfrak{R}$ will denote this

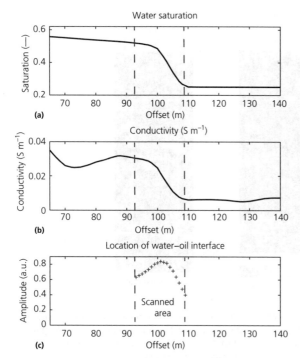

Figure 6.11 Determination of the position of the oil water interface using a set of beamforming points at the same depth that cross the interface. **a)** Spatial distribution of water saturation (snapshot #3) showing the position of the saturation front. **b)** Electrical conductivity distribution. **c)** Source intensity as a function of offset. This shows that the strongest seismoelectric conversions are generated at the position of the saturation front.

scanning domain). The entire procedure is summarized in the chart shown in Figure 6.13.

There are two main ways in which we can implement our procedure in practice:

1 First, we could consider a broad distribution of sources located at all locations in the boreholes. These sources are characterized by different phase delays but are synchronized such that the generated acoustic energy focuses at a specified location in the subsurface. Every focus location requires a different physical experiment with sources characterized by different phase delays. The effort required by this experimental setup is proportional to the number of scanning points in the subsurface.

2 Second, we could consider a single source located at different points in the boreholes. We simulate waves

propagating separately from every point in the borehole and linearly recombine the resulting wave fields and electrical potential fields to reflect delays that would focus energy at a given point in the subsurface. In this case, the experimental effort is proportional to the number of sources in the boreholes, since we can consider the computational effort to recombine the electrical potential fields negligible.

In the numerical experiment shown in the following, we use an even simpler method. We loop over all the points located in the scanning area and inject an identical pulse (Ricker wavelet) to generate wave fields propagating in the medium. We assume that we know the velocity model with sufficient accuracy in order to predict correctly the kinematics in the subsurface. Figure 6.14 shows an example of wave propagation from a scanning point to the receivers located in the wells. The seismic wave field encodes the model heterogeneities and carries this information to the boreholes surrounding the scanning point. Then, we record these simulated seismic signals at the receivers located in the wells (see the seismograms in Figure 6.15), and we reinject the recorded seismograms into the formation in order to focus the seismic energy back at the chosen scanning point (Figure 6.16). As indicated earlier, the recorded seismograms (see Figure 6.15) could be generated by linear superposition of seismograms generated separately from all points in the boreholes.

The backpropagating waves (Figure 6.16) generate seismoelectric sources at every location where rapid changes of physical properties occur in the medium (Figure 6.17). The corresponding normalized electric potential distribution is shown in Figure 6.18. The reason for normalization of the potential map is that the source we use is of arbitrary strength, and this in turn means that the virtual electrode obtained by focusing is characterized by a proportional potential. However, since the potential normalization is done globally, that is, for all time steps, the relative relationships between the potentials reconstructed at different times and positions are preserved. This simulation, shown in Figures 6.14, 6.15, 6.16, 6.17, and 6.18, illustrates the main difficulty of using seismoelectric methods to investigate the subsurface: relatively weak sources of energy are located everywhere in the medium at all times. However, our method addresses this issue. Instead of observing the seismoelectric conversions at all times, we concentrate on only one moment, that is, the time when the injected wave field

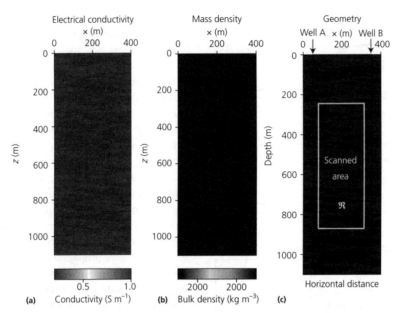

Figure 6.12 Model used for the scanning simulations. **a), b)** Heterogeneous distribution of the electrical conductivity and mass density of the formations. **c)** Geometry of the acquisition array. The porous material is bimodal with a larger correlation length in the horizontal direction. The seismic receivers are located in two wells (A and B). The area scanned through focusing is indicated by the yellow rectangle (scanning region $\mathfrak{R}$). (*See insert for color representation of the figure.*)

Table 6.3 Material properties used for the simulations. The viscosity of the pore water is taken equal to 10^{-3} Pa s.

Property	Reservoir rock	Nonreservoir rock
Bulk modulus K (Pa)	22×10^9	7×10^9
Skempton coefficient B (—)	0.65	0.85
Density ρ (kg m^{-3})	2300	1900
Log (water-phase permeability, k, m^2)	-12	-16

focuses on the desired scanning point. In this case, we know precisely the source of electric current position and time, and we also maximize the strength of the source. The number of focus points is determined from the wavelength of the wave. Typically, we use two focus points per wavelength.

We repeat the process outlined earlier and record the electric potential at a small number of electrodes located in the two wells. These electrodes record the electrical potential over time, but we only use the voltage observed at the focusing time. This is mainly because the seismoelectric sources prior to the focus time are distributed everywhere in the medium. This wide distribution of

seismoelectric sources does not allow us to associate an observed electric potential with a given position in the subsurface. This causes interpretation ambiguity, which can be resolved by restricting our attention to the focus time. Figure 6.19 shows the potential observed at a given electrode by focusing at all locations in the target area. Although the potential is measured at a single point, we record all of the measurements from the corresponding seismic energy focus positions. The accumulation of the stimulated seismoelectric responses at each electrode during the scanning process generates a seismoelectric map of the subsurface. This map highlights the locations characterized by hydroelectric contrasts. Also, this map is subject to the effects of "illumination" from a given electrode. This causes the potential to weaken with distance from the measurement electrode, as well as a function of the distribution of conductivity in the medium. This means that different electrodes are sensitive to different portions of the subsurface. Such maps are shown for various electrodes in Figure 6.19. A cumulative map of the electrical potential is shown in Figure 6.20. This "seismoelectric image" illuminates the position of all heterogeneities in the scanning area located between the boreholes.

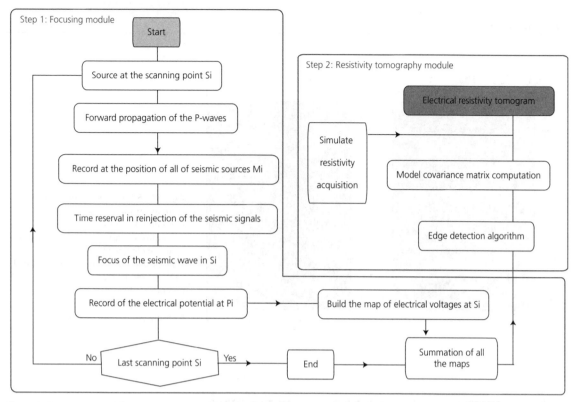

Figure 6.13 Description of the two-step procedure for obtaining a high-definition electrical resistivity tomogram between two wells. We scan all the points for a given region $\mathfrak{R}$ between the two wells. The density of points depends on the wavelength of the seismic wave used to scan the formations. At each scanning point, we place a seismic source (e.g., a Ricker wavelet) and record the seismograms at the positions of all the seismic sensors located in the wells. We then reinject the recorded seismograms in the formation in order to focus at the position of the scanning point (time reversal). The voltage is recorded at the focusing time at a set of electrodes, Pi (the reference of the voltage is considered to be at infinity). This voltage is reported at the focus position; a map of these voltages obtained for a collection of scanning points generates a "seismoelectric image" usable as a physical constraint for electrical tomography (Step 2).

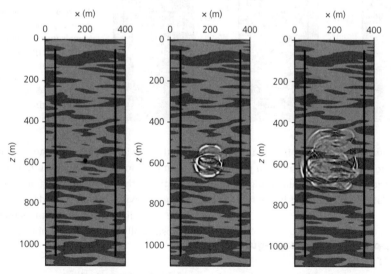

Figure 6.14 Seismic wave fields from a seismic source located at one of the scanning locations belonging to the scanning domain $\mathfrak{R}$. The seismic wave field spreads throughout the medium and reaches the seismic receivers located in the two wells. The corresponding seismograms are recorded at these seismic receivers (see Figure 6.15).

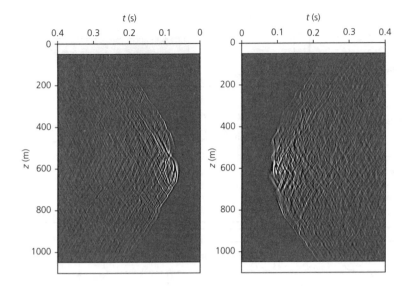

Figure 6.15 The propagating wave field recorded at the seismic receivers in the two boreholes located in the vicinity of the scanning area. Similar seismograms could be obtained by linear superposition of seismograms generated from individual locations in the boreholes, after appropriate phase delay corresponding to the desired focusing point in the region $\mathfrak{R}$.

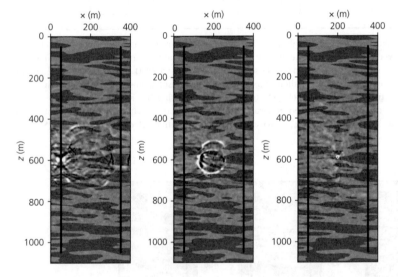

Figure 6.16 Seismic wave field obtained by backpropagating the recorded seismic signals into the medium in order to focus seismic energy at the location of the original scanning point (time reversal). In this case, the seismograms do not need to be delayed since they are generated through modeling from the desired focusing point.

In the next section, the structural information contained in this seismoelectric image is used to guide the inversion of electrical resistivity data using current sources and potential receivers located in the two wells.

6.3.2 Step 2: application of image-guided inversion to ERT

In this section, we consider a set of electrodes located in the two wells with a spacing of $L = 10$ ms between the electrodes. Using the map of electric potential defined

in the previous section, we can proceed to the next step, that is, to invert for the distribution of electrical conductivity in the medium using apparent resistivity data acquired with a set of electrodes located in the two wells. We show in this section that this method may improve substantially the resolution of cross-well ERT.

6.3.2.1 Edge detection

First, we filter the seismoelectric image to detect the boundaries of the different formations. Among the

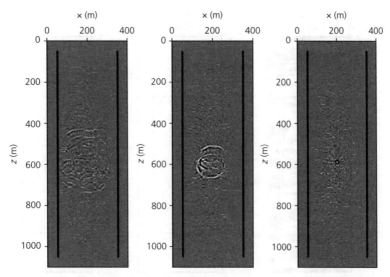

Figure 6.17 The backpropagating wave fields (for the three snapshots shown in Figure 6.16) generate seismoelectric current sources at many locations characterized by appropriate contrasts of physical properties. The seismoelectric source is thus distributed in space and time, which creates a strong interpretation ambiguity. However, at the focusing time, the seismoelectric current source is mostly localized in space, and it is stronger than at other times. We don't plot the scale of the seismoelectric current source here because they depend on the intensity of the seismic source.

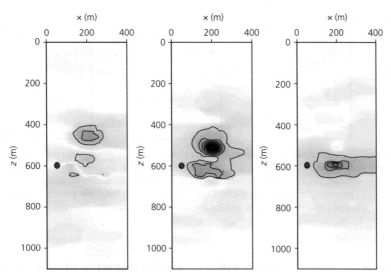

Figure 6.18 Electrical potential observed as a function of time for the seismoelectric sources depicted in Figure 6.17, corresponding to the three snapshots shown in Figure 6.16. Interpretation of this potential is ambiguous, given that the seismoelectric source is distributed in space and time. In our method, only the potential seen in the rightmost frame is relevant, and its interpretation is not ambiguous since in this case we deal with a known position and time of the source. We don't plot the scale of the electrical potential because they depend on the intensity of the seismic source. The only thing that matters is that the strength of the electrical potential is recordable at the electrodes located in the two wells. The gray-filled circle corresponds to the position of the electrodes at which the potential is recorded at the time of focusing.

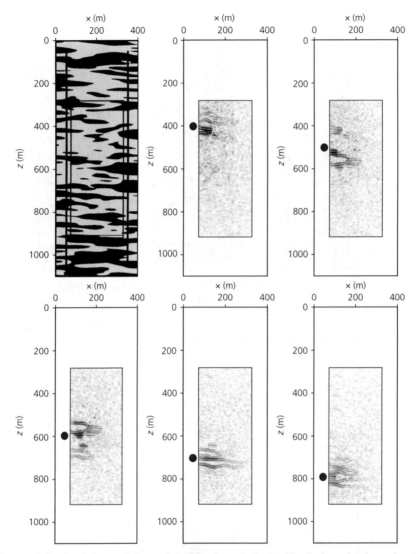

Figure 6.19 Distribution of the electrical potential, recorded at the electrode depicted by the large dot, for each point in the scanning area. Each point corresponds to the electrical potential recorded at the electrode at the time of focusing (the reference electrode is supposed to be at infinity). These electrical potential maps illuminate the heterogeneities in the vicinity of the particular electrode of interest here. Seismoelectric responses that are too far from this particular electrode are too weak (in a relative sense) to help image those relatively distant heterogeneities.

numerous techniques available to detect the edge, in this work, we used a simple approach of image sharpening (Figure 6.21). Once the boundaries are identified, we discretize the domain between the two wells in $L \times L$ m cells (based on the electrodes separation of L). All cells that fall within the boundaries are considered to belong to the same geological unit.

6.3.2.2 Introduction of structural information into the objective function

We consider again our two vertical wells, A and B, located 300 m apart with a depth of 1050 m. Within each borehole, we have electrodes with $L = 10$ m spacing (105 electrodes per borehole; see Figure 6.22). We simulate bipole–bipole data (one injection source and one

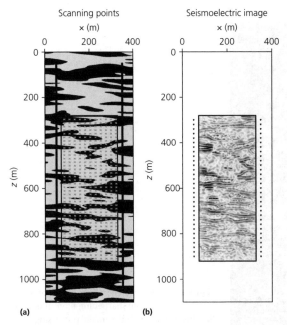

Figure 6.20 Seismoelectric image of the electrical potential using a set of electrodes located in the two wells for the domain $\mathfrak{R}$. **a)** Position of the scanning (focusing) points. The total number of scanning points is actually $(N_x = 128) \times (N_z = 320)$, that is, a total of more than 40,960 points. Only a fraction of these points is shown here. **b)** High-definition voltage map with the position of the electrodes used in the wells to scan the electrical potential. We see that the voltage map contains a lot of structural information regarding the position of the heterogeneities between the wells in the scanning area. The voltages are opposite in sign on each side of the heterogeneities. (*See insert for color representation of the figure.*)

measurement electrode in each borehole) with a total of 19,292 measurements.

The next step is to invert these apparent resistivity data. In electrical impedance (resistivity) tomography, we traditionally use a least-square minimization of the following objective function, $P^\beta(X)$, using the L2 norm to find an optimal resistivity model, X^*:

$$X^* = \arg \min P^\beta(X), \tag{6.1}$$

$$P^\beta(X) = [G(X) - d]^T C_d^{-1} [G(X) - d] + \beta [X - X_0]^T C_m^{-1} [X - X_0], \tag{6.2}$$

where $P^\beta(X)$ is the sum of a data misfit function (first term) and a model objective function (second term), which is used for the regularization of the inverse

problem. In this equation, X denotes the 2.5D resistivity model (vector of model parameters; note that we estimate the log-conductivity distribution to enforce positivity of the resistivity of each cell), X_0 can represent a prior resistivity model (we will not use such a prior hereinafter); G is the nonlinear forward operator used to map the predicted data, given a resistivity model; X, d signifies the vector of measured apparent resistivity (measured data vector), C_d^{-1} stands for the data covariance matrix (taken diagonal and based on the variance on the measurements); β designates the regularization parameter that balances the data misfit function and the model objective function, and C_m^{-1} symbolizes the smoothing matrix. In the classical least-square inversion, C_m^{-1} is a square $n \times n$ matrix corresponding to the Laplacian operator of the model (n denotes the number of cells). The classical inversion of the apparent resistivity data with no structural information (isotropic smoothness) is shown in Figure 6.23. We see that this approach yields a very smooth resistivity image because of the lack of resolution of the resistivity method far from the electrodes. In addition, the true values of the electrical resistivity are not recovered very well.

The question now is to see what would be the effect of incorporating into the resistivity data the structural information contained in the seismoelectric image. Since we are able to identify the boundaries with the seismoelectric focusing approach, we assign weights to the neighbors of a pixel only if they belong to the same material property unit (Figure 6.24). This way, we directly incorporate structural information into the objective function by changing only the matrix C_m^{-1}. A detailed approach on how this structural information is incorporated in C_m^{-1} can be found in a number of papers (e.g., Farquharson, 2008; Hale, 2009a, b, c; Lelièvre & Farquharson, 2013; Zhou et al., 2014).

6.3.2.3 Results
In the following, we study the effect of the structural information on the recovered model. The L-curve method can be used, in principle, to find the optimal value of the regularization parameters. In our approach, we enforce structural information; therefore, finding a value with an L-curve may not necessarily be the optimal way to proceed. For our tests, we considered the following values of the regularization parameter β: 0.1 (light), 1 (medium), 10 (hard), and 100 (extreme) (see Figure 6.25). A typical

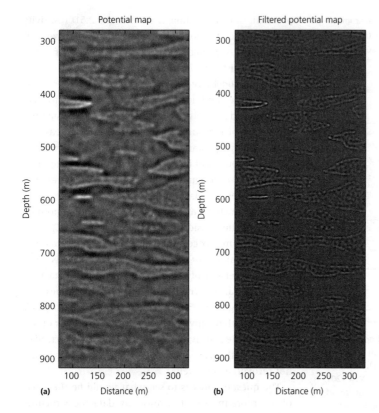

Figure 6.21 Edge detection from the potential map using an image sharpening filter in domain $\mathfrak{R}$. This is done with the image processing toolbox of MATLAB (function imsharpen). The structural information contained in the filtered potential map is used to impose textural information onto the electrical resistivity tomogram.

value for the L2 scheme is on the range of 0.01–0.1 using the L-curve approach. For our image-guided scheme, we enforce a constant value for all iterations. Larger values of the Lagrange parameter, β, enforce structural similarity within each layer (conductivity values show less variation within each formation) and have a negative effect on the data RMS. We see clearly from Figure 6.25 that the texture of the resistivity tomogram is improved and that the resistivity values are closer to the true resistivity value by comparison with the tomogram obtained in Figure 6.23. Such an improvement is very important in order to be able to reliably apply petrophysical models to be able to transform resistivity or complex conductivity into relevant parameters such a saturation, salinity, porosity, or permeability.

6.3.3 Discussion

We have shown that by focusing seismic waves at set of points, we can image the structural heterogeneities of formations. However, we have not discussed the order of

magnitude of the electrical field produced by the seismoelectric coupling for three fundamental reasons:

1 We did not care about the amplitude of the electrical potentials when we built a seismoelectric image. Indeed, the only things that matter in building the seismoelectric image displayed in Figure 6.20 are the local variations of the electrical potentials that map heterogeneities. Geometrical spreading could be accounted for, but this was not done in building Figure 6.20.

2 The conversions need only to be higher than the background noise. Typically, the electrical noise is on the order of $1\,\mu V\,m^{-1}$ for the electrical field but can decrease down to $1\,nV\,m^{-1}$ at depth.

3 If electrical noise is high, stacking may be required. One way to stack seismoelectric signals would be to repeat the same seismic (Ricker wavelet) sources. Another possibility would be to inject a harmonic pressure source and to stack the resulting harmonic electrical field over the number of cycles needed to have a good signal-to-noise ratio. Therefore, in

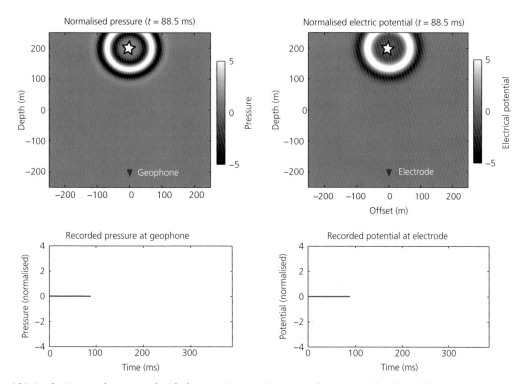

Figure 1.21 Synthetic example computed with the acoustic approximation to demonstrate the effect of the coseismic electrical field. A seismic source generates a pressure wave in a homogeneous material. We see here the normalized pressure (left panel) and the normalized electrical potential (right panel) at time 88.5 ms. The star represents the seismic source.

The Seismoelectric Method: Theory and Applications, First Edition. André Revil, Abderrahim Jardani, Paul Sava and Allan Haas.
© 2015 John Wiley & Sons, Ltd. Published 2015 by John Wiley & Sons, Ltd.

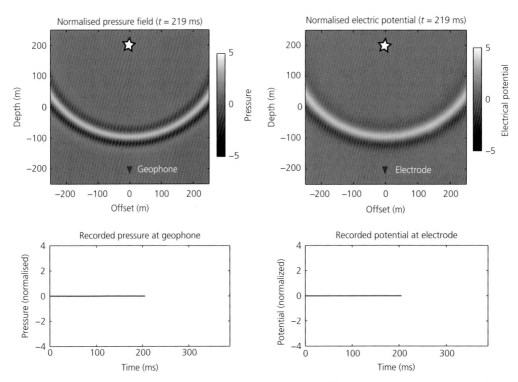

Figure 1.22 Same as Figure 1.18 for time 219.0 ms. The traveling P-wave is located in between the source and the position of the recording geophone located with an electrode (reference electrode to infinity). We see no disturbances at the position of the geophone and electrode.

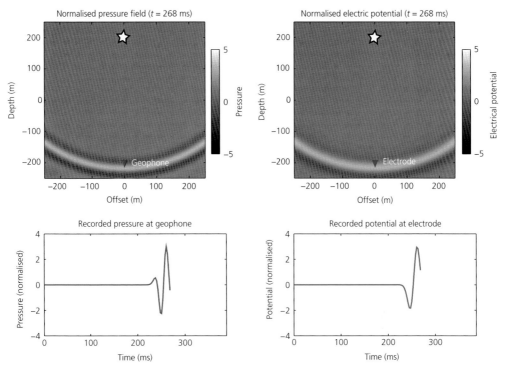

Figure 1.23 Same as Figure 1.18 for time 268.0 ms. The traveling (P-wave) seismic disturbance has reached ith an electrode (reference electrode to infinity). The geophone and the electrode record a pressure and an associated electrical disturbance. The electrical field corresponding to the electrical disturbance is called the coseismic field.

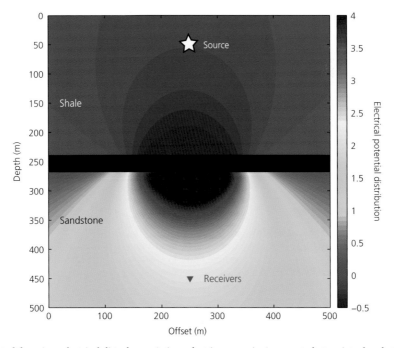

Figure 1.24 Snaphot of the seismoelectrical disturbance (seismoelectric conversion) generated at an interface between two media of different electrical conductivities (here a conductive shale and a less conductive sandstone) during the passage of a seismic P-wave. The star represents the seismic source. The radiation occurs in the first Fresnel zone of the seismic wave.

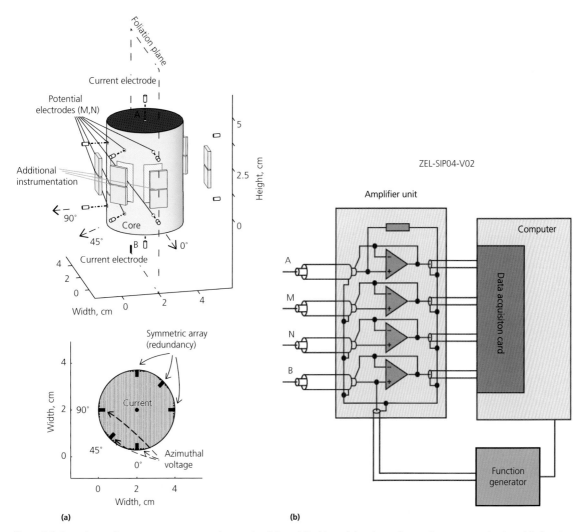

Figure 2.7 Experimental setup to measure complex conductivity. **a)** Position of the electrodes on the core sample. A and B denote the current electrodes used to inject a harmonic current, while the electrodes M and N denote the voltage electrodes used to record the harmonic electrical field. **b)** ZEL-SIP04-V02 impedance meter built by Egon Zimmermann in Germany (Zimmermann et al., 2008). The data acquisition system operates in the frequency range from 1 mHz to 45 kHz with a phase accuracy close to 0.1 mrad below 1 kHz. This instrument can be used to measure the complex conductivity (amplitude and phase) in the frequency range 1 mHz to 45 kHz.

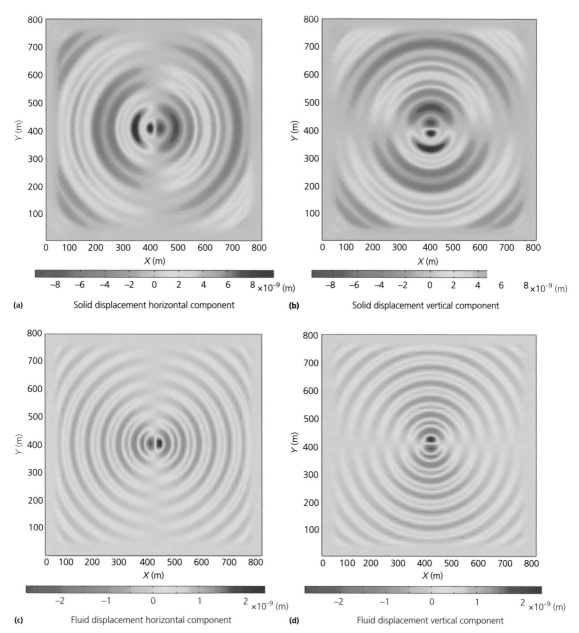

(a)   Solid displacement horizontal component

(b)   Solid displacement vertical component

(c)   Fluid displacement horizontal component

(d)   Fluid displacement vertical component

Figure 4.1 Horizontal and vertical components of the solid displacement vector and relative fluid displacement in the frequency domain (real components). **a)** Horizontal component of the solid displacement vector. **b)** Vertical component of the solid displacement vector. **c)** Horizontal component of the fluid displacement vector. **d)** Vertical component of the fluid displacement vector. Note the efficiency of the C-PML approach, at the boundaries, in attenuating the seismic waves. The use of the C-PML effectively eliminates reflections from the outer boundaries of the model.

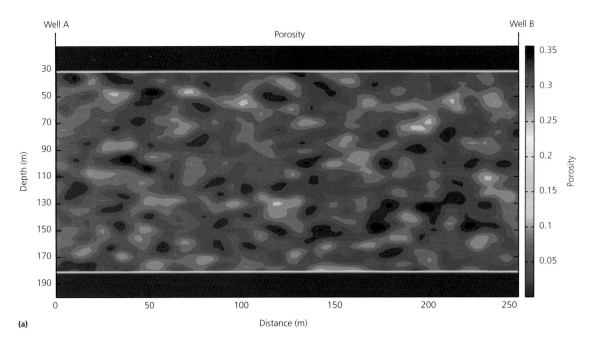

(a)

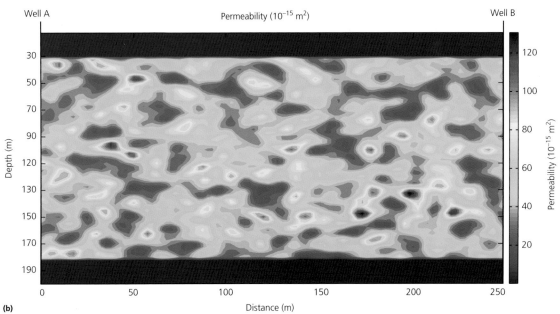

(b)

Figure 4.3 Porosity and permeability map of a NAPL contaminated aquifer between two wells for a water-flood simulation. **a)** Porosity distribution. **b)** Permeability distribution.

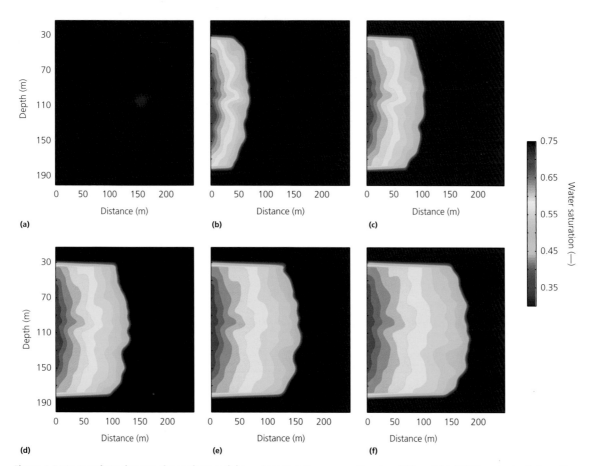

Figure 4.4 Six snapshots showing the evolution of the water saturation s_w over time in a 150-m-thick NAPL contaminated aquifer. The initial water saturation in the aquifer is equal to the irreducible water saturation $s_r = 0.25$ (which correspond to a NAPL saturation of 0.75). In this study the NAPL is considered to be the non-wetting phase. **a)** Reference snapshot T1. **b)** Snapshot T2 at 200 days. **c)** Snapshot T3 at 400 days. **d)** Snapshot T4 at 600 days. **e)** Snapshot T5 at 800 days. **f)** Snapshot T6 at 1000 days.

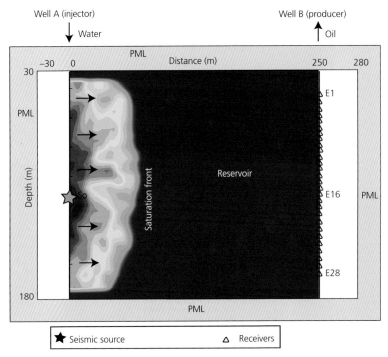

Figure 4.5 Sketch of the domain used for the modeling. The total modeling domain is a 410 m × 250 m rectangle. Injector Well A is used, located at position *x* = 0 m, and is also used for the seismic source. Production and recording Well B is located at *x* = 250 m. The discretization of the domain comprises a finite-element mesh of 205 × 125 rectangular cells. 28 receivers are located in Well B, approximately 30 m away from the nearest PML boundary (the PML boundary layers are shown in gray).

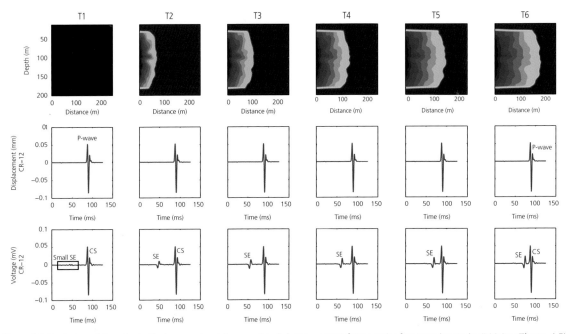

Figure 4.6 Evolution of the seismic displacement and the associated electric potential time series from receiver point E12 (see Figure 4.5) due to changes in the position of the oil/water encroachment front during snapshots T1 to T6. The arrival of the P-wave is identified in the seismic time series. SE denotes the seismoelectric conversions occurring at any electrical and mechanical heterogeneity in the aquifer (especially at the oil/water encroachment front), while CS denotes the coseismic electrical field associated with the P-wave.

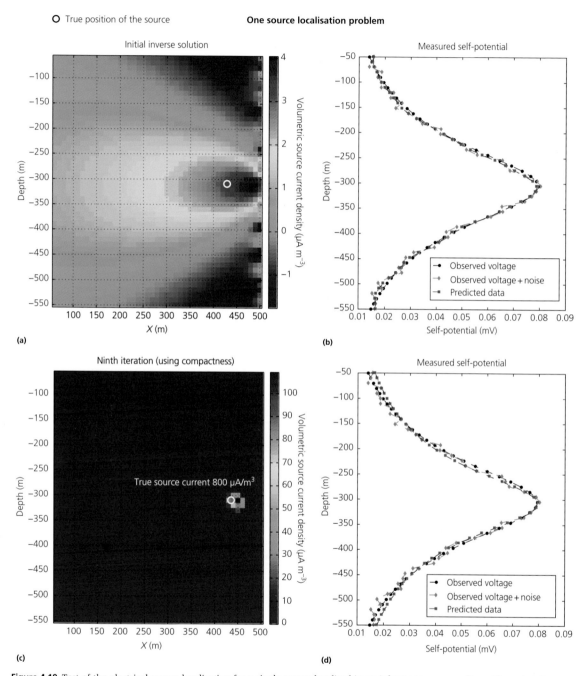

Figure 4.19 Test of the electrical source localization for a single source localized in an inhomogeneous medium. The resistivity distribution is given in Figure 4.15a (see Table 4.4). A Gaussian noise with a standard deviation equal to 10% of the computed data mean has been added to the data. The sources are analyzed in terms of volumetric source current distributions. **a)** Source current distribution at the first iteration. **b)** Comparison between the true self-potential distribution and the one determined from the inverted source current density distribution. **c)** Source current distribution using compactness. **d)** Comparison between the true self-potential distribution and the one determined from the inverted source current density distribution using compactness.

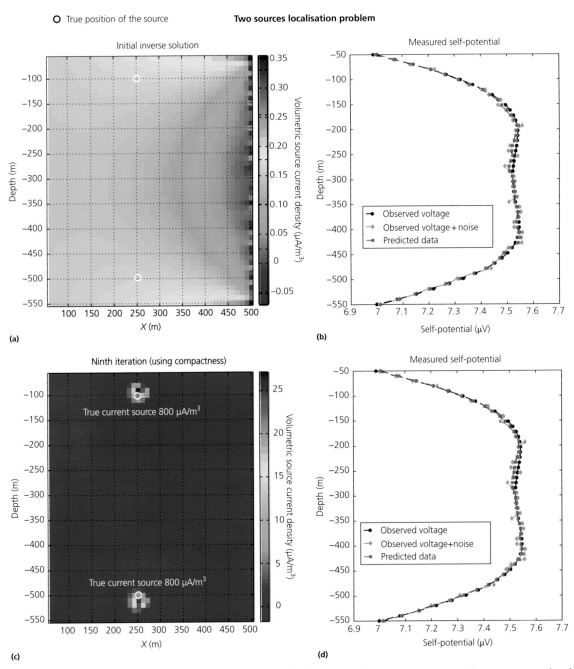

Figure 4.20 Test of the electrical source localization for two sources localized in an inhomogeneous medium. The sources are analyzed in terms of volumetric source current distributions. **a)** Source current distribution at the first iteration. **b)** Comparison between the true self-potential distribution and the one determined from the inverted source current density distribution. **c)** Source current distribution using compactness. **d)** Comparison between the true self-potential distribution and the one determined from the inverted source current density distribution using compactness.

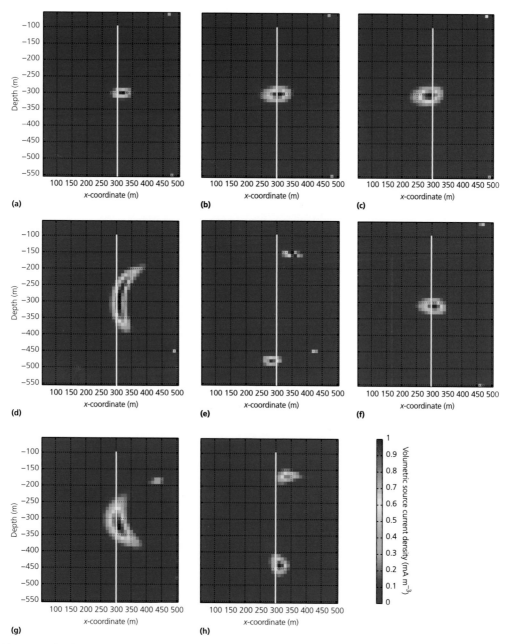

Figure 4.22 Source localization for the electrical potential distributions shown in Figure 4.21 (shot #3). The sources are analyzed in terms of volumetric source current distributions. The white line indicates the position of the discontinuity between the units U1 and U2. **a)** Source current distribution at time 184.96 ms. **b)** Source current distribution at time 187.77 ms. **c)** Source current distribution at time 190.58 ms. **d)** Source current distribution at time 193.38 ms. **e)** Source current distribution at time 196.19 ms. **f)** Source current distribution at time 210.23 ms.

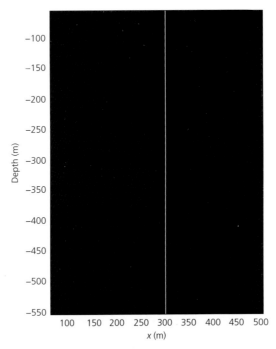

Figure 4.23 An aggregate of all the source distributions for the 5 shots using 6 characteristic times for each shot for case study #1. The white line indicates the true position of the discontinuity.

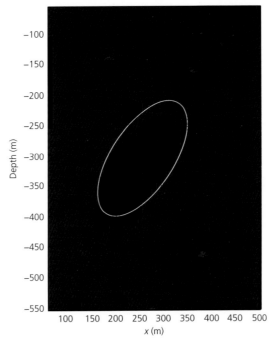

Figure 4.24 Aggregate of all the source current distributions for the 5 shots using 6 characteristic times for each shot for case study #2. The white line indicates the true position of the discontinuity between the domains U1 and U2.

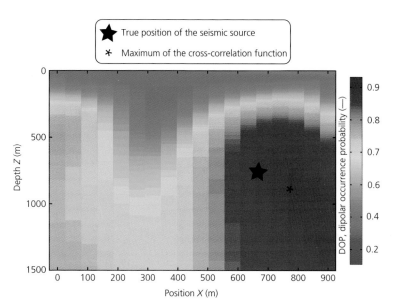

Figure 5.7 Dipole occurrence probability (DOP) of the source using a normalized cross-correlation of a dipole source with the signals obtained at the time of the source at the eight stations located at the ground surface (each pixel is 50 m in x by 1.5 m in z). The maximum of the DOP (denoted by the star symbol, $*$) can be used to provide a prior localization of the source assuming that the subsurface has a constant (unknown) electrical conductivity. The filled star, $\star$, denotes the true position of the seismic source.

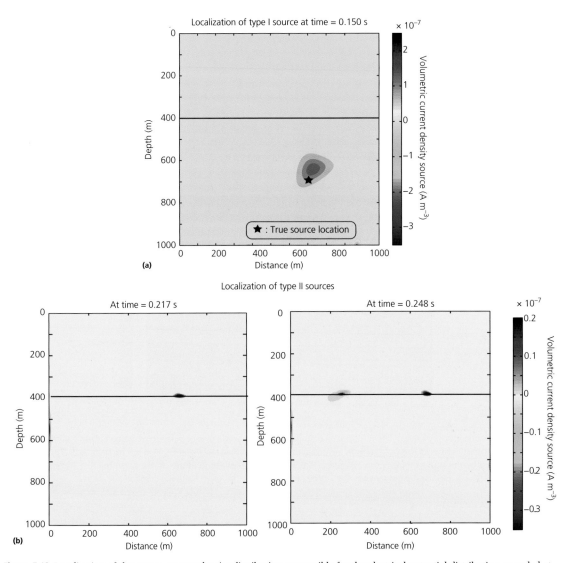

Figure 5.12 Localization of the source current density distribution responsible for the electrical potential distribution recorded at the surface stations (5% noise). **a)** Type I anomaly associated with the seismic source. **b)** Type II anomalies (seismoelectric conversions) (iteration #8, data root mean square error: 3%). The horizontal line denotes the interface between L1 and L2 where seismoelectric conversions take place.

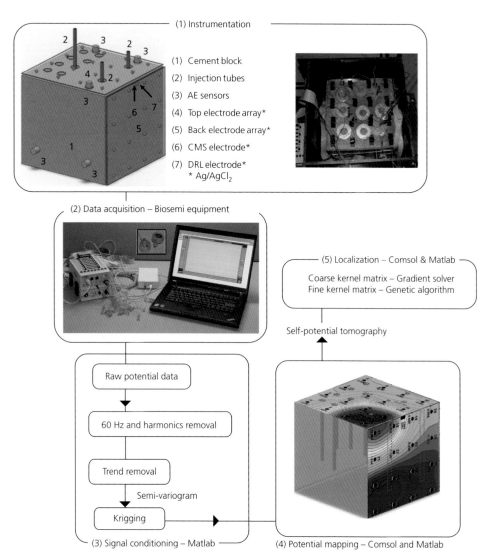

Figure 5.16 Flowchart for the processing of the electrical potential data. (1) Instrumentation of the porous block. (2) Data acquisition showing the BioSemi EEG system and the laptop computer. (3) Signal conditioning of the raw data. (4) Mapping the voltage response using ordinary kriging. (5) Localization of the causative sources in the block.

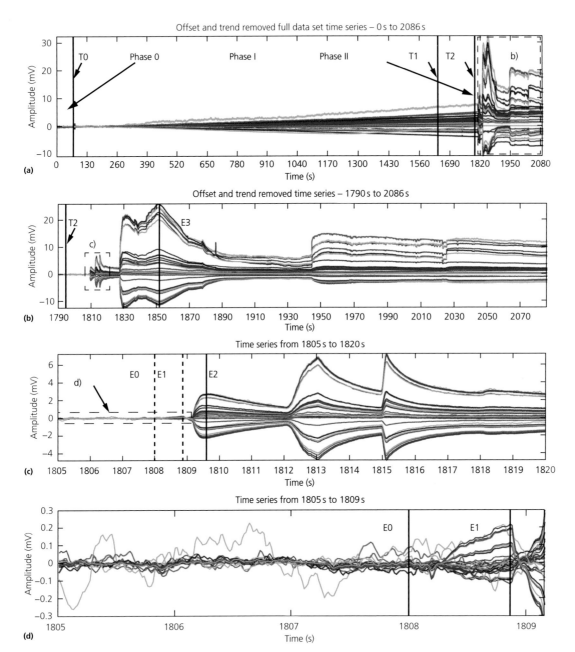

Figure 5.17 Self-potential time series related to Hole #9 saline water injection. **a)** Full time series data set showing the different fluid injection time periods during data acquisition (T0, T1, and T2). Note the significant change in electrical response after T2 that is bounded by region **b)**. **b)** Zoom highlighting the background normalized electrical response showing distinct electrical impulses related to the start of constant flow injection at T2 with selected peak events. **c)** Zoom in region **c)** highlighting the first series of impulsive signals with selected peak Events E2, E3, and E4 with temporal reference to E0 and E1. **d)** Zoom in region **d)** showing the temporal noise leading up to Event E1, with a voltage background time slice at E0. Note the change in potential after E1.

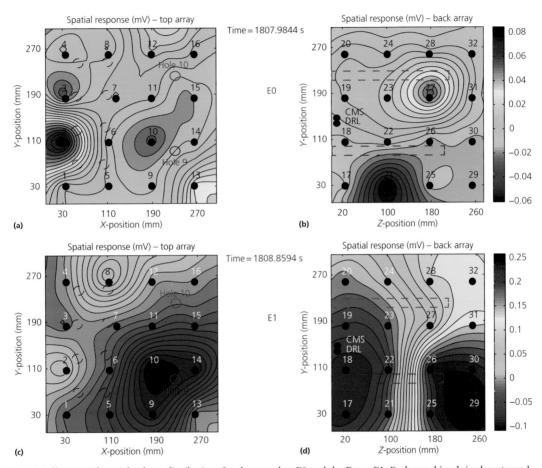

Figure 5.18 Self-potential spatial voltage distributions for the snapshot E0 and the Event E1. Each panel is a kriged contoured distribution of the electrical potential on the top and back panels (ordinary kriging). The black dots denote the position of the electrode positions. **a)** and **b)** Spatial electrical potential distribution for snapshot E0 showing the spatial variations associated with the background noise. Note the very small color bar voltage scale (+0.08 to −0.06 mV). **c)** and **d)** Voltage distribution for Event E1 showing the burst of the electrical field associated with the first hydraulic pulse taking place during constant flow injection. Note the voltage polarities in the spatial distributions and the much larger color bar voltage scale (+0.25 to −0.1 mV).

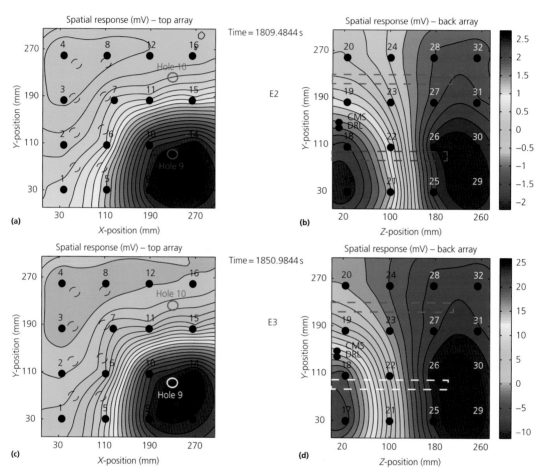

Figure 5.19 Self-potential spatial voltage distributions of Events E2 and E3. **a)** and **b)** Event E2 voltage distribution showing the peak voltage associated with the second hydraulic pulse during constant flow injection. Note the voltage polarities in the spatial distributions and the color bar voltage scale (+2.5 to −2.0 mV). Event E2 represents the first of a series of electrical field bursts. Panels **a)** and **b)** show a reversal in polarity and increase in peak magnitude relative to figure panels **c)** and **d)**. **c)** and **d)** Event E3 spatial voltage distribution showing the peak voltage associated with the highest magnitude pulse during constant flow injection.

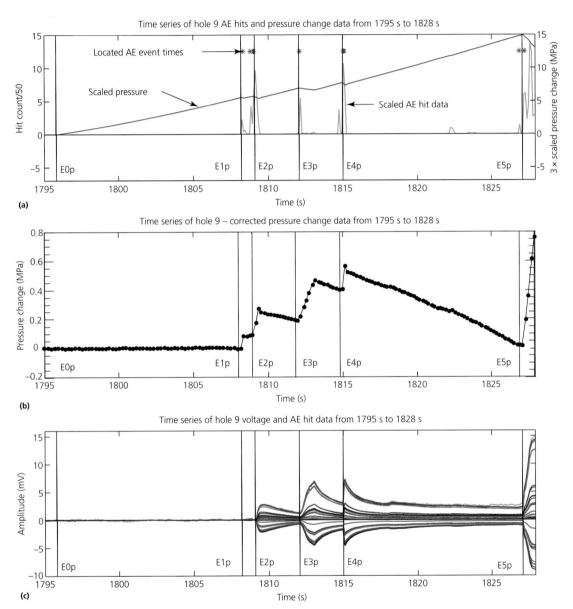

Figure 5.20 Fluid pressure, acoustic emissions, and electrical potential changes during a given time window. **a)** The acoustic emission (AE) data and pressure change correlation. The red asterisks denote the located AE events. **b)** The trend removed pressure changes. **c)** The voltage responses at all 32 measurement points. The events labeled E0p–E5p represent the same events in each panel of the figure. The AE hits, the pressure changes, and the voltage data all align, indicating that there is a relationship between these phenomena.

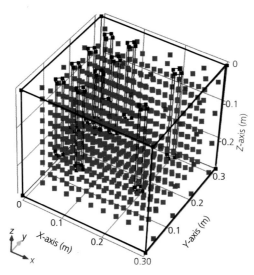

Figure 5.21 COMSOL model showing the distribution of points used to compute the coarse kernel matrix (blue points, 729 positions) and the voltage measurement points (red points, 32 positions). The model geometry includes each of the 10 holes that were drilled into the block (black cylinders) considered to be perfect insulators. The arrows labeled *x*, *y*, and *z* indicate the positive directions of the corresponding axes in the figure for the Cartesian coordinate system.

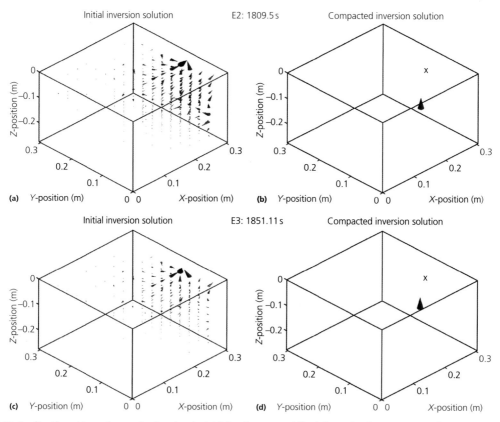

Figure 5.22 Gradient based inversion results showing the initial and compacted dipole inversion for Events E2 and E3. In these cases, the compact inversion of Events E2 and E3 yields one dipole that dominates the response through a higher magnitude. Note that Event E3 is localized at a shallower depth than Event E2, showing that the causative source is moving closer to the top surface of the block in the vicinity of Hole 9. **a)** Event E2 results of the source distribution after inversion. **b)** Same after 40 iterations in compacting the support of the source. **c)** and **d)** Same for Event E3.

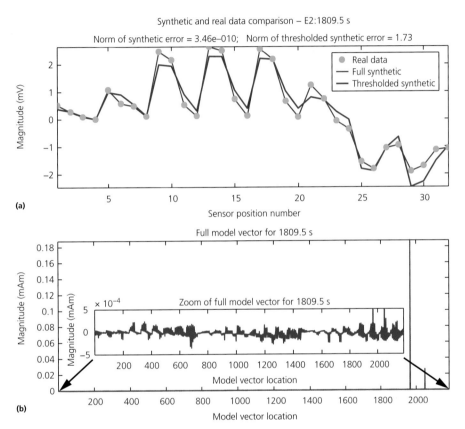

Figure 5.23 E2 compacted inversion results and forward modeled comparison with a thresholded forward model. **a)** Comparison of the real data with the forward model calculations using the full inverted model and a thresholded model derived from the full inversion model. **b)** The full model vector that was found through inversion and compaction. The inset shows the multitude of minor dipoles spread throughout the model vector that were generated by the inversion and compaction process.

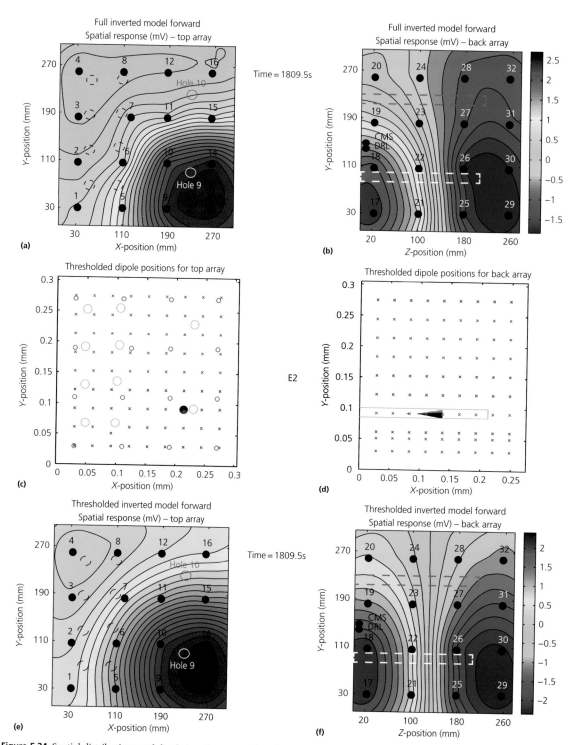

Figure 5.24 Spatial distributions and dipole location comparison for inverted and forward modeled results for Event E2. **a)** and **b)** Forward computation results using the full model vector (all the minor dipoles included). **c)** and **d)** Thresholded model vector dipole positions (there are actually two dipoles in these panels) showing the orientation and position of the main dipole (visible). **e)** and **f)** Forward computation results using the thresholded model vector.

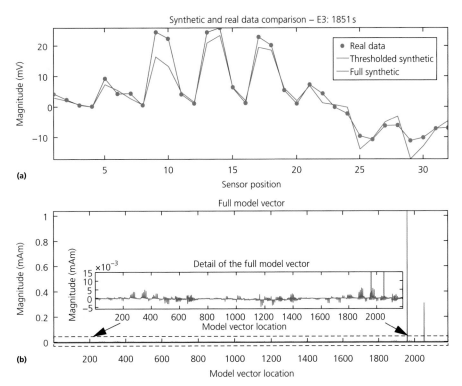

Figure 5.25 E3 compacted inversion results and forward modeled comparison with a thresholded forward model. **a)** Comparison of the real data with the forward model calculations using the full inverted model and a thresholded model derived from the full inversion model. **b)** Full model vector found through inversion and compaction. The inset shows the multitude of minor dipoles spread throughout the model vector that were generated by the inversion and compaction process. The main figure shows the thresholded model with zero values everywhere except at the model locations where the dipoles were localized. Note that the located dipoles are in close proximity to each other.

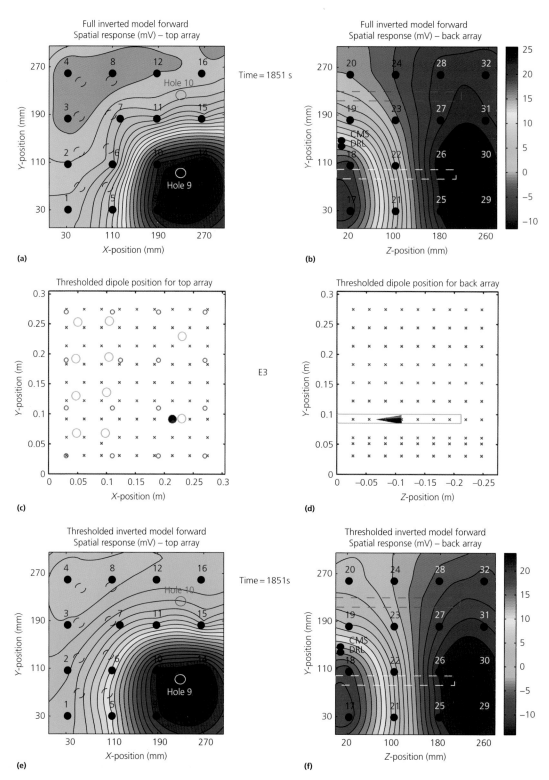

Figure 5.26 Spatial distributions and dipole location comparison for inverted and forward modeled results for Event E3. **a)** and **b)** Forward computation results using the full model vector (all the minor dipoles included). **c)** and **d)** Thresholded model vector dipole positions (there are actually two dipoles in these panels) showing the orientation and position of the main dipole (visible). **e)** and **f)** Forward computation results using the thresholded model vector.

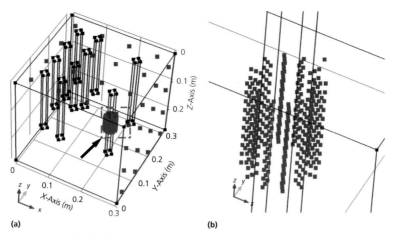

(a) (b)

Figure 5.27 COMSOL geometry used for the fine geometry kernel matrix computations (360 positions). **a)** Finer resolution cylindrical kernel matrix point distribution along with the measurement points used with the genetic algorithm. **b)** Close-up of the kernel matrix point distribution.

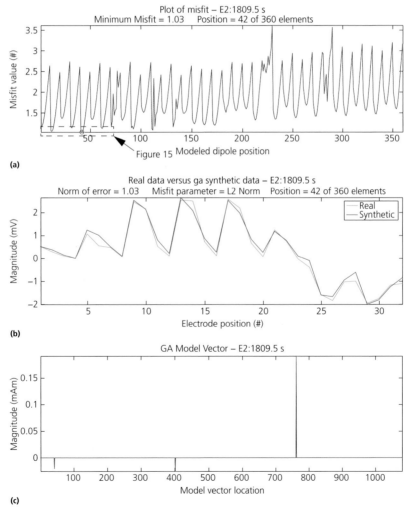

(a)

(b)

(c)

Figure 5.28 a) Shows the E2-related genetic algorithm localization misfit values for each position in the kernel matrix. The red circle represents the minimum value and therefore the best fit for the measured data, given the inversion constraints. **b)** Comparison of the real data and the genetic algorithm-based model vector forward computation. **c)** The genetic algorithm-based best fit model vector. Note that there are three points on the vector where there are values that are greater than zero. These points are exactly 360 elements apart, representing a single dipole moment orthogonal components.

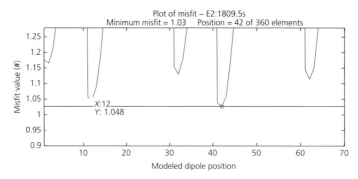

Figure 5.29 Zoom in misfit for all the observations of the genetic algorithm search algorithm of Event E2. This figure shows that there are dipole positions that are close to the minimum of the misfit function, which is shown by the small open circles. The position marked by x = 12 represents another possible location that is only slightly greater than the minimum found at 42. There are other points that are spatially adjacent to 42, including 11, and 41.

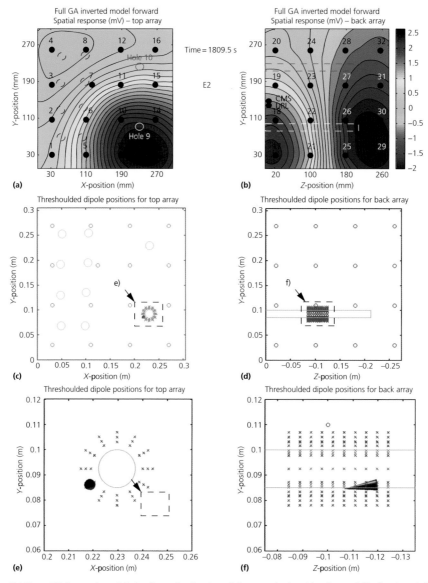

Figure 5.30 a) and **b)** Event E2 forward modeled voltage distribution of the genetic algorithm located dipole. **c)** and **d)** Spatial location of the dipole within the concrete block. **e)** and **f)** Close-up of the dipole location showing the off-vertical orientation of the dipole moment.

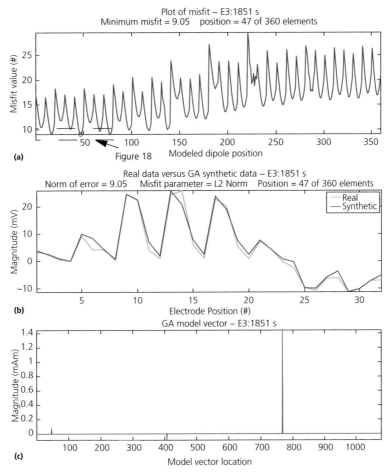

(a)

Figure 18

(b)

(c)

Figure 5.31 **a)** E3-related genetic algorithm localization misfit values for each position in the kernel matrix. The red circle represents the minimum value and therefore the best fit for the measured data, given the inversion constraints. **b)** Comparison of the real data and the genetic algorithm-based model vector forward computation. **c)** The genetic algorithm-based best fit model vector. Note that there are three points on the vector where there are values that are greater than zero. These points are exactly 360 elements apart, representing a single dipole moment orthogonal components.

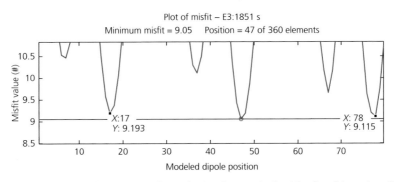

Figure 5.32 Zoom in the values of the misfit parameter obtained using the genetic algorithm-based inversion of Event E3. This figure shows that there are several dipole solutions that are close to the minimum of the data misfit function. In this figure, there are several other possible solutions, and they are all close to the point found at position 47, the minimum. These additional possible locations corresponds to dipole positions 17, 18, 48, 77, and 78.

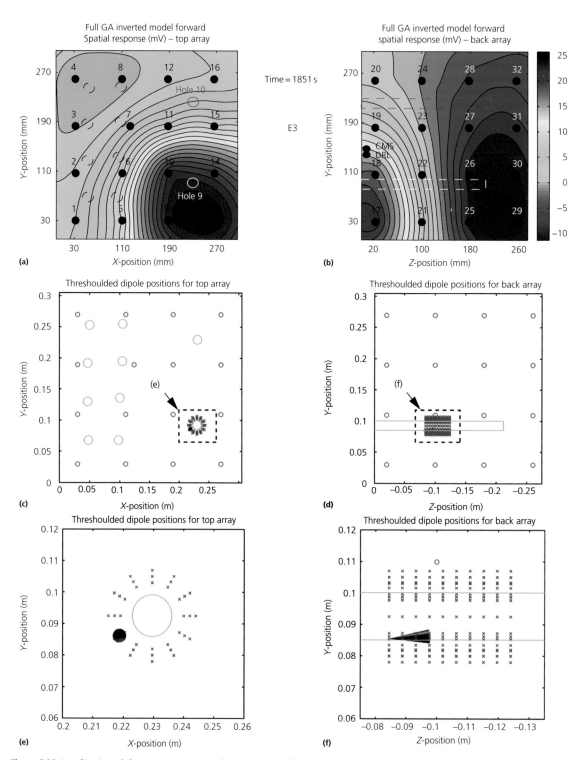

Figure 5.33 Localization of the causative source of current. **a)** and **b)** E3 forward modeled voltage distribution of the genetic algorithm located dipole. **c)** and **d)** Spatial location of the dipole within the concrete block. **e)** and **f)** Close-up of the dipole location showing the vertical orientation of the dipole moment.

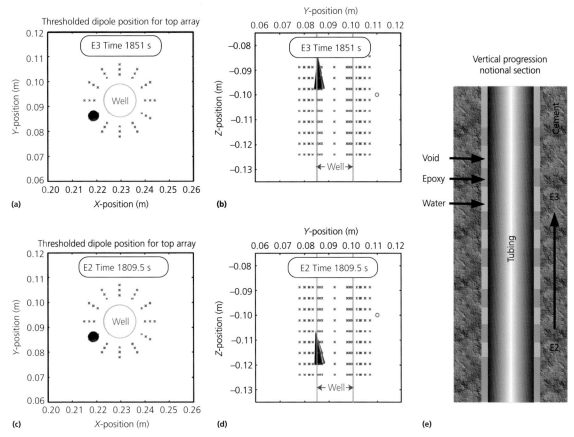

Figure 5.34 Proposed explanation of the electrical disturbances observed in the experiment. Events E2 and E3 are shown together with their spatial placement near Hole 9 in panels **a)** through **d)**. The blue x's in panels **a)** through **d)** are the dipole point positions used in the genetic algorithm source inversion process. The green circle in panels **a)** and **c)** represents the position of Hole 9 as viewed from the top, and the vertical green lines in panels **b)** and **d)** represent the position of Hole 9 as seen from the side. The red circles in panels **b)** and **d)** are the position of one of the measurement electrodes. It can be clearly seen in this figure that the first event, E2, occurs lower in the model representation (and by correlation, in the test specimen) than the later event, E3. The figure in panel *e* represents a conceptualization of the movement of the fluid through multiple epoxy barriers, filling voids in between each the barriers with water. The voids represent points of open contact with the cement that allow the fracturing fluid to make direct contact, generating the observed pressure fluctuations and electrical impulses. **e)** Sketch of the tubing showing the progression of the Events E2 and E3 along the well over time.

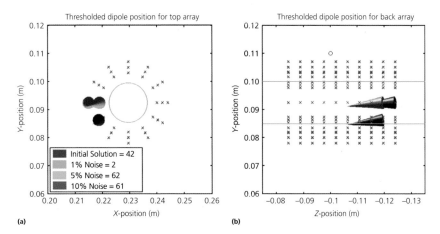

Figure 5.35 Localization of the causative source of current and noise analysis. **a)** and **b)** These figures show the spatial positions of the dipoles found during the localization uncertainty test. Note that the solutions cluster near the initial solution found during the inversion process. The bias in the +y and −z directions may indicate that the true solution may be between the solutions with noisy data and the solution found initially.

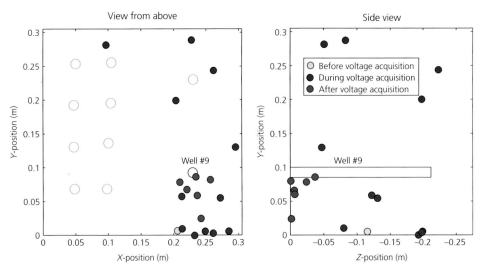

Figure 5.36 Localization of the acoustic emissions with respect to the time window shown in Figure 5.20. The events localized far from Well #9 are probably associated with the reactivation of small cracks. Note that only a tiny fraction of the AE hits shown in Figure 5.20 are localizable.

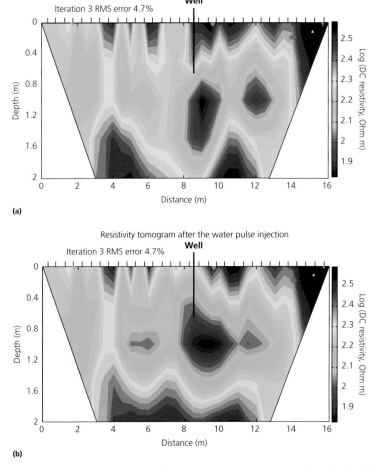

Figure 5.44 Inverted DC resistivity section using the Gauss–Newton method. The data are collected using 32 electrodes with 50 cm spacing. **a)** Tomogram prior the pulse injection. **b)** Tomogram after the pulse injection.

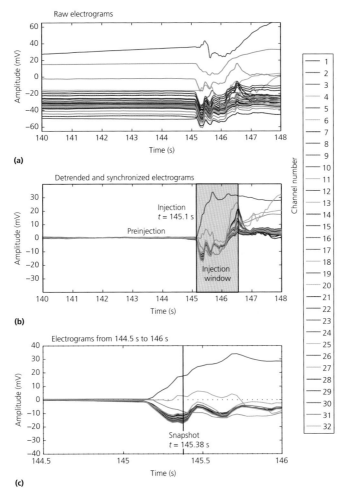

(a)

(b)

(c)

Figure 5.45 Self-potential time series from the water injection experiment. **a)** Raw electrograms. **b)** Detrended and background corrected electrograms. **c)** Magnification of the electrograms showing the preinjection and injection time windows.

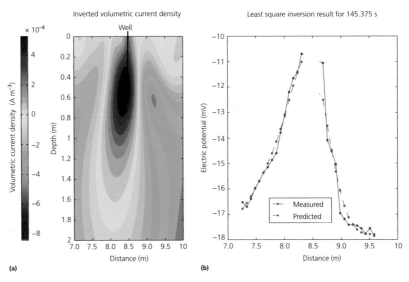

(a)

(b)

Figure 5.46 Source localization of a single snapshot at time 145.38 s the localization of the self-potential anomaly (positive pole associated with the pulse injection) is located very close to the end of the open well. The electrical resistivity has been accounted for this inversion. Results shown are at the third iteration.

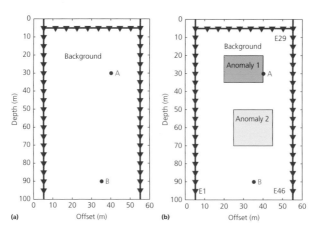

Figure 6.1 Geometry used for the beamforming problem. The medium consists of a homogeneous background model (reference model, fully saturated) plus two anomalies termed Anomaly 1 and Anomaly 2. These anomalies correspond to areas that are unsaturated (see Table 6.3). The survey area is surrounded by two vertical wells located on each side. The triangles correspond to the location of the seismic sources/geophones/electrodes. The spacing between two consecutive sensors is 5 m. The two boreholes have 19 dipoles of electrodes each, and sets of sensors are located close to the ground surface (5 m deep). The two red-filled circles correspond to the focusing points used for our numerical experiments. Ei corresponds to the position of electrode i. There are 46 set of sensors in total with E1 and E46 at the bottom of the two wells. **a)** Reference model (without the two heterogeneities). **b)** Model used for the numerical simulation with the two heterogeneities.

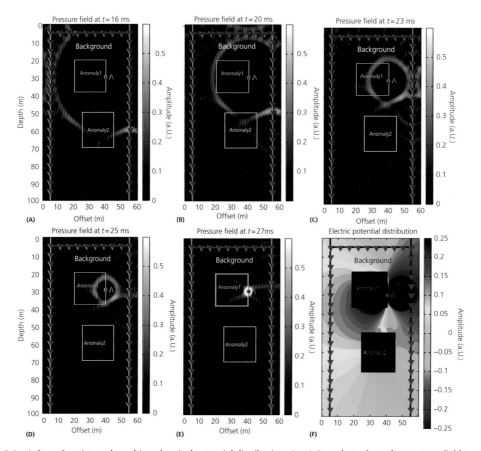

Figure 6.3 Seismic beamforming and resulting electrical potential distribution. **(a–e)** Snapshots show the pressure field (coming from the 46 seismic sources shown in Figure 6.1) focusing at point A. At $t = 27$ ms, the wave field interferes constructively at point A producing a strong seismoelectric response that is recorded at the receiver electrodes. **f)** The distribution of the electrical potential corresponds to the case where the seismic energy is focused at point A. The seismoelectric conversion at the time of focusing is characterized by a strong dipolar behavior.

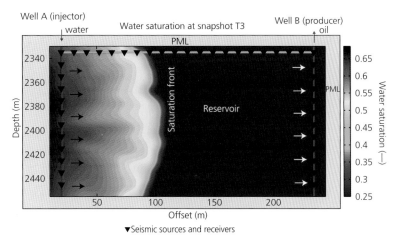

Figure 6.6 Sketch of the domain used for modeling. The total modeling domain is a 410 m × 250 m rectangle. Injector Well A, located at position $x = 0$ m, is also used for the seismic source. Production and recording Well B is located at $x = 250$ m. The discretization of the domain comprises a finite-element mesh of 205 × 125 rectangular cells. Twenty-eight receivers are located in Well B, approximately 30 m away from the nearest PML boundary (the PML boundary layers are shown in gray).

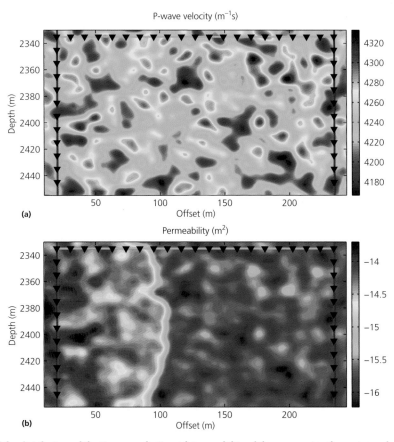

Figure 6.7 Sketch of the distribution of the P-wave velocity and permeability of the pore water phase at snapshot 3. Note that the saturation front is characterized by a sharp contrast in permeability. **a)** Velocity distribution for the compressional wave. **b)** Distribution of the permeability.

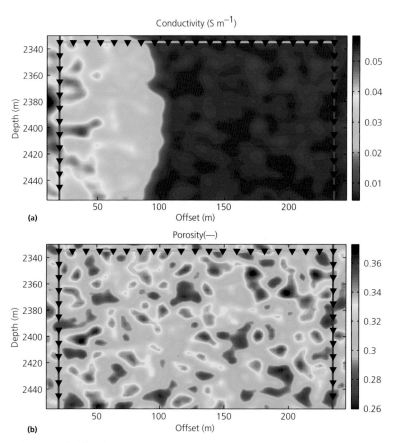

Figure 6.8 Sketch of the distribution of the electrical conductivity and porosity at snapshot 3. Note that the saturation front is characterized by a sharp contrast in electrical conductivity. **a)** Electrical conductivity distribution. **b)** Porosity distribution.

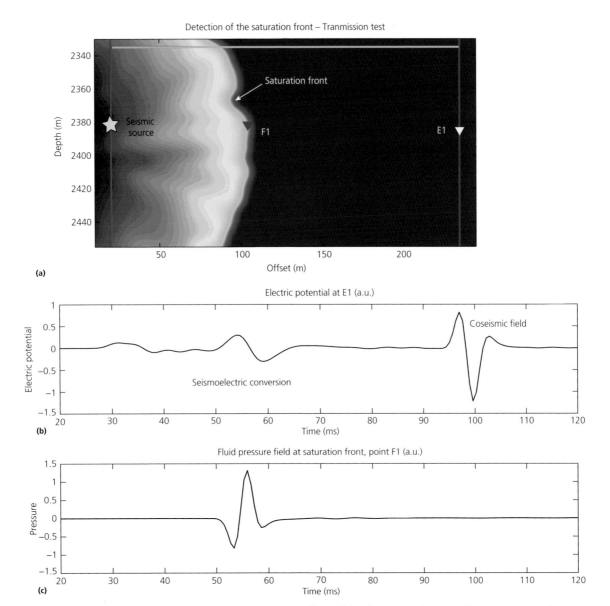

Figure 6.9 Transmission experiment with the seismic source in Well A and the electrical receiver in Well B. **a)** Geometry of the test. **b)** Time series for the electrical potential showing the seismoelectric conversion occurring at the interface and the coseismic field. **c)** Fluid pressure field at a point located at the saturation front at the center of the Fresnel zone.

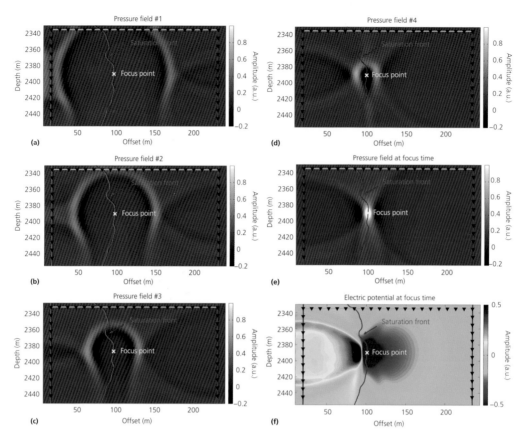

Figure 6.10 Seismic beamforming at the saturation front and resulting electrical potential distribution. **(a–e)** Evolution of the confining pressure $P(r, t)$ for 5 snapshots. The propagating wave fields interfere constructively at the focus point. **f)** Electrical potential distribution corresponding to the seismic energy focused at the saturation front (snapshot e).

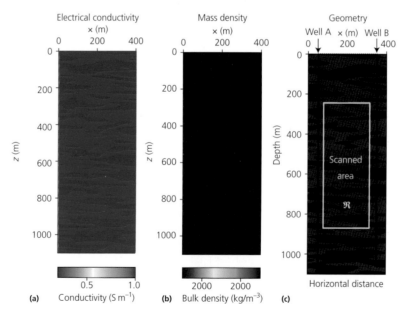

Figure 6.12 Model used for the scanning simulations. **a), b)** Heterogeneous distribution of the electrical conductivity and mass density of the formations. **c)** Geometry of the acquisition array. The porous material is bimodal with a larger correlation length in the horizontal direction. The seismic receivers are located in two wells (A and B). The area scanned through focusing is indicated by the yellow rectangle (scanning region $\Re$).

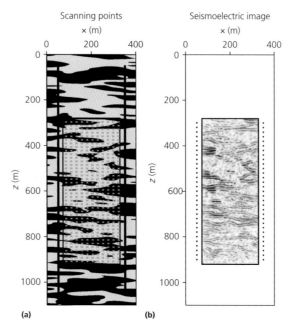

Figure 6.20 Seismoelectric image of the electrical potential using a set of electrodes located in the two wells for the domain $\mathfrak{R}$.
a) Position of the scanning (focusing) points. The total number of scanning points is actually $(N_x = 128) \times (N_z = 320)$, that is, a total of more than 40,960 points. Only a fraction of these points is shown here. **b)** High-definition voltage map with the position of the electrodes used in the wells to scan the electrical potential. We see that the voltage map contains a lot of structural information regarding the position of the heterogeneities between the wells in the scanning area. The voltages are opposite in sign on each side of the heterogeneities.

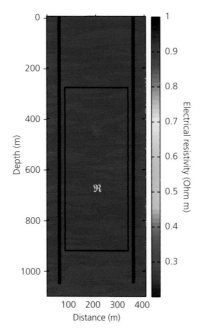

Figure 6.22 Position of the electrodes (black circles, 105 electrodes per well, spacing 10 m) for the cross-well resistivity tomography and discretization of the domain (the size of the cells is 10 × 10 m). The resistivity image denotes the true resistivity image used to simulate the acquisition of the apparent resistivity data. We are interested in recovering the resistivity information in domain $\mathfrak{R}$.

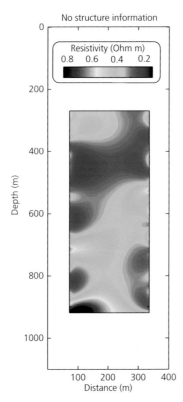

Figure 6.23 Result of a classical cross-well resistivity tomography in domain 𝕽 using the least-square method without structural information to constrain the model covariance matrix. Classical cross-well resistivity tomography is unable to image the formations between the wells due to a lack of resolution far away from the electrodes. In addition, the values of the resistivity are smoother than the true values, which mean that the application of a petrophysical model to the recovered resistivity would lead to a misinterpretation of the petrophysical parameters of interest.

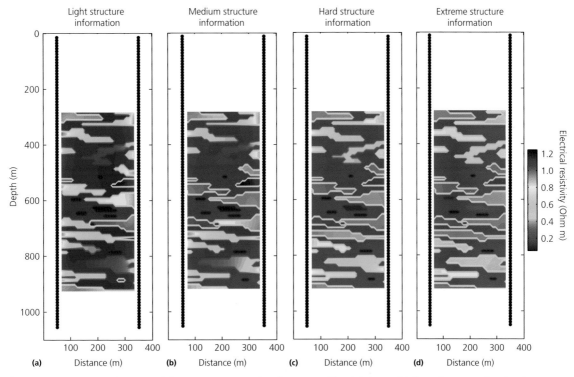

Figure 6.25 Importance of structure information in recovering the cross-well resistivity model through cross-well resistivity tomography. We consider the following value of the regularization parameter β: 0.1 (light), 1 (medium), 10 (hard), and 100 (extreme).

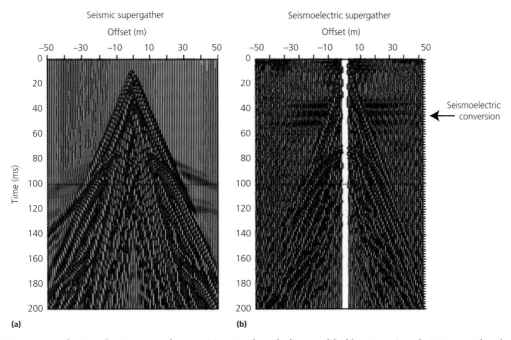

Figure 7.3 Seismic and seismoelectric supergathers (position 128 along the line, modified from Dupuis et al., 2007, reproduced with the authorization from the SEG and the authors). **a)** Seismic supergather offset records. **b)** Seismoelectric supergather offset records.

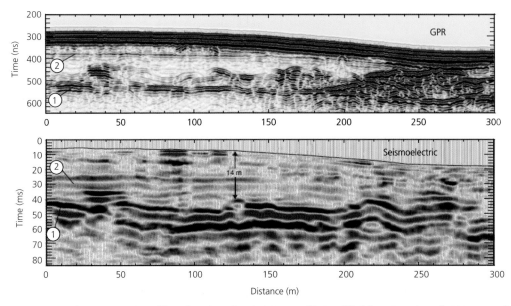

Figure 7.4 Comparison between a GPR profile and a seismoelectric 300 m profile (modified from Dupuis et al., 2007, reproduced with the authorization from the SEG and the authors). Variable elevation static delays were applied to the data to account for the effect of elevation. Anomaly 1 corresponds to the water table, while anomaly 2 likely corresponds to a unit that might have a high water content.

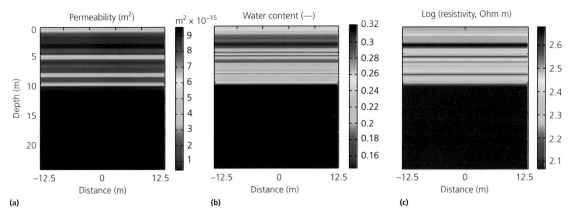

Figure 7.11 Material properties distribution constructed for modeling. **a)** Permeability. **b)** Volumetric water content. **c)** Electrical resistivity distribution. In addition, we use a constant porosity for the domain of interest (Table 7.1).

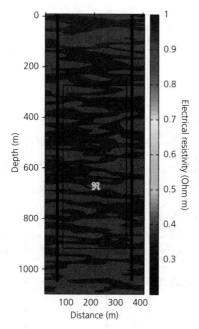

Figure 6.22 Position of the electrodes (black circles, 105 electrodes per well, spacing 10 m) for the cross-well resistivity tomography and discretization of the domain (the size of the cells is 10 × 10 m). The resistivity image denotes the true resistivity image used to simulate the acquisition of the apparent resistivity data. We are interested in recovering the resistivity information in domain $\mathfrak{R}$. (*See insert for color representation of the figure.*)

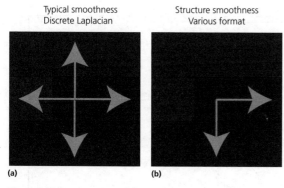

(a)   (b)

Figure 6.24 Computation of the model covariance matrix using the detection of boundaries. **a)** The classical discrete Laplacian correlates each pixel with all the neighbor ones in an isotropic way. **b)** The structural constrains correlate only neighbor pixel that belong to the same unit. In our approach, the structural constraints are imposed from the seismoelectric image.

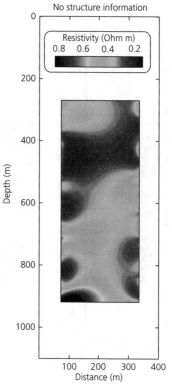

Figure 6.23 Result of a classical cross-well resistivity tomography in domain $\mathfrak{R}$ using the least-square method without structural information to constrain the model covariance matrix. Classical cross-well resistivity tomography is unable to image the formations between the wells due to a lack of resolution far away from the electrodes. In addition, the values of the resistivity are smoother than the true values, which mean that the application of a petrophysical model to the recovered resistivity would lead to a misinterpretation of the petrophysical parameters of interest. (*See insert for color representation of the figure.*)

principle, there should be no issues in building the seismoelectric image required to perform the image-guided inversion of ERT data.

6.4 Spectral seismoelectric beamforming (SSB)

We can beamform a pulse but we can also beamform a harmonic wave field for a range of frequencies. We call this approach "spectral beamforming" since the resulting electrical response can be measured frequency by frequency and the result will be a spectral seismoelectric

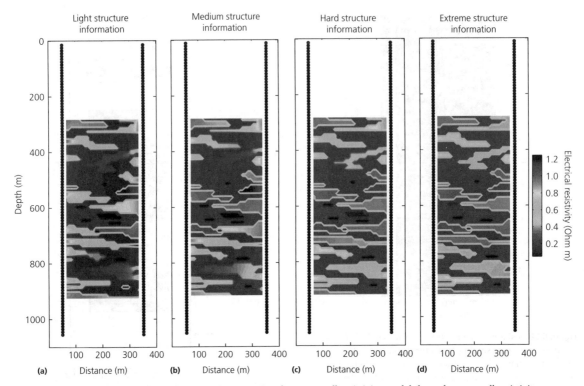

Figure 6.25 Importance of structure information in recovering the cross-well resistivity model through cross-well resistivity tomography. We consider the following value of the regularization parameter β: 0.1 (light), 1 (medium), 10 (hard), and 100 (extreme). (*See insert for color representation of the figure.*)

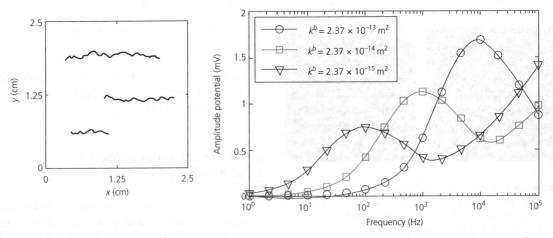

Figure 6.26 In the presence of harmonic stimulations of a porous material containing fractures, the electrokinetic response shows resonance effects that depend on the permeability of the fractures (Jougnot et al., 2013). The figure on the right side shows the electrical response for a stimulation at different frequencies from 1 to 10^5 Hz.

response for a range of frequencies. If the target is a fractured material, we can expect to see some resonance effects associated with the focusing of the seismic waves on the fractures (Figure 6.26). Jougnot et al. (2013) have shown, indeed, that we could expect specific electrical resonance effects when a crack is vibrated at particular frequencies. Therefore, the spectral beamforming method could be used to detect and characterize fractures in a reservoir.

6.5 Conclusions

We have demonstrated that virtual electrodes can be used, at known positions and times, to image a heterogeneous medium. This is achieved through the focusing of acoustic waves at specific coordinates using multiple acoustic receivers located in the vicinity of the target. The acoustic energy at the focus time is the largest relative to all other times, thus insuring the strongest possible seismoelectric response at the target position. This methodology can be used to investigate the electric and hydraulic properties of a medium by surrounding the area of investigation with multiple electrodes at known positions. The dense virtual electrode distribution has the potential to increase the robustness and improve the resolution of ERT. This methodology can also be employed to infer hydraulic parameters, for example, permeability, through a controlled seismoelectric procedure.

Based on the beamforming approach, we can map heterogeneities in resistivity between a set of wells. The method we discussed in this chapter is based on scanning, through seismic focusing, electrokinetic conversion, and measurement of the electrical potential between two wells. The focusing is done through the time reversal of seismic waves using a set of seismic sources located in two wells. The obtained "seismoelectric image" contains structural information regarding the heterogeneities in the scanned volume. The resulting seismoelectric-based heterogeneity image can then be used to constrain resistivity tomography using image-guided inversion to impose structural constraints in the model covariance matrix. This approach opens a new perspective on the imaging of a fine resistivity structure between wells and then, using the derived resistivity structure, to convert the data into petrophysical properties of interest.

This method could also be applied to complex conductivity, which is known to contain information that can be used to determine the permeability structure between wells.

References

Farquharson, C.G. (2008) Constructing piecewise-constant models in multidimensional minimum-structure inversions. *Geophysics*, **73**, K1–K9.

Hale, D. (2009a) Structure-oriented smoothing and semblance. CWP Report, 635, http://inside.mines.edu/~dhale/papers/Hale09StructureOrientedSmoothingAndSemblance.pdf [accessed on December 3, 2014].

Hale, D. (2009b) Image-guided blended neighbor interpolation. CWP Report, 634, http://inside.mines.edu/~dhale/papers/Hale09ImageGuidedBlendedNeighborInterpolation.pdf [accessed on December 3, 2014].

Hale, D. (2009c) Image-guided blended neighbor interpolation of scattered data. *79th Annual International. Meeting, SEG, Expanded Abstracts*, **28**, 1127–1131.

Jardani, A., Revil, A., Slob, E. & Söllner, W. (2010) Stochastic joint inversion of 2D seismic and seismoelectric signals in linear poroelastic materials. *Geophysics*, **75**(1), N19–N31, doi:10.1190/1.3279833

Jougnot, D., Rubino J.G., Rosas Carbajal M., Linde N. & Holliger K. (2013) Seismoelectric effects due to mesoscopic heterogeneities. *Geophysical Research Letters*, **40**, 2033–2037, doi:10.1002/grl.50472.

Lelièvre, P.G. & Farquharson, C.G. (2013) Gradient and smoothness regularization operators for geophysical inversion on unstructured meshes. *Geophysical Journal International*, **178**, 623–637, doi: 10.1111/j.1365-246X.2009.04188.x.

Revil, A. & Cathles, L.M. (1999) Permeability of shaly sands. *Water Resources Research*, **35**(3), 651–662, 1999.

Revil, A. & Mahardika, H. (2013) Coupled hydromechanical and electromagnetic disturbances in unsaturated porous materials. *Water Resources Research*, **49** (2), 744–766, doi:10.1002/wrcr.20092.

Saunders, J.H., Jackson, M.D. & Pain, C.C. (2008) Fluid flow monitoring in oilfields using downhole measurements of electrokinetic potential. *Geophysics*, **73**, E165–E180.

Sava, P. & Revil, A. (2012) Virtual electrode current injection using seismic focusing and seismoelectric conversion. *Geophysical Journal International*, **191**, 3, 1205–1209, doi: 10.1111/j.1365-246X.2012.05700.x.

Zhou, J., Revil, A., Karaoulis, M., Hale, D., Doetsch, J. & Cuttler, S. (2014) Image-guided inversion of electrical resistivity data. *Geophysical Journal International*, **197**, 292–309, doi:10.1093/gji/ggu001

CHAPTER 7

Application to the vadose zone

In this chapter, we discuss how seismoelectric data can be acquired to characterize the vadose zone, that is, the unsaturated, usually shallow, portion of the ground. Then, we present a case study in the United Kingdom, performed by B. Kulessa and L. J. West, in which seismoelectric signals are correlated with the water content of the vadose zone. Finally, we apply the numerical model discussed in Chapter 4 to this case study with the objective of reproducing the field data using the seismoelectric theory we developed earlier in this book that is valid in unsaturated conditions. We show that we can reproduce the field data fairly well, and from that, we can potentially infer the water content of the vadose zone using the seismoelectric method.

7.1 Data acquisition

In this section, we describe seismoelectric data acquisition systems that have been used to characterize vadose zone phenomena where seismoelectric effects occur relatively close to the network of electrodes used to record the seismoelectric signals. Dupuis et al. (2007) published an interesting case study showing clear seismoelectric signals in the vadose zone of a sandy aquifer. Their electrical data acquisition system consisted of 26 electrodes with a takeoff separation of 4 m. These electrodes were arranged to form 24 end-to-end dipoles except for a 4 m shot gap at the center (Figure 7.1). The seismic acquisition system itself consisted of three 12-channel 24-bit

Geometrics Geode system. Four shot points spaced 1 m apart were placed in the shot gap offset laterally, by about 2 m from the acquisition line. The seismic source was a 40 kg accelerated weightdrop source. Three to five impacts were recorded at each point, and the data were staked. The total length of the profile was 300 m, and the acquisition was done using the methodology shown in Figure 7.1.

The electrodes used for the investigation of Dupuis et al. (2007) were stainless steel electrodes. A better choice would be to use nonpolarizable electrodes (see Figure 7.2), which are known to be much more stable over time (Perrier et al., 1997, 1998; Petiau, 2000). The Petiau nonpolarizable electrode is a two-compartment electrode that is composed of a metal wire that is connected to an ionic reservoir containing a dissociated, saturated salt solution of the same metal cation (electrolyte). The first compartment in the electrode is a fluid reservoir containing the electrolyte, and the second compartment is filled with a salt solution (incorporating the electrolyte) saturated clay mineral (such as kaolinite). The last and critical part of the electrode, the part that makes contact with the Earth, is the diffusion membrane, and it is usually made of wood or ceramic. The combination of salt and clay in the second compartment maintains ionic continuity throughout the electrode while slowing down fluid loss from diffusion. The two compartments of the electrode are connected to each other through a small diameter hole that is able to supply fluid to the second compartment as needed to replace lost fluid.

The Seismoelectric Method: Theory and Applications, First Edition. André Revil, Abderrahim Jardani, Paul Sava and Allan Haas.
© 2015 John Wiley & Sons, Ltd. Published 2015 by John Wiley & Sons, Ltd.

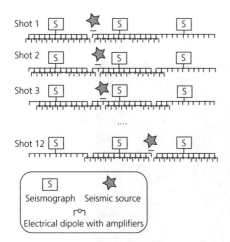

Figure 7.1 Seismoelectric array geometry and shooting progression (modified from Dupuis et al., 2007, reproduced with the authorization from the SEG and the authors).

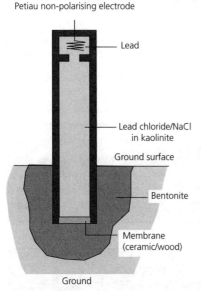

Figure 7.2 Petiau nonpolarizing electrodes (see Petiau, 2000). These electrodes are composed of a metal in contact with its own salt (e.g., Cu in $CuSO_4$, Ag in AgCl, or Pb in $PbCl_2$). This salt is usually in a saturated solution (can be solid too, as in one form of the Ag–AgCl system) and typically fully saturates a clay material (usually kaolinite) contained inside the electrode to slow down the fluid diffusion processes. The contact between the electrode and the ground is made though a membrane with micropores (some woods or ceramics can be used for this purpose). A bentonite mud cake can be placed between the electrode and the ground to decrease the contact resistance of the electrode and keep the moisture content constant during the time of the survey. Note that these electrodes have a thermal drift of 0.2 mV per °C.

Nonpolarizability is maintained through the diffusion of small amounts of the electrolyte through the membrane into the soil immediately outside of the electrode.

The electrical measurements made during seismoelectric data acquisition are made with a multichannel digital voltmeter that makes a series of time-synchronized measurements. To make accurate voltage measurements, the input impedance of the voltmeter has to be at least ten times higher than the impedance of the ground between the two electrodes in order to avoid errors caused by input bias currents in the voltmeter. A voltmeter with an internal impedance of 100 MOhm would be high enough for most applications. However, working over very resistive materials (>10,000 Ohm m, e.g., ice, permafrost, crystalline rocks) may imply the use of a voltmeter with much higher internal impedance (some voltmeters can be made with an internal impedance of 10^{12}–10^{14} Ohm). The voltmeter, like all instrumentation in geophysics, has to be calibrated regularly against known resistances to check its accuracy over a broad range of resistance values. This accuracy needs to be known if we want to use the amplitude of the recorded voltages, especially in some applications like the one described in the succeeding text designed to connect the seismoelectric signals to the water content of the vadose zone. Note, however, that in a number of the applications we described in the previous chapters, we were more interested in the timing of the seismic excitation relative to the resulting seismoelectric disturbances and known electrode positions. Importantly, amplifiers can be used to boost the voltages that are recorded. The amplified and recorded voltages will require correction of the amplitude of the measured data with respect to the magnitude of the source of the phenomena causing the signal.

An example of a seismic and seismoelectric supergathers obtained by Dupuis et al. (2007) is shown in Figure 7.3. We can clearly see the seismoelectric conversions on this seismoelectric supergather. The final result of their survey is shown in Figure 7.4 and is compared to a georadar (GPR) cross section. The two surveys clearly show the position of the water table and possibly some pockets of high water content in the vadose zone. This confirms that the seismoelectric method can be used for shallow surveys to potentially characterize the vadose zone in terms of water content distribution. If the measurements are repeated over time, it is possible to monitor the progress of unsaturated flow in the vadose

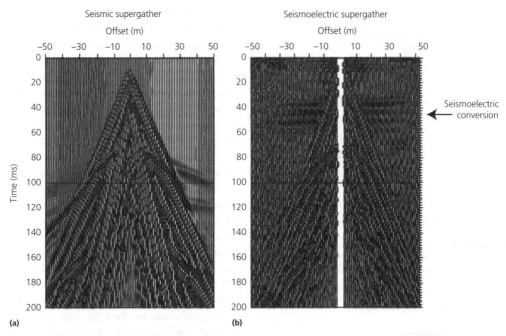

Figure 7.3 Seismic and seismoelectric supergathers (position 128 along the line, modified from Dupuis et al., 2007, reproduced with the authorization from the SEG and the authors). **a)** Seismic supergather offset records. **b)** Seismoelectric supergather offset records. (*See insert for color representation of the figure.*)

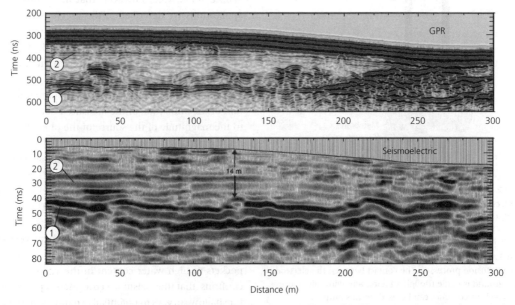

Figure 7.4 Comparison between a GPR profile and a seismoelectric 300 m profile (modified from Dupuis et al., 2007, reproduced with the authorization from the SEG and the authors). Variable elevation static delays were applied to the data to account for the effect of elevation. Anomaly 1 corresponds to the water table, while anomaly 2 likely corresponds to a unit that might have a high water content. (*See insert for color representation of the figure.*)

zone. In the next section, we will show how seismo-electric data from the vadose zone can be used to quantitatively determine its water content.

7.2 Case study: Sherwood sandstone

7.2.1 Experimental results

This case study describes work that was performed in a sand quarry near the village of Great Heck, Yorkshire (United Kingdom), by B. Kulessa and L. J. West. A summary of previous works at this site can be found in West and Truss (2006). This quarry was chosen because the glacial till cover was absent, so that the top of the 14.5 m thick vadose zone of the Sherwood Sandstone was directly accessible. Furthermore, spatially highly resolved (0.25 m intervals), volumetric water content-depth profiles down to 11.25 m were available from a borehole using time-domain reflectometry (TDR). Detailed lithological profiles derived from sandstone core and gamma-ray borehole measurements (West & Truss, 2006, their Figure 5) were also available. The vadose zone at this site consists mostly of horizontally stratified deposits of fine- to coarse-grained sandstones typically 0.3–1.5 m thick. Volumetric water content, θ, ranges from ~0.15 to 0.33 (Figure 7.5). The existence of these water content variations is facilitated by substantial differences in the grain sizes of the sandstone units and, as such, results in variations of the capillary entry pressure (West & Truss, 2006) as a function of sandstone unit composition. The volumetric water content shown

in the left panel of Figure 7.5 is the arithmetic mean of nine repeated borehole logs acquired over a 10-month period from December 2002 to October 2003.

In this case study, we report on seismoelectric soundings conducted in July 2003 within 25 m of each other at four locations centered on the axis spanned by two boreholes, purpose-drilled 15 m apart to 30 m depth (Figure 7.5). At each of the four seismoelectric sounding locations, ten sets of data were recorded. The commercially available Groundflow EKS GF2500™ was used for the seismoelectric data acquisition. The seismic source consisted of sledgehammer blows on an aluminum plate. This source demonstrated good shot-to-shot repeatability. Two 0.5 m long copper rods, spaced 2 m apart, at 0.5 and 2.5 m distance from and in-line with the shot point, served as the two receiving electrodes. There are therefore eight electrodes in total (4 dipoles, 2 on each side of the shot point). This system was also used by Kulessa et al. (2006). Since seismoelectric conversions are focused within the first Fresnel zone of the downgoing seismic wave (see Chapter 4), the survey setup used in this investigation is favorable to the detection of the seismoelectric conversions.

The seismoelectric experiment was complemented by two types of seismic surveys. Split-spread surface seismic data were recorded simultaneously with the seismoelectric data, using 24 (40 Hz) vertical geophones connected to a Geometrics Geode seismograph. Geophone spacing was 1 m, and the seismic array was centered on the respective seismoelectric shot point, aligned parallel to the axis spanned by the seismoelectric array at a distance

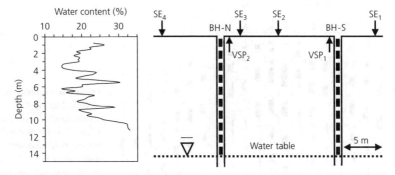

Figure 7.5 Mean volumetric water content-depth profile in the vadose zone (left panel; from West & Truss, 2006) and present field acquisition geometries (right panel). The four seismoelectric sounding locations (SE_1–SE_4) were complemented by VSP surveys using borehole geophone spacing of 0.5 m down to the water table at 14.5 m. Hammer-and-plate shot points at VSP_1 (for geophone deployment in the northern borehole (BH-N)) and VSP_2 (for geophone deployment in the southern borehole (BH-S)) were used as the seismic source. Only selected geophone positions are indicated for figure clarity.

of ~1 m. Test surveys confirmed that cross-talk between the electrodes and the geophones and the cables was not of concern. This innovative approach allowed us to simultaneously record seismoelectric and seismic data sets in response to the same hammer blow. Vertical seismic profiles (VSP) of the vadose zone were also acquired using a three-component Geostuff BHG-2 borehole geophone with a hammer-and-plate source. The geophone sonde was lowered at 0.5 m intervals down the northern borehole (labeled BH-N in Figure 7.5), and the plate was hit at the surface at a distance of 14.5 m immediately adjacent to the southern borehole (shot point labeled VSP_1 in Figure 7.5), and vice versa for geophone deployment in the southern borehole (BH-S) and shots adjacent to the northern borehole (VSP_2).

7.2.2 Results

The seismoelectric traces, recorded with the two dipoles of electrodes located on either side of the shot point, typically had maximum absolute amplitudes of several tens of millivolts (Figure 7.6a). Strong, longer-period components commonly dominated the first 15–20 ms of the measured signals, onto which many shorter-period, inphase (e.g., between ~2 and 3, 5 and 6, and 10 and 11 ms in Figure 7.6a), and out-of-phase (e.g., between 11 and 20 ms in Figure 7.6a) components were superimposed. After ~35 ms a strong 50 Hz, powerline-related noise dominated the signals until the end of the recording period (Figure 7.6a). The 50 Hz noise component appears to be inphase at the two dipoles on either side of the shot point. The associated split-spread seismograms were

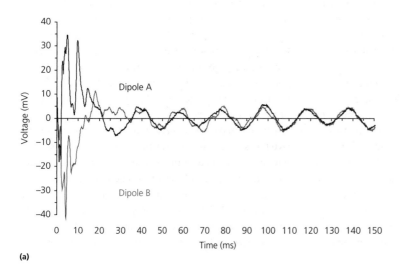

(a)

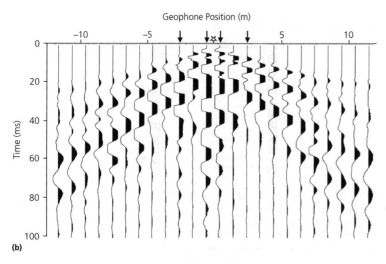

(b)

Figure 7.6 Recorded seismoelectric and seismic data. **a)** Raw seismoelectric sounding data recorded at SE_3 (see Figure 7.4) for electrodes A and B, respectively, spanned by two electrodes at distances of 0.5 m and 2.5 away from, and on opposite sides of, the shot point. **b)** Corresponding split-spread seismogram using 24 40 Hz vertical geophones spaced 1 m. The locations of the seismoelectric electrodes are indicated by arrows, and the shot point is indicated by a star.

dominated by direct arrivals and surface waves, while reflected or refracted arrivals are not readily apparent (Figure 7.6b). The first four to six seismic channels were typically saturated with surface-wave energy, confirming good shot coupling with the ground surface. While good energy transmission is favorable for the generation of seismoelectric conversions in the subsurface, unfortunately, it is also likely that the coseismic electrical field is very strong, since the first six geophones are colocated with the seismoelectric dipoles (Figure 7.5). Therefore, we expect possible seismoelectric conversions to be superimposed on top of strong coseismic energy and powerline noise. We processed the data to minimize the noise influence (see the following section).

Assuming straight-ray seismic propagation and horizontal stratification of sandstone units, seismic velocities are readily determined from first-arrival VSP data and averaged for the two boreholes to estimate a seismic velocity-depth profile for the vadose zone. Velocities have a mean of ~811 m s^{-1} and decrease from maximum values of >850 m s^{-1} at around 1.5–2 m depth to relatively low values of <800 m s^{-1} at around 8–9 m depth, whence increasing again to above-average values closer to the water table (located at 14.5 m; Figure 7.7). Many smaller fluctuations with meter-scale spatial amplitudes are superimposed on this general velocity trend. The absolute value of the sharp near-surface velocity

gradient, apparent as >100 m s^{-1} within the first 2 m of the ground surface (Figure 7.7), is debatable owing to the straight-ray assumption. It is, nonetheless, consistent with the expectation of poor very near-surface cementation of sandstone deposits owing to previous quarrying operations (see Section 7.2).

7.2.3 Interpretation
7.2.3.1 Seismoelectric signal preprocessing
Our initial concern is the elimination of powerline and other environmental noise from the recorded seismoelectric signals. Seismoelectric conversions and coseismic arrivals received by two dipoles on opposite sides of a shot point (s_1, s_2) are expected to be 180° out of phase (Butler et al., 1996), while powerline noise typically appears inphase (Beamish, 1999). The inphase powerline noise is, indeed, readily apparent in the raw seismoelectric data (Figure 7.6a). Kulessa et al. (2006) proposed a simple gradient-based scheme that exploits the contrasting phase relationships of seismoelectric conversions and coseismic arrivals as compared with powerline and other distant environmental noise. Initially, the phase of one of the raw signals is reversed so that seismoelectric conversions and coseismic arrivals are now inphase, while noise components are out of phase. Next, the average gradient, s_{grad}, is calculated from the two recorded signals $\dot{s}_1$ and $\dot{s}_2$:

$$s_{grad} = \frac{1}{2}(\dot{s}_1 + \dot{s}_2) \quad \text{if } \text{sgn } \dot{s}_1 = \text{sgn } \dot{s}_2$$
$$s_{grad} = 0 \quad \text{if } \text{sgn } \dot{s}_1 \neq \text{sgn } \dot{s}_2 \tag{7.1}$$

Although Equation (7.1) is useful in removing noise from seismoelectric records, it is desirable to recreate full waveform noise-minimized seismoelectric signals for the purposes of modeling and petrophysical analyses. In this study, we have therefore reconstructed the average components of the noise-minimized seismoelectric signals (s_{ek}) received by the two dipoles by reintegrating s_{grad} using

$$s_{ek}(t) = \int_{0}^{t_{max}} s_{grad} dt \tag{7.2}$$

and considering that $s_{ek}(0) = 0$ is a boundary condition. We have sequentially applied Equations (7.1) and (7.2) to all individual seismoelectric shots at the four survey locations (Figure 7.5). To minimize random noise, we

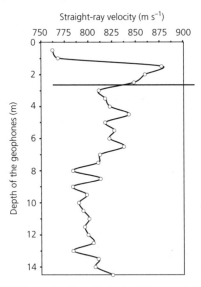

Figure 7.7 Velocity-depth profile derived from vertical seismic profiling (VSP).

followed common seismic practice and stacked the resulting seismoelectric signals, initially separately for the four locations, and then for all seismoelectric signals from all locations. This step is justified because spatial changes in surface topography of, the horizontal stratification of the sandstone units within the vadose zone, and the depth of the water table were all negligible within the study area. The resulting noise-minimized and stacked seismoelectric signal (hereafter "preprocessed seismoelectric signal") (Figure 7.8) is characterized by only a minor energy return from the water table, which we expect at ~17.5 ms one-way seismic travel time (marked by an arrow in Figure 7.8). This expectation is based on the travel time-depth conversion using the VSP data (Figure 7.7). Indeed, while there was little seismoelectric energy received from the saturated zone (>17.5 ms in Figure 7.8), we observe that strong seismoelectric energy returns are received from the vadose zone. These returns are characterized by shorter-term fluctuations, typically several milliseconds long, superimposed on a longer-term positive background trend (<17.5 ms in Figure 7.8).

7.2.3.2 Seismoelectric–water content relationship
Along with the transient pressure disturbances forced by a downgoing seismic wave, seismoelectric signals are sensitive to energy loss due to spherical spreading. We have therefore spherically corrected the preprocessed seismoelectric signal and smoothed it with a 0.25 m running mean that corresponds to the depth interval over

which the available volumetric water contents are integrated (Figure 7.5). We also removed the data from the first 1.25 m below the ground surface since they were unreliable given the assumption of straight seismic-ray travel (see preceding text). Comparison of the preprocessed seismoelectric signal (Figure 7.9a) with the volumetric water content-depth profile (Figure 7.9b) readily reveals an inverse relationship. For instance, the seismoelectric amplitudes increase over the first ~1–4 m as volumetric water content decreases, followed by a series of predominantly inverse major fluctuations from ~4 to 9 m, and, respectively, the final decrease or increase in seismoelectric amplitudes and volumetric water contents between ~9 and 11 m (Figure 7.9).

To quantify these observations, we picked the principal peak-to-peak amplitudes of the preprocessed seismoelectric signal (Figure 7.8) and the water content profile (Figure 7.5). A cross-plot of these changes (Figure 7.10) confirms a statistically significant ($R^2 = 0.81$) linear relationship between these amplitudes. Note that the least-squares regression line was forced through the origin assuming that seismoelectric energy cannot be generated in the absence of water content changes. Allowing for some natural and statistical variability, we can therefore infer that seismoelectric energy returns are produced in close vertical succession within the vadose zone. We emphasize that this relationship is robust within the combined study areas of West and Truss (2006) and that reported here, which were colocated within some 50 m

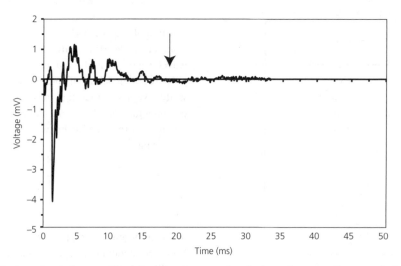

Figure 7.8 Processed seismoelectric data revealing seismoelectric conversions in the vadose zone. The arrow indicates the position of the groundwater table at ~17.5 ms.

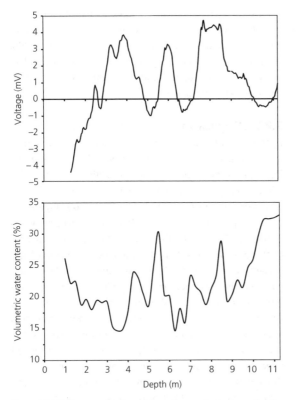

Figure 7.9 Voltage and volumetric water content versus depth. **a)** Seismoelectric conversions from depths of 1.25–11.25 m following the application of the model described in the main text with spherical correction applied and smoothed with a 0.25 m running mean. **b)** Volumetric water content-depth profile from TDR measurements.

of each other. The relationship is robust because volumetric water contents were averaged over several boreholes and several measurement periods (West and Truss, 2006), while the preprocessed seismoelectric signals were averaged over some ten repeat shots at four different sites. We can therefore conclude that judiciously processed seismoelectric sounding data can be used to determine the volumetric water content in the vadose zone of unconfined aquifers. We will confirm in the following text, using a numerical test, this feature. We will show that the seismoelectric signals are dominated by electrokinetically induced seismoelectric conversions at times later than ~4 ms, after which coseismic energy returns become negligible.

7.2.4 Empirical modeling

Accepting such dominance of seismoelectric conversions pending confirmation in Section 7.3, we develop an empirical petrophysical framework. We expect that we can explain the statistically significant and robust inferences from the framework we develop here and thus can facilitate the inversion of seismoelectric conversions for subsurface hydrological properties. We can model a seismoelectric conversion as the response to a transient streaming potential and can connect the seismically induced pressure disturbances propagating through the vadose zone with the unsaturated electrokinetic coupling coefficient (e.g., Guichet et al., 2003; Revil & Cerepi, 2004; Kulessa et al., 2012). The streaming potential generated by differential fluid pressure imposed on a saturated medium is given by (see Chapter 1)

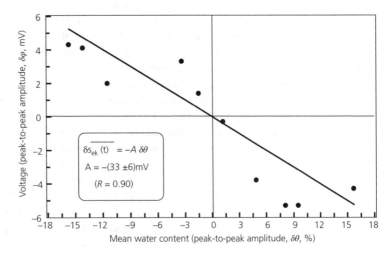

Figure 7.10 Statistical relationship between principal spatial amplitudes (picked from Figure 7.9) of volumetric water content against those of seismoelectric conversions.

$$\nabla \varphi(t) = C \nabla p(t), \qquad (7.3)$$

where C is the streaming potential coupling coefficient described in detail in Chapter 3. This assumes that induced flow is laminar, that all pore radii are much larger than the thickness of electrical double layer at the fluid–solid interface, and that surface conductivity is negligible compared to that of the bulk fluid. We use the simplified relationship discussed in Chapter 3:

$$C(s_W) = C(s_W = 1)s_W, \qquad (7.4)$$

$$C(s_W = 1) = -\frac{\hat{Q}_V^0 k}{\eta_W \sigma}. \qquad (7.5)$$

In these equations, $\hat{Q}_V^0$ (expressed in C m^{-3}) denotes the effective charge density of the diffuse layer per unit pore volume that is dragged by the flow of the pore water, k the permeability (in m^2), σ the electrical conductivity of the porous material (in S m^{-1}), and η_w the dynamic viscosity of the pore water (in Pa s). Combining Equations (7.3) and (7.4), and by replacing s_w with the volumetric water content, θ, where $s_w = \theta/\phi$ (ϕ denotes porosity), we have

$$\nabla \psi(t) = C(s_W) \nabla p(t) = -\frac{\hat{Q}_V^0 k}{\eta_W \sigma \phi} p_0(t) \frac{1}{r} \theta H(t), \qquad (7.6)$$

assuming the equation

$$\nabla p(t) = p_0(t) \frac{1}{r} H(t), \qquad (7.7)$$

corresponds to the differential fluid pressure caused by the propagating seismic wave, where $p_0(t)$ is the pressure pulse generated by the hammer impact at the ground surface, r is radial distance from the shot point, and $H(t)$ denotes the Heaviside (step) function. We can substitute $s_{ek}(t) = \nabla \varphi(t) r$ as the spherically corrected seismoelectric conversion and assign $\overline{s_{ek}(t)}$ and $\overline{p_0(t)}$, respectively, as the stacked seismoelectric conversion signal and stacked pressure pulse. If $\nabla \overline{s_{ek}}(t)$ and $\nabla \theta$ are the peak-to-peak positive or negative change in seismoelectric conversion and volumetric water content, respectively, as considered in Figure 7.10, we obtain

$$\delta \overline{s_{ek}(t)} = -A \, \delta \theta, \qquad (7.8)$$

$$A \equiv \frac{\hat{Q}_V^0 k}{\eta_W \sigma \phi} \overline{p_0(t)}. \qquad (7.9)$$

Assuming that vertical changes in parameter A are negligible compared to changes in the volumetric water content, Equation (7.8) describes a linear relationship, with slope A, between the magnitude of the seismoelectric conversions, $\delta \overline{s_{ek}}(t)$, and volumetric water content changes, $\delta \theta$, with depth.

7.2.5 Discussion

In the present case, we have no reason to expect significant spatial changes to occur in any of the fluid or interfacial properties. The range of porosities (0.33 to 0.36; West & Truss, 2006) is smaller than the range of volumetric water contents (~0.15 to 0.33; see Figure 7.9b) and, as such, supports a linear relationship according to Equation (7.7). Notwithstanding, the spatial variation of the porosity, albeit small, could explain some of the variability in Figure 7.10. The inverse relationship between the seismoelectric conversions and water content changes (Figure 7.10) must imply that $\hat{Q}_V^0$ is positive (negative zeta potential, ζ) because all other parameters affecting A in Equation (7.7) are positive. This is not surprising, however, since the zeta potential is indeed negative in most naturally occurring environments (see Revil et al., 1999a, b and Chapter 1). In the present case, A has a value of ~0.33 (Figure 7.10).

Haines and Pride (2006) demonstrated that layers as thin as 1/20th of the dominant seismic wavelength can be responsible for observable seismoelectric conversions. In this case study, the mean seismic velocity in the vadose zone was 811 m s^{-1}, and the dominant seismic frequency was 61 Hz, implying that layers as thin as ~0.6 m should generate detectable seismoelectric conversions. Vadose zone volumetric water content typically varies on this scale (Figures 7.5 and 7.9b), effectively representing vertically successive thin layers, each producing a strong electrical dipole. The inference of a temporally continuous seismoelectric conversion (Figure 7.9a), produced at the surface by the superimposed effect of these dipoles, is therefore consistent with Haines and Pride (2006).

Since Figure 7.10 effectively represents a calibration of seismoelectric conversions for volumetric water content, Equation (7.7) could thus be used to either complete the volumetric water content-depth profile between 11.25 m depth and the water table at 14.5 m or to estimate volumetric water content-depth profiles at some distance from this study site where no borehole control is available. Alternatively, if the pressure pulse ($P_0(t)$) was measured in field surveys and the zeta potential (ζ) or $\hat{Q}_V$ could be estimated (e.g., from core samples), depth profiles of either porosity or selected fluid properties could be derived using Equation (7.7). This could be of interest where porosity estimates are not available or, for example, fluid properties change significantly at depth such as at contaminated land sites.

The longer-period seismoelectric conversion components (Figures 7.6a and 7.8) are most likely manifestations of the direct field, as documented previously by Kulessa et al. (2006). These components are most likely caused by transient streaming potentials generated by ground compression owing to the impact of the hammer on a poorly cemented top sandstone layer. This could generate transient water movement toward the two dipoles and act effectively as the superposition of a series of electrical dipoles to produce the direct field.

7.3 Numerical modeling

In this section, we present a numerical experiment that simulates the field study discussed earlier. The goal of this exercise is not to reproduce exactly the field seismoelectric data, but to place the inferred empirical model, relating seismoelectric conversions to volumetric water contents, on a solid physical grounding. We begin with a summary of the relevant theory developed in Revil and Mahardika (2013) by first describing the propagation of seismic waves in poroelastic media before providing the equations that govern electrokinetic processes in unsaturated conditions, accounting for variability of fluid saturation.

7.3.1 Theory

Assuming an $e^{-j\omega t}$ time dependence for the mechanical disturbances (j being the pure imaginary number, $j^2 = -1$, ω is the angular frequency, and t is time) while using the displacement of the solid phase and the pore pressure as unknowns instead of the solid and fluid phase displacement vectors, the poroelastodynamic equations of motion, in the frequency domain, are given by (Chapter 2)

$$-\omega^2 \rho_\omega^S \mathbf{u} + \theta_\omega \nabla p = \nabla \cdot \hat{\mathbf{T}}, \qquad (7.10)$$

$$\hat{\mathbf{T}} = (\lambda \nabla \cdot \mathbf{u})\mathbf{I} + G(\nabla \mathbf{u} + \nabla \mathbf{u}^{\mathrm{T}}), \qquad (7.11)$$

$$\frac{p}{M} + \nabla \cdot [k_\omega(\nabla p - \omega^2 \rho_f \mathbf{u})] = -\alpha \nabla \cdot \mathbf{u}, \qquad (7.12)$$

where the superscript "T" denotes "Transpose." These three equations correspond to (i) a macroscopic momentum conservation equation for the solid phase, (ii) a constitutive equation for the effective stress tensor, $\hat{\mathbf{T}}$, and (iii) a momentum conservation equation for the pore fluid. In these equations, $\mathbf{I}$ is the identity matrix, $\mathbf{u}$ represents the averaged solid displacement vector (in m), $\mathbf{w}$ denotes the averaged fluid–solid relative displacement vector (in m), M symbolizes one of the Biot coefficients (described in the following text), and p is the average fluid pressure in the pore water phase. The relationship between the effective stress tensor, $\hat{\mathbf{T}}$, and total stress tensor, $\mathbf{T}$ (in Pa), is

$$\mathbf{T} = \hat{\mathbf{T}} - \alpha p \mathbf{I}. \qquad (7.13)$$

The Biot coefficients M and α are given by (Revil & Mahardika, 2013; see Chapter 3)

$$\frac{1}{M} = \theta\left(\frac{1 + \Delta}{K_f}\right), \qquad (7.14)$$

$$\alpha = 1 - K/K_S, \qquad (7.15)$$

where $\theta = s_w \phi$ represents the volumetric water content (dimensionless, equal to the porosity at saturation), $\Delta = (K_f/\phi K_S^2)[(1-\phi)K_S - K_{fr}]$, and K_s and K_f are the bulk moduli of the solid and fluid phases, respectively. Two mechanical properties enter into these equations: λ is the drained Lamé coefficient (in Pa) defined by $\lambda = K - (2/3)G$ where $K = K_{fr}$ and $G = G_{fr}$ symbolize the bulk modulus and the shear modulus of the solid frame (index "fr," solid skeleton without any fluid inside), respectively. The density, ρ_ω^S, in Equation (7.10) represents an apparent mass density of the solid phase. The coupling term in Equation (7.1), θ_ω, denotes a hydromechanical coupling coefficient, k_ω, and (k_ω (in m^6 s^2 kg^{-1}) in Equation (7.12) is a permeability-related constant. These three coefficients are given by (see Jardani et al., 2010, and Chapter 2)

$$k_\omega = \frac{1}{\omega^2 \tilde{\rho}_f + i\omega b}, \qquad (7.16)$$

$$\rho_\omega^S = \rho - \omega^2 \rho_f^2 k_\omega, \qquad (7.17)$$

$$\theta_\omega = \alpha - \omega^2 \rho_f k_\omega, \qquad (7.18)$$

$$\tilde{\rho}_f = F\rho_f, \qquad (7.19)$$

where $b = \eta_f/k_0$ and k_0 (in m^2) is the permeability for quasistatic flow. In the following,

$$\rho = (1-\phi)\rho_S + \phi\rho_f, \tag{7.20}$$

(in kg m^{-3}) denotes the bulk density of material, ϕ represent the connected porosity, ρ_S and ρ_f are the mass densities of the solid and pore fluid phases, respectively.

Here, we provide the equations describing the electrokinetic theory in unsaturated conditions using the theory developed by Revil and Mahardika (2013). In the following, the electrical field is given by $\mathbf{E} = -\nabla\varphi$ in the quasistatic limit of the Maxwell equations, and σ (in S m^{-1}) represents the bulk electrical conductivity of the porous medium. The electric potential ψ (in V) is obtained by solving the following elliptic equations (see details in Jardani et al., 2010):

$$\nabla\cdot(\sigma\nabla\psi) = \mathfrak{I}, \tag{7.21}$$

where the volumetric source current density is given by $\mathfrak{I} = \nabla\cdot\mathbf{J}_S$. As discussed in Chapter 2, the dynamic streaming current density, $\mathbf{J}_S$ (in A m^{-2}), is related to the properties of the seismic waves by

$$\mathbf{J}_S = \frac{\hat{Q}_v^0}{s_w}\dot{\mathbf{w}} = -i\omega\frac{\hat{Q}_v^0}{s_w}k_\omega\left(\nabla p - \omega^2\rho_f\mathbf{u}\right). \tag{7.22}$$

In Equation (7.22), $\hat{Q}_v^0$ is the effective charge per unit pore volume that is dragged by the flow of the pore water relatively to the solid phase, and $\dot{\mathbf{w}}$ symbolizes the Darcy velocity (see Chapters 1 and 2). The charge density at saturation, $\hat{Q}_v^0$, is related to the low-frequency permeability at saturation, k_0, by (Figure 1.11)

$$\log_{10}\hat{Q}_v^0 = -9.2349 - 0.8219\log_{10}k_0. \tag{7.23}$$

This relationship avoids the introduction of additional parameters other than those used to describe the seismic wave propagation part of the overall formulation.

The last effect to account for is the variability of fluid saturation. The bulk electrical conductivity, σ, is determined by (Revil, 2013a, b)

$$\sigma = \frac{1}{F}\sigma_w s_w^n + \sigma_s s_w^{n-1} \tag{7.24}$$

where s_w is the water saturation, n (≥ 1, dimensionless) symbolizes the second Archie's exponent (or the

saturation exponent), σ_w denotes the electrical conductivity of the pore water (in S m^{-1}), and σ_s stands for the surface conductivity (in S m^{-1}) associated with conduction in the electrical double layer coating the surface of the grains. The dimensionless formation factor, F, is related to the connected porosity, ϕ (dimensionless), by Archie's law, $F = \phi^{-m}$ (Archie, 1942), where m (≥ 1, dimensionless) is usually called the cementation exponent and more precisely the porosity exponent or first Archie's exponent. In the following example, we use $m = n = 1.8$, and the surface conductivity is reported in Table 7.1.

The effect of the saturation on the other properties is given by (see Chapter 3)

$$\rho_f = (1-s_w)\rho_g + s_w\rho_w, \tag{7.25}$$

$$\frac{1}{K_f} = \frac{1-s_w}{K_g} + \frac{s_w}{K_w}, \tag{7.26}$$

$$\eta_f = \eta_g\left(\frac{\eta_w}{\eta_g}\right)^{s_w}, \tag{7.27}$$

$$\tilde{\rho}_f = \frac{1}{F}s_w^n\rho_f, \tag{7.28}$$

$$k_0(s_w) = k_r(s_w)k_S = s_w^{n+2}k_S. \tag{7.29}$$

The model parameters used for the simulation are given in Table 7.1.

Table 7.1 Material properties used in the numerical modeling.

Parameter	Symbol	Value	Units
Matrix density grains	ρ_S	2650	kg m^{-3}
Water density	ρ_w	1000	kg m^{-3}
Matrix bulk modulus	K_s	36.5	GPa
Frame bulk modulus	K_{fr}	0.65	GPa
Shear modulus	G	0.09	GPa
Water bulk modulus	K_w	0.20	GPa
Water viscosity	η_w	1×10^{-3}	Pa s
Permeability	k_0	1×10^{-14}	Pa s
Water conductivity	σ_w	6.8×10^{-2}	S m^{-1}
Surface conductivity	σ_s	1×10^{-5}	S m^{-1}
Air density	ρ_g	1.205	kg m^{-3}
Air bulk modulus	K_g	0.140	MPa
Air viscosity	η_g	2×10^{-5}	Pa s

7.3.2 Description of the numerical experiment

We use the volumetric water content given in Figure 7.5, the VSP velocity data given in Figure 7.7, and the additional material properties given in Table 7.1 to construct the water saturation, electrical resistivity, and seismic velocity distributions shown in Figure 7.11. The area of interest corresponds to a 25 m × 25 m 2D subsurface model that matches the seismic split-spread acquisition (Figure 7.12b) and incorporates the seismoelectric and borehole VSP geometry. We place a seismic source on the top surface at position S_0 (0 m, 0 m). The seismic source is a Gaussian function with a magnitude of 120 N m that typifies a hammer blow on a plate (Keisswetter & Steeples, 1995), and the dominant frequency is f_c =500 Hz. Simulating the geometry of the field experiment, the numerical experiment used 24 geophones in split-spread configuration (Figure 7.6) and four electrodes making up two sets of electrical dipole receivers (Figures 7.5 and 7.6a). The separation between the geophones is 1 m, and the two electrodes for each dipole receiver are spaced 2 m apart at 0.5 m and 2.5 m distances from and in-line with the shot point.

We solve the poroelastodynamic wave equations and the Poisson equation for the electrical potential in the frequency domain using COMSOL Multiphysics 4.2a. We compute the poroelastic and electric property distributions given the porosity, fluid permeability, and saturation profiles, before solving for the displacement of the solid phase and the pore fluid pressure. Then, we compute the electrical potential by solving the Poisson equation coupled to the poroelastodynamic problem for a frequency range of 10–1000 Hz that is characteristic of seismic and seismoelectric measurements. We then compute the inverse fast Fourier transform (FFT^{-1}) to get the time series of the seismic displacements, u_x and u_z, and the time series of the electrical potential response (see Jardani et al., 2010). The meshing option for the whole computation is 0.25 m × 0.25 m grid squares that are smaller than smallest wavelength of the seismic wave, which corresponds to the smallest mesh for which the solution of the partial differential equation is mesh independent. At the four external boundaries of the domain, we apply a 2.5 m thick convoluted perfectly matched layer (C-PML; see Chapter 4 for further details on the implementation).

7.3.3 Model application and results

The application of this numerical model follows a four-step strategy (steps I to IV as described below):

Step I. We compute the full 24-channel seismograms and electrograms (Figure 7.12), dominated by the seismic direct wave as well as two seismic reflections (annotated in Figure 7.12a) that correlate with major changes in VSP-derived seismic velocities centered at depths of ~4 and ~7 m (Figure 7.7). The associated electrogram is characterized by corresponding coseismic returns from the direct wave and the two reflections and by a series of seismoelectric conversions with a direct field contribution (annotated in Figure 7.12b).

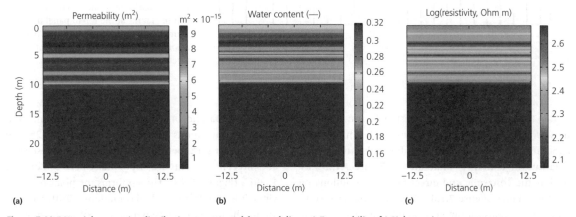

Figure 7.11 Material properties distribution constructed for modeling. **a)** Permeability. **b)** Volumetric water content. **c)** Electrical resistivity distribution. In addition, we use a constant porosity for the domain of interest (Table 7.1). (*See insert for color representation of the figure.*)

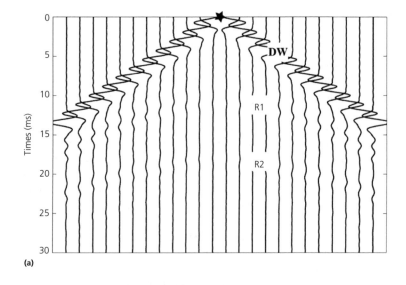

(a)

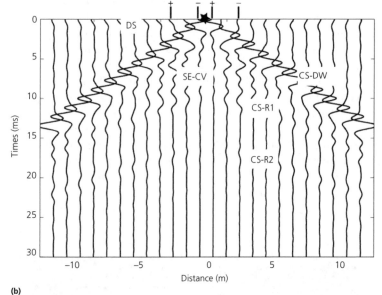

(b)

Figure 7.12 Numerical modeling results. **a)** Split-spread seismogram, showing direct wave (DW) and two reflections (R1, R2). **b)** Corresponding electrogram, showing direct field (DF), coseismic direct wave (CS-DW), two coseismic reflections (CS-R1, CS-R2), and seismoelectric conversions (SE-CV). The star marks the location of the shot point, and the marks at the top of panel b in this figure indicate the locations of the two dipoles used in the field experiment together with polarity conventions.

Step II. We simulate the total seismoelectric signals as they would be received by the two dipoles in the field experiment (Figure 7.13a; see dipole locations and polarity conventions as annotated in Figure 7.12b), as well as the coseismic and direct field contributions in the total signal (Figure 7.13b). The total modeled seismoelectric signals for dipoles 1 and 2 are characterized by strong, longer-period components that dominate the first 4 ms, followed by many shorter-period signals

(Figure 7.13a and b). Given the dipole locations and polarity conventions annotated in Figure 7.12b, we confirm that seismoelectric conversions and coseismic and direct field events are out of phase between dipoles 1 and 2 (Figures 7.13). This confirms the polarity behaviors inferred from field observations (Butler et al., 1996), the theory (Haartsen & Pride, 1997), and numerical computation (Haines & Pride, 2006), which are first explained conceptually in Butler et al. (1996).

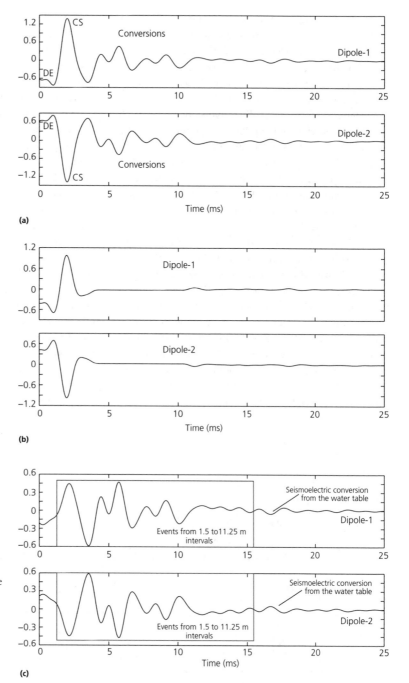

Figure 7.13 Numerical modeling results: computed signals for dipoles 1 and 2. **a)** Total seismoelectric signals, including direct field effect (DE), coseismic response (CS), and seismoelectric conversions. **b)** Coseismic signals only. **c)** Seismoelectric conversion signals, obtained by subtracting coseismic signals **b)** from total signals. The boxes outline seismoelectric conversions used in Figure 7.14.

While the total seismoelectric signals at the two dipoles are dominated by the direct field and strong coseismic energy at early times (<4 ms; annotated in Figure 7.13a), seismoelectric conversions are dominant later in time (>4 ms). Numerical simulation of the coseismic and direct field contributions only (Figure 7.13b) confirms that coseismic and direct field contributions are strong only at times less than ~4 ms, with minor contributions from coseismic energy at ~11 and ~17 ms caused by the two seismic reflections.

Step III. The modeled coseismic and direct field contributions (Figure 7.13b) are subtracted from the total modeled seismoelectric signal (Figure 7.13a), emphasizing the contribution of the seismoelectric conversions to the latter (Figure 7.13c). The conversions are revealed to have considerable amplitudes between the start of the time series and ~11 ms and to continue toward later times with smaller amplitudes (>>20 ms) (Figure 7.13c). Similar to the field data (Figure 7.10), the modeled seismoelectric conversion (Figure 7.13c) at the water table is small compared to conversions in the vadose zone above it. It is highly significant that strong seismoelectric conversions occur within the vadose zone at times (>4 ms) when the coseismic and direct field contributions (Figure 7.13a and b) are small or negligible. This behavior readily explains the statistically significant relationship between the principal amplitudes of the seismoelectric conversions and vadose zone volumetric water contents inferred from the field data (Figure 7.10), which used data from times outlined by the red box in Figure 7.13c. We may thus hypothesize that modeled seismoelectric conversions (Figure 7.13c) should likewise correlate well with vadose zone volumetric water contents.

Step IV. The modeled seismoelectric conversions (Figure 7.13c) are processed in the same way as the field data, using Equations (7.1) and (7.2) together with spherical correction and smoothing with a 0.25 m running mean (Figure 7.14a). Processed seismoelectric conversions (Figure 7.14a) are characterized by the same inverse relationship with volumetric water content (Figure 7.14b) as the field data. The computed seismoelectric conversions similarly increase in amplitude over the first ~1–4 m as volumetric water content decreases, followed by a series of predominantly inverse major fluctuations from ~4 to 9 m, and the final decrease or increase in seismoelectric conversions and volumetric water content between ~9 m and 11 m (Figure 7.14), respectively. This allows the principal peak-to-peak amplitudes of the seismoelectric conversions to be picked and plotted against those of vadose zone volumetric water contents (Figure 7.15), in exactly the same way as we did for the field data. We obtain a statistically significant relationship ($R^2 = 0.76$) between the conversion and water content amplitudes (Figure 7.15), which confirms our hypothesis. Numerical modeling is therefore able to place the empirically inferred seismoelectric conversion–volumetric water content relationship on a solid physical grounding, as anticipated in the previous chapters of this book.

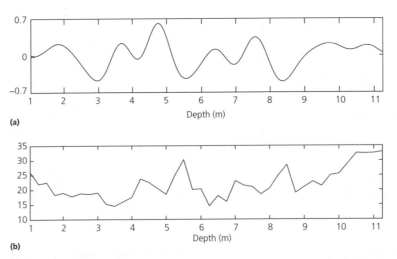

Figure 7.14 Computed seismoelectric conversions and measured volumetric water contents versus depth. **a)** Seismoelectric conversions from depths of 1.25–11.25 m using spherical correction applied and smoothed with a 0.25 m running mean. **b)** Volumetric water content-depth profile from TDR measurements as it was indicated from the measured data.

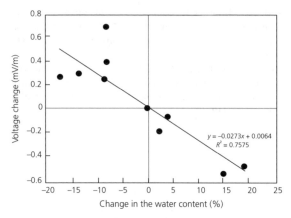

Figure 7.15 Statistical relationship between principal spatial amplitudes (picked from Figure 7.14) of volumetric water content against those of modeled seismoelectric conversions.

7.4 Conclusions

We have shown that (1) vadose zone seismoelectric conversions can be much stronger than those originating at the water table or within the saturated zone, (2) vertical changes in vadose zone volumetric water contents produce seismoelectric conversions that generate a continuous seismoelectric conversion signal at the ground surface (linear regression can calibrate seismoelectric conversion amplitudes in terms of those of volumetric water contents), and (3) seismoelectric numerical modeling from fundamental principles can reproduce the empirically inferred seismoelectric conversion–volumetric water content relationship, thus placing it on a solid physical grounding by confirming that it arises from transient streaming potentials generated by seismically induced pressure disturbances, as governed by the unsaturated electrokinetic coupling coefficient. Under favorable circumstances and pending additional works, seismoelectric sounding at the ground surface shows considerable potential to serve as a nonintrusive vadose zone water content sensor. Going one step further, this approach could be used to monitor the water content in the vadose zone and therefore nonintrusively estimate its flow properties.

References

Archie, G.E. (1942) The electrical resistivity log as an aid in determining some reservoir characteristics. *Transactions of the American Institute of Mining, Metallurgical and Petroleum Engineers*, **146**, 54–62.

Beamish, D. (1999) Characteristics of near-surface electrokinetic coupling. *Geophysical Journal International*, **137**, 231–242.

Butler, K.E., Russell, R.D., Kepic, A.W. & Maxwell, M. (1996) Measurement of the seismoelectric response from a shallow boundary. *Geophysics*, **61**, 1769–1778.

Dupuis, J.C., Butler, K.E. & Kepic, A.W. (2007) Seismoelectric imaging of the vadose zone of a sand aquifer. *Geophysics*, **72**(**6**), A81–A85, doi: 10.1190/1.2773780.

Guichet, X., Jouniaux, L. & Pozzi, J.P. (2003) Streaming potential of a sand column in partial saturation conditions. *Journal of Geophysical Research*, **108**, doi:10.1029/2001JB001517.

Haartsen, M.W. & Pride, S.R. (1997) Electroseismic waves from point sources in layered media. *Journal of Geophysical Research*, **102**(**B11**), 24745–24769.

Haines, S. & Pride, S. (2006) Seismoelectric numerical modeling on a grid. *Geophysics*, **71**, N57–65.

Jardani, A., Revil, A., Slob, E. & Sollner, W. (2010) Stochastic joint inversion of 2D seismic and seismoelectric signals in linear poroelastic materials. *Geophysics*, **75**(**1**), N19–N31, doi: 10.1190/1.3279833.

Keiswetter, D.A. & Steeples, D.W. (1995) A field investigation of source parameters for the sledgehammer. *Geophysics*, **66**, 1051–1057.

Kulessa, B., Murray, T. & Rippin, D. (2006) Active seismoelectric exploration of glaciers. *Geophysical Research Letters*, **33**, L07503, doi:10.1029/2006GL025758.

Kulessa, B., Chandler, D.C., Revil, A. & Essery, R. (2012) Theory and numerical modelling of electrical self-potential (SP) signatures of unsaturated flow in melting snow, *Water Resources Research*, **48**, W09511, doi:10.1029/2012WR012048.

Perrier, F., Petiau, G., Clerc, G., et al. (1997) A one-year systematic study of electrodes for long period measurements of the electric field in geophysical environments. *Journal of Geomagnetism and Geoelectricity*, **49**, 1677–1696.

Perrier, F., Trique, M., Lorne, B., Avouac, J.-P., Hautot, S. & Tarits, P. (1998) Electrical potential variations associated with yearly lake level variations. *Geophysical Research Letters*, **25**, 1955–1959.

Petiau, G. (2000) Second generation of lead-lead chloride electrodes for geophysical applications. *Pure and Applied Geophysics*, **157**, 357–382.

Revil, A. (2013a) Effective conductivity and permittivity of unsaturated porous materials in the frequency range 1 mHz–1GHz. *Water Resources Research*, **49**, doi:10.1029/2012WR012700.

Revil, A. (2013b) On charge accumulations in heterogeneous porous materials under the influence of an electrical field. *Geophysics*, **78**(**4**), D271–D291, doi: 10.1190/GEO2012-0503.1.

Revil, A. & Cerepi, A. (2004) Streaming potentials in two-phase flow conditions. *Geophysical Research Letters*, **31**, doi:10.1029/2004GL020140.

Revil, A. & Mahardika, H. (2013) Coupled hydromechanical and electromagnetic disturbances in unsaturated clayey materials. *Water Resources Research*, **49**, doi:10.1002/wrcr.20092.

Revil, A., Pezard, P.A. & Glover, P.W.J. (1999a) Streaming potential in porous media. 1. Theory of the zeta-potential. *Journal of Geophysical Research*, **104**(**B9**), 20,021–20,031.

Revil, A., Schwaeger, H., Cathles, L.M. & Manhardt, P. (1999b) Streaming potential in porous media. 2. Theory and application to geothermal systems. *Journal of Geophysical Research*, **104**(**B9**), 20,033–20,048.

West, L.J. & Truss, S.W. (2006) Borehole time domain reflectometry in layered sandstone. Impact of measurement technique on vadose zone process identification. *Journal of Hydrology*, **319**, 143–162.

CHAPTER 8

Conclusions and perspectives

In the previous chapters, we developed the main concepts behind the seismoelectric theory, including its petrophysical foundations, the macroscopic field equations, their implementation in a finite-element package to simulate forward problems, and some algorithms to perform inverse problems. The theory has been described in water-saturated conditions and extended to the partially saturated case and two-phase flow conditions. This allows the theory described in this book to be very general and versatile regarding geophysical applications. That said, the applications of the seismoelectric methods in environmental geosciences are still in their infancy, and very few commercially available equipment exist to democratize the method among geophysicist and non-geophysicists. We discuss below eight challenges that will need to be investigated in the next years:

1 The first challenge is related to the definitions of clear protocols and equipment sensitivity to perform field measurements. Seismoelectric conversion can be measured in the field. This has been demonstrated in a number of published case studies that were briefly outlined in this book. Because the seismoelectric method is an active method, stacking can be used to improve the signal-to-noise ratio. There is, however, the need to define clear operational protocols to record seismoelectric signals in the field and to define what type of equipment is required to get clear signals. Once such protocols are available, there is a need to further explore the seismoelectric method with various near-surface applications. We choose to work on near-surface applications first, because the seismoelectric conversions are expected to be relatively strong where the conversions are taking place in the vicinity of the receivers (<100 m). Any application in environmental geosciences that studies electrical double layer properties remotely could benefit from seismoelectric-based investigations. These include contaminant plumes, CO_2 sequestration, and the use of specific tracers to follow flow paths in shallow aquifers. This will require further developments in time-lapse seismoelectric investigations using, for instance, the active-time constrained approach recently applied to a number of geophysical methods or the fully coupled inversion approach. In the first case, the objective function to minimize contains a time-dependent regularizer. The regularization is done in time like it is classically done in space using deterministic algorithms. In the second case, we need first to be able to predict geophysical data using the modeling of the process we wish to monitor. Then we use the geophysical data to optimize some of the fundamental parameters controlling these processes.

2 The second challenge is the separation of different contributions of the seismoelectric phenomena that appear in real, field-based, electrical potential (electromagnetic) time series data. These contributions include the signals associated with the seismic source, the coseismic effects, and the seismoelectric conversions. We have shown in this book that all of these contributions can be properly modeled, and therefore,

The Seismoelectric Method: Theory and Applications, First Edition. André Revil, Abderrahim Jardani, Paul Sava and Allan Haas.
© 2015 John Wiley & Sons, Ltd. Published 2015 by John Wiley & Sons, Ltd.

full waveform inversion is possible. An alternative is to enhance the seismoelectric conversion effects through the beamforming approach discussed in Chapter 6. Additionally, various strategies can be used to separate the coseismic effects from the seismoelectric conversions using, for instance, k–f (wave number–frequency) filters.

3 The seismoelectric theory needs to be extended to anisotropic formations especially in the case of transverse isotropy. In principle, such an extension should be rather straightforward since the Biot–Frenkel theory has already been extended in anisotropic conditions. Also, actual electrical conductivity data and models (including the effect of surface conductivity and quadrature conductivity) are available for anisotropic formations. The same can be said about formation permeability. It remains unclear if the charge per unit volume used extensively in this book should be considered as a scalar or a second-order symmetric tensor as electrical conductivity and permeability are. As a first step, the seismoelectric theory based on the acoustic approximation should be easily amenable to the anisotropic case. Then it would be interested to see what valuable information the seismoelectric method could bring on the table to characterize anisotropy.

4 Because both the seismoelectric method and the induced polarization method are sensitive to the electrical double layer, there is likely some interest in formulating time-lapse joint inversion approaches for these two methods. In addition, their spatial sensitivities are different and complementary. The petrophysics of these methods is now fairly well established, so we can formulate the fully coupled geophysical inversion that involves reactive transport modeling codes.

5 The present book is focused on the seismoelectric method. We did not discuss the symmetric effect, the so-called electroseismic effect. Here again, there are very few undisclosed field trials of the electroseismic effect. The use of an electrical current provides a highly reproducible seismic source of electroosmotic nature (the drag of the excess of charge in the pore water provides a fluid pressure source). Thanks to Onsager reciprocity relations in the constitutive equations of transport, the theory discussed in the present book can easily be used to produce an electroseismic theory in unsaturated conditions or in two-phase flow conditions. The application of the electroseismic approach to near-surface applications is nearly unheard of, for example, in environmental-related applications. The use of high-precision lasers can be used to monitor extremely small fluctuations of the displacement of the ground surface due to electroseismic influences. A lot of recent research regarding the inversion of ground displacement in terms of mechanical sources (for the monitoring of active volcanoes) could be used to interpret electroseismic effects. There is a lot of work to do in this direction.

6 While the present book has been focused on the occurrence of electrical fields, the seismoelectric method can be based on the magnetic field as well. Similarly, electrical resistivity tomography is usually based on the record of the electrical field, but the magnetic field associated with the flow of the current in the subsurface can be measured as well (magnetoresistivity). We know that the magnetic field is measurable in the laboratory, but it remains uncertain if the magnetic field of electrokinetic nature can be measured in field conditions. If so, what is the value of using the magnetic field alone or in combination with the electric field? Once generated, the magnetic field is not sensitive to the heterogeneous distribution of the electrical conductivity of the subsurface. In fact, the relative permittivity of the subsurface is usually one or very close to one (except in some iron-based deposits such as magnetite and other similar minerals). This implies that it is, in principle, easier to invert the magnetic field fluctuations relative to the inversion of the electrical field fluctuations. However, the magnetic field strength of magnetic dipoles decreases with distance faster (proportional to $1/r^3$) than the electric field strength of electric dipoles (proportional to $1/r^2$). This means that there may be an effective range of use for magnetic field-based applications that is different from the electric field effective range of use. Even so, it is very likely that the joint inversion of electromagnetic fields can provide more information relative to the inversion of just the electrical field alone. There is much to explore here as well.

7 We have seen in Chapter 6 that the seismoelectric approach can be used to produce spectral seismoelectric signals. The presence of cracks should have a very specific signature on these spectra. In other words, we could foresee the use of the seismoelectric methods to detect cracks and to determine their preferential orientations and possibly their hydraulic conductances.

8 We know that disseminated ores generate a strong frequency-dependent electrical conductivity due to induced polarization effects. This implies in turn that the seismoelectric coupling coefficient should reflect this frequency dependence. In other words, a spectral seismoelectric beamforming approach (such as underlined in Chapter 6) could be used to localize ore bodies and to image their extensions with a resolution far better than the methods presently used. We could foresee huge applications of such a method for the mining industry and especially for the exploration and mapping of ore bodies and in situ (bio)leaching.

Glossary: The Seismoelectric Method

Acoustic emission: Acoustic emissions describe the generation of elastic waves in a porous material, for instance, associated with a change in the stress state (and the formation of microcracks or the reactivation of existing cracks) or with the rapid movement of the meniscus between two immiscible fluids in the pore space of a porous material.

Adaptive Metropolis algorithm (AMA): AMA describes a specific choice of a proposal distribution for Markov chain Monte Carlo (McMC) sampling methods. This distribution is crucial factor for the convergence of the stochastic algorithm. AMA uses a Gaussian proposal distribution, which is updated along the sampling process using the full information cumulated so far. Due to the adaptive nature of the process, AMA is non-Markovian in nature.

Bayesian approach: In geophysics, the Bayesian inference approach is a method of inference in which Bayes' rule is used to update the probability density estimate for a model vector as additional information (e.g., geophysical data) is acquired.

Beamforming: Geophysical method allowing to focus seismic energy at a desired location in space and at a known time. It requires some knowledge or assumptions regarding the seismic velocity model.

Capillary pressure curve: The capillary pressure denotes the difference in pressure across the interface between two immiscible fluid phases in a porous material, the nonwetting phase and the wetting phase (e.g., oil and water in a siliciclastic material). It is caused by interfacial tension between the two phases that must be overcome to initiate flow. The capillary pressure curve denotes the evolution, with respect to the saturation of the wetting phase, of the capillary pressure when the nonwetting phase replaces the wetting phase (drainage) or when the wetting phase replaces the nonwetting phase (imbibition).

Cation exchange capacity (CEC): Amount of exchangeable cations in a porous material reported per unit mass of grains (solid). Very often, the CEC is reported in equivalent charge per unit mass of solid.

Complex conductivity: Electrical conductivity of a porous material written as a complex number to account for the phase lag between the current and the electrical field. The inphase (real) component characterizes the ability of the porous material to conduct electrical current, while the quadrature (imaginary) component describes the ability of the porous material to store reversibly electrical charges.

Coseismic effect: The coseismic effect describes the electrical and magnetic fields associated with the passage of a seismic wave through an electromagnetic receiver. This type of electromagnetic disturbances travels with the seismic wave itself.

Diffuse layer: External part of the electrical double layer coating the surface of the minerals in contact with water. If the surface of the mineral is negatively charged, the diffuse layer carries an excess of cations and has a deficiency of anions to counterbalance the charge of the mineral surface and the Stern layer. The thickness of the diffuse layer is given by twice the Debye length.

Double layer: Generic term describing the disturbances in the ionic concentrations in the vicinity of the mineral surface in contact with water. The electrical double layer comprises the Stern layer of sorbed ions and the diffuse layer.

Drained: A drained mechanical behavior describes the behavior of a porous material when the fluid is free to flow through the connected porosity. In this case, the fluid does not resist to the deformation of the elastic skeleton (frame) of the porous material.

Drainage: Replacement of the wetting fluid by the nonwetting fluid in a porous material.

Electrical resistivity tomography: Geophysical imaging technique using the injection of electrical current in the ground and the measurements of the electrical field on another set of electrodes. Through inversion, an image of the electrical resistivity distribution can be obtained.

Electroencephalography: Medical technique based on the passive measurement of the electrical field on the scalp and the localization of the causative source of electrical current in the brain of a patient.

Electrogram: An electrogram is a record of the ground electrical field at a measuring station as a function of time. Electrograms are typically recorded in two Cartesian axes along the ground surface (the vertical component is usually zero).

Electrokinetic coupling: Electrokinetic coupling mechanism denotes a set of phenomena involving the relative displacement between a charged solid and the pore water including a fraction of the electrical diffuse layer. The plane of zero velocity is called the shear plane and characterizes the position

The Seismoelectric Method: Theory and Applications, First Edition. André Revil, Abderrahim Jardani, Paul Sava and Allan Haas.
© 2015 John Wiley & Sons, Ltd. Published 2015 by John Wiley & Sons, Ltd.

where the zeta potential is estimated. Very often, this shear plane is considered to be between the Stern layer and the diffuse layer.

Electroseismic effect: Geophysical method in which a current is injected in the ground and the ground motion is measured. The coupling is electrokinetic in nature.

Filtration displacement: Relative displacement between the fluid localized in the pore space of a porous material and the displacement of the solid phase.

First Fresnel zone: A Fresnel zone is one of a number of concentric ellipsoids defining volumes in the radiation pattern of a seismic wave. The cross section of the first Fresnel zone is a circle.

Gassmann substitution formula: The Gassmann equation allows to connect the bulk moduli of the same porous material saturated by two different viscous fluids. For the fluid velocities, Gassmann equations correspond essentially to the lower frequency limit of Biot theory of poroelasticity.

Haines jump: Haines jumps denote jumps in the position of the meniscus between two immiscible fluids during drainage or imbibition. These jumps happen so quickly that they generate recordable acoustic waves that are audible.

Helmholtz decomposition: Decomposition of a vector field as the sum of the gradient of a scalar potential and the curl of a vector potential.

Hooke's law: Linear constitutive equations between the stress and the strain in linear elasticity. The material property involved in this linear constitutive law is the stiffness tensor, which is represented by a matrix of $3 \times 3 \times 3 \times 3 = 81$ components, and because of the symmetry of the stress and strain tensors, only 21 elastic coefficients are independent (the stiffness tensor is a fourth-order tensor).

Hydraulic fracturing: Hydraulic fracturing (or fracking) denotes the formation of a fracture in a rock due to the increase of the fluid pressure. They can happen naturally, or they can be induced from a well to increase the permeability of a reservoir.

Imbibition: Replacement of the nonwetting fluid by the wetting fluid in a porous material.

Impedance meter: Instrument that can be used to measure the impedance of a porous material. It can be used to determine the complex conductivity of a porous material for a range of frequencies.

Interface response: Electromagnetic effect associated with the conversion of hydromechanical to electromagnetic energy at the interface between two porous media.

Inverse modeling: Modeling consisting in retrieving a model from observed data. The inverse problem is intrinsically non-unique in geophysics. It can be solved with deterministic or stochastic techniques.

Magnetogram: A magnetogram is a record of the ground magnetic field at a measuring station as a function of time. Magnetograms are typically recorded in the three Cartesian axes.

Markov chain Monte Carlo (MCMC) sampler: MCMC algorithms are used to sample a probability distribution based on constructing a Markov chain that has the desired distribution as its equilibrium distribution (e.g., Gaussian). The state of the chain after a number of realizations is then used as a sample of the desired probability distribution on the model parameters. The quality of the sampling improves as a function of the number of realizations. Therefore, after statistical convergence of the chain, the algorithm is used to determine the posterior probability distribution on each component of the model vector.

Maxwell fluid: A Maxwell fluid is a class of viscoelastic fluid having the properties of both elastic and viscous materials.

Maxwell–Wagner polarization: The Maxwell–Wagner polarization occurs at the inner dielectric boundary layers of a porous material and is related to a separation of electrical charges. These charges are often separated over a considerable distance relative to the distance involved in true dielectric polarization at the molecular scale. This polarization mechanism is due to the discontinuity of the displacement current at the interface between different phases of a porous material.

NAPL: Nonaqueous phase liquids that are not easily miscible with water.

Nonpolarizable electrode: A nonpolarizable electrode is characterized by faradic current can that freely pass (without polarization) between the two sides of the electrical double layer between the electrode and a porous material or an electrolyte. Examples correspond to the Ag/AgCl or $Pb/PbCl_2$ electrodes.

Objective function: In geophysics, an optimization problem consists of minimizing an objective function. Optimization includes finding optimal values of the objective function given a set of constraints, including a variety of different terms in the objective functions and different types of domains. For instance the objective function is the sum of a misfit term and a regularizer balanced by a regularization parameter.

Oil wet: A porous material is oil wet if the oil has a preferential wettability (affinity) for the solid phase. For instance, carbonate reservoirs are often oil wet, while siliciclastic reservoirs are usually water wet.

Outer Helmholtz plane: Plane separating the Stern layer from the diffuse layer in the electrical double layer coating the surface of the mineral grains.

Perfectly matched layer (PML): A PML is a synthetic absorbing layer used in the numerical modeling of wave propagation. It is used to simulate material with open boundaries for the propagation of the waves (for instance, in the finite-element method).

P-wave: Compressional or primary elastic wave. The motion occurs in the direction of propagation. This type of wave is faster than the S-wave.

Quadrature conductivity: Imaginary part of the complex electrical conductivity. It corresponds to the reversible storage of electrical charges at low frequencies in porous rocks. This component is measured in geophysics through the induced polarization method.

Richards equation: Field partial differential equation for the pressure of the water phase in unsaturated condition. The validity of this equation is based on the assumption that the nonwetting phase is very compressible and at constant (atmospheric) pressure.

Reynolds number: Dimensionless ratio between the inertial and viscous term in the Navier–Stokes equation.

Seismogram: Record of the ground motion at a measuring station as a function of time. Seismograms typically record ground motions in three Cartesian axes. The seismic source can be an artificial source or a natural source such as an earthquake.

S-wave: Shear or secondary elastic wave characterize by a motion normal to the direction of propagation. This wave does not diverge (the divergence of the displacement is equal to zero). This type of wave does not travel through a viscous fluid.

Seismoelectric effect: Geophysical method in which seismic waves propagate in the ground and the associated electromagnetic effects are measured. The coupling is electrokinetic in nature.

Self-potential: Passive geophysical method consisting in measuring the electrical field at the ground surface (possibly in boreholes) in order to retrieve the causative source of current in the subsurface.

Skeleton: The skeleton denotes the solid frame of a porous material.

Stern layer: Inner portion of the electrical double layer corresponding to ions that are weakly or strongly sorbed on the mineral surface.

Streaming current: Source current generated by the flow of the pore water and the drag of the excess of electrical charges contained in the pore water because of the electrical diffuse layer.

Tortuosity: The tortuosity of the pore space of a porous material describes the relative path in a given direction imposed by the topology of the pore network with respect to the straight direction. Different tortuosities can be defined for different transport properties because of the local boundary condition problem defining these properties.

Undrained: An undrained mechanical behavior describes the behavior of a porous material when the fluid has no time or cannot flow through the connected pore space. In this case, the fluid resists to the deformation of the elastic skeleton (frame) of the porous material.

Vadose zone: Unsaturated portion of the ground comprised between the ground surface and the water table of an unconfined aquifer. The saturation of the water phase is smaller than one.

Water wet: A porous material is water wet if the water has a preferential wettability (affinity) for the solid phase.

Water table: Interface between the vadose zone and an unconfined aquifer where the porous material is at saturation and the water at atmospheric pressure.

Zeta potential: Microscopic electrical potential of the electrical double layer at the position of the shear plane, typically between the Stern layer and the diffuse layer.

Index

The Seismoelectric Method: Theory and Applications, First Edition. André Revil, Abderrahim Jardani, Paul Sava and Allan Haas.
© 2015 John Wiley & Sons, Ltd. Published 2015 by John Wiley & Sons, Ltd.